Lecture Notes
in Control and Information Sciences 222

Editor: M. Thoma

W0245715

Springer-Verlag London Ltd.

A. Stephen Morse (Ed.)

Control Using Logic-Based Switching

 Springer

ISBN 978-3-540-76097-9   ISBN 978-3-540-40943-4 (eBook)
DOI 10.1007/978-3-540-40943-4

British Library Cataloguing in Publication Data
Control using logic-based switching. - (Lecture notes in
    control and information sciences ; 222)
    1.Logic circuits 2.Control theory
    I.Morse, A. Stephen
    621.3'95

ISBN 978-3-540-76097-9

Library of Congress Cataloging-in-Publication Data
A catalog record for this book is available from the Library of Congress

Typesetting: Camera ready by author
69/3830-543210 Printed on acid-free paper

Preface

By a *logic-based switching controller* is meant a controller whose subsystems include not only familiar dynamical components such as integrators, summers, gains, etc., but event-driven logic and switches as well. More often than not the predominately logical component within such a system is called a supervisor, a mode changer, a gain scheduler, a reference governor or something similar. Within the past few years there has been a growing interest in determining what might be gained by utilizing "hybrid" controllers of this type. Toward this end, a workshop was held in the fall of 1995 on Block Island, Rhode Island with the aim bringing together a small number of individuals to discuss research of common interest within the area. The invited speakers were as follows:

Michael S. Branicky	Sanjoy K. Mitter
Roger W. Brockett	Felipe M. Pait
Michael Chang	Njal B. O. L. Pettit
Peter E. Caines	Kameshwar Poolla
Edward J. Davison	Peter J. Ramadge
John Guckenheimer	Michael G. Safonov
Robert P. Kurshan	Jeff S. Shamma
Jorgen Malmborg	Mark W. Spong
N. Harris McClamroch	George Zames
Daniel E. Miller	

This volume is an outgrowth of that workshop. It encompasses not only much of what transpired on Block Island, but also a broad spectrum of related topics contributed by individuals who were not at the workshop.

The first eight papers, authored by Branicky through Teel, discuss a number of issues within the burgeoning area of *hybrid dynamical systems* [1, 2, 3]. The next seven papers, written by Ledyaev through Praly, address various problems concerned with the control of hard-bound constrained and nonlinear systems. Automotive problem involving switching control are covered next in two papers authored by Butts through Kolmanovsky, as well as in the paper by Sangiovanni-Vincentelli. The volume's final six papers, by Shamma through Kassab, discuss a variety ways to control systems in the face of large-scale modeling errors.

Cover: The computer flow diagram appearing cover describes a simple switching logic proposed in [4] for stabilizing the nonholonomic integrator.

References

[1] *Hybrid Systems I*, Grossman, Ravn, Rischel and Nerode, editors, Springer Lecture Notes in Computer Science 736, 1993

[2] *Hybrid Systems II*, Antsaklis, Kohn, Nerode and Sastry, editors, Springer Lecture Notes in Computer Science 999, 1995

[3] *Hybrid Systems III*, Alur, Sontag and Henzinger, editors, Springer Lecture Notes in Computer Science 999, 1996

[4] J. P. Hespanha, "Stabilization of the Nonholonomic Integrator Via Logic-Based Switching," *Proceedings of the 1996 IFAC Congress*, San Francisco, June, 1996, to appear

Acknowledgments

Financial support for the Block Island Workshop on "Control Using Logic-Based Switching" was provided by the National Science Foundation, the U. S. Army Research Office, the Yale University Faculty of Engineering, and the Princeton University Department of Electrical Engineering.

A. S. Morse
New Haven, June 14, 1996

Contents

Hybrid Dynamical Systems, or HDS: The Ultimate Switching Experience

Michael S. Branicky*

Laboratory for Information and Decision Systems
Massachusetts Institute of Technology

Abstract

In previous work I have concentrated on formalizing the notion of a hybrid system as switching among an indexed collection of dynamical systems. I have also studied in some depth the modeling, analysis, and control of such systems. Here, I give a quick overview of the area of hybrid systems. I also briefly review the formal definition and discuss the main approaches taken in the study of hybrid systems. Finally, I elucidate issues in each of the research areas in light of previous results.

1 Introduction

Many complicated control systems today (e.g., those for flight control, manufacturing systems, and transportation) have vast amounts of computer code at their highest level. More pervasively, programmable logic controllers are widely used in industrial process control. We also see that today's products incorporate logical decision-making into even the simplest control loops (e.g., embedded systems). Thus, virtually all control systems today issue continuous-variable controls and perform logical checks that determine the mode—and hence the control algorithms—the continuous-variable system is operating under at any given moment. As such, these "hybrid control" systems offer a challenging set of problems.

Hybrid systems involve both continuous-valued and discrete variables. Their evolution is given by equations of motion that generally depend on both. In turn these equations contain mixtures of logic, discrete-valued or **digital** dynamics, and continuous-variable or **analog** dynamics. The continuous dynamics of such systems may be continuous-time, discrete-time, or mixed (sampled-data), but is generally given by differential equations. The discrete-variable dynamics of hybrid systems is generally governed by a **digital automaton**, or input-output transition system with a countable number of states. The continuous and discrete dynamics interact at "event" or "trigger" times when the continuous state hits certain prescribed sets in the continuous state space. See Figure 1(a). *Hybrid control systems* are control systems that involve both continuous and discrete dynamics and continuous and discrete controls. The

---

*Supported by Army Research Office and Center for Intelligent Control Systems, grants DAAL03-92-G-0164/-0115. MIT, 35-415, Cambridge, MA 02139. branicky@lids.mit.edu

continuous dynamics of such a system is usually modeled by a controlled vector field or difference equation. Its hybrid nature is expressed by a dependence on some discrete phenomena, corresponding to discrete states, dynamics, and controls. The result is a system as in Figure 1(b).

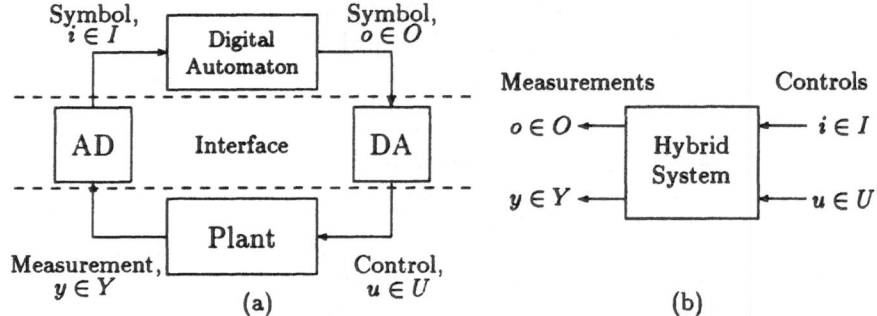

Figure 1: (a) Hybrid System. (b) Hybrid Control System.

Hybrid systems research may be broken down into four broad categories:

- **Modeling**: formulating precise models that capture the rich behavior of hybrid systems. i.e., *How do we "fill in the boxes" in Figure 1? What is their dynamics? How can we classify their rich structure and behavior— and sort through the myriad hybrid systems models appearing?*

- **Analysis**: developing tools for the simulation, analysis, and verification of hybrid systems. i.e., *How do we analyze systems as in Figure 1(a)? What does continuity mean? What is their complexity? How do they differ from continuous dynamical systems? How do we test their stability? or analyze examples?*

- **Control**: synthesizing hybrid controllers —which issue continuous controls and make discrete decisions—that achieve certain prescribed safety and performance goals for hybrid systems. i.e., *How do we control a plant as in Figure 1(b) with a controller as in Figure 1(b)? How can we synthesize such hybrid controllers?*

- **Design**: conceiving new schemes and structures leading to easier modeling, verification, and control.

In previous work [8, 10, 11, 12, 14, 17] I have concentrated on formalizing the notion of a hybrid system sketched above and the subsequent answering of the above questions for that model. Here, I briefly review the formal definition. I then discuss the main approaches taken in the study of of hybrid systems. Finally, I elucidate issues in the above research areas in light of previous results.

2   Related Work

Hybrid systems are certainly pervasive today. But they have been with us at least since the days of the relay. In control theory, there has certainly been a lot of related work in the past, including variable structure systems, jump linear systems, systems with impulse effect, impulse control, and piecewise deterministic processes. More recently, in computer science there has been a successive build-up in the large formal verification literature toward verification of systems that include both continuous and discrete variables. In control theory we have witnessed a resurgence in examining quantization effects [22, 29, 39, 40] and recent progress in analyzing switched [11, 31, 38, 45] and discretely-controlled continuous-variable systems [20, 30, 35, 34].

3   Hybrid Dynamical Systems

Briefly, a hybrid dynamical system is an indexed collection of dynamical systems along with some map for "jumping" among them (switching dynamical system and/or resetting the state). This jumping occurs whenever the state satisfies certain conditions, given by its membership in a specified subset of the state space. Hence, the entire system can be thought of as a sequential patching together of dynamical systems with initial and final states, the jumps performing a reset to a (generally different) initial state of a (generally different) dynamical system whenever a final state is reached.

Formally, a **controlled general hybrid dynamical system (CGHDS)** is a system $H_c = [Q, \Sigma, \mathbf{A}, \mathbf{G}, \mathbf{V}, \mathbf{C}, \mathbf{F}]$, with constituent parts as follows.

- Q is the set of **index states**, or **discrete states**.

- $\Sigma = \{\Sigma_q\}_{q \in Q}$ is the collection of controlled dynamical systems, where each $\Sigma_q = [X_q, \Gamma_q, f_q, U_q]$ (or $\Sigma_q = [X_q, \Gamma_q, \phi_q, U_q]$) is a controlled dynamical system. Here, the X_q are the **continuous state spaces** and ϕ_q (or f_q) are called the **continuous dynamics**.

- $\mathbf{A} = \{A_q\}_{q \in Q}$, $A_q \subset X_q$ for each $q \in Q$, is the collection of **autonomous jump sets**.

- $\mathbf{G} = \{G_q\}_{q \in Q}$, where $G_q : A_q \times V_q \to S$ is the **autonomous jump transition map**, parameterized by the **transition control set** V_q, a subset of the collection $\mathbf{V} = \{V_q\}_{q \in Q}$; they are said to represent the **discrete dynamics**.

- $\mathbf{C} = \{C_q\}_{q \in Q}$, $C_q \subset X_q$, is the collection of **controlled jump sets**.

- $\mathbf{F} = \{F_q\}_{q \in Q}$, where $F_q : C_q \to 2^S$, is the collection of **controlled jump destination maps**.

Thus, $S = \bigcup_{q \in Q} X_q \times \{q\}$ is the **hybrid state space** of H. The case where the sets U_q and **G** through **F** above are empty is simply a **general hybrid dynamical system (GHDS)**: $H = [Q, \Sigma, \mathbf{A}, \mathbf{G}]$,

A CGHDS can be pictured as an automaton as in Figure 2. There, each node is a constituent dynamical system, with the index the name of the node. Each edge represents a possible transition between constituent systems, labeled by the appropriate condition for the transition's being "enabled" and the update of the continuous state (cf. [26]). The notation ![condition] denotes that the transition *must* be taken when enabled. The notation ?[condition] denotes an enabled transition that *may be taken* on command; ":∈" means reassignment to some value in the given set.

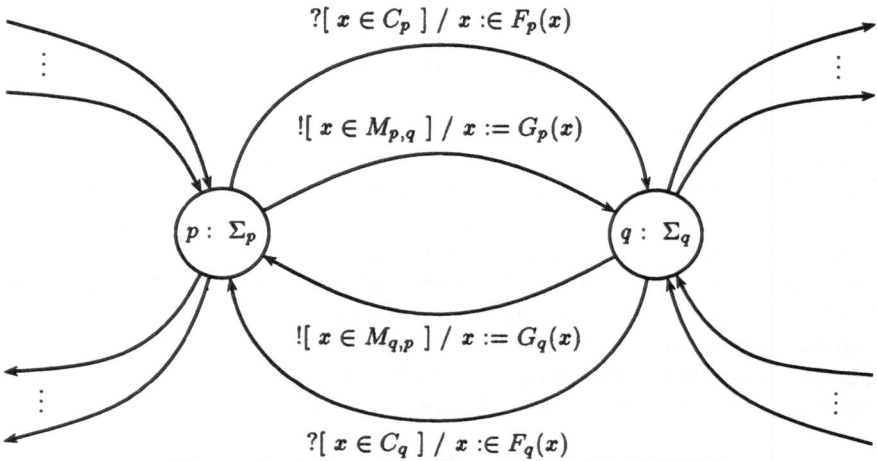

$$?[\, x \in C_p \,] \,/\, x :\in F_p(x)$$

$$![\, x \in M_{p,q} \,] \,/\, x := G_p(x)$$

$$p : \Sigma_p \qquad q : \Sigma_q$$

$$![\, x \in M_{q,p} \,] \,/\, x := G_q(x)$$

$$?[\, x \in C_q \,] \,/\, x :\in F_q(x)$$

Figure 2: Automaton Associated with CGHDS.

Roughly, the dynamics of H_c are as follows. The system is assumed to start in some hybrid state in $S \backslash A$, say $s_0 = (x_0, q_0)$. It evolves according to $\phi_{q_0}(\cdot, \cdot, u)$ until the state enters—if ever—either A_{q_0} or C_{q_0} at the point $s_1^- = (x_1^-, q_0)$. If it enters A_{q_0}, then it *must* be transferred according to transition map $G_{q_0}(x_1^-, v)$ for some chosen $v \in V_{q_0}$. If it enters C_{q_0}, then we *may* choose to jump and, if so, we may choose the destination to be any point in $F_{q_0}(x_1^-)$. Either way, we arrive at a point $s_1 = (x_1, q_1)$ from which the process continues. See Figure 3.

The following are some important notes about CGHDS:

Dynamical Systems. GHDS with $|Q| = 1$ and **A** empty recover all these.

Hybrid Systems. The case of GHDS with $|Q|$ finite, each X_q a subset of $\mathbb{R}^n$, and each $\Gamma_q = \mathbb{R}$ largely corresponds to the *usual* notion of a hybrid system, viz. a coupling of finite automata and differential equations [13, 14, 25]. Herein, a **hybrid system** is a GHDS with Q countable, and with $\Gamma_q \equiv \mathbb{R}$ (or $\mathbb{R}_+$) and $X_q \subset \mathbb{R}^{d_q}$, $d_q \in \mathbb{Z}_+$, for all $q \in Q$: $[Q, [\{X_q\}_{q \in Q}, \mathbb{R}_+, \{f_q\}_{q \in Q}], \mathbf{A}, \mathbf{G}]$, where

f_q is a vector field on $X_q \subset \mathbb{R}^{d_q}$.[1]

Changing State Space. The state space may change. This is useful in modeling component failures or changes in dynamical description based on autonomous—and later, controlled—events which change it. Examples include the collision of two inelastic particles or an aircraft mode transition that changes variables to be controlled [33]. We also allow the X_q to overlap and the inclusion of multiple copies of the same space. This may be used, for example, to take into account overlapping local coordinate systems on a manifold [4].

Refinements. We may refine the concept of, say, GHDS H by adding:

- inputs, including control inputs, disturbances, or parameters (see *controlled* HDS below).

- outputs, including **state-output** for each constituent system as for dynamical systems [12, 42] and **edge-output:** $H = [Q, \Sigma, \mathbf{A}, \mathbf{G}, O, \eta]$, where $\eta : A \rightarrow O$ produces an output at each jump time.

- $\Delta : A \rightarrow \mathbb{R}_+$, the **jump delay map**, which can be used to account for the time which abstracted-away, lower-level transition dynamics actually take.[2]

- Marked states (including initial or final states), timing, or input and output for any constituent system.

Other Notes. (1) **Nondeterminism** in transitions may be taken care of by partitioning ?[condition] into those which are controlled and uncontrolled (cf. DEDS) **Disturbances** (and other nondeterminism) may be modeled by partitioning U, V, and C into portions that are under the influence of the controller or nature respectively. Systems with state-output, edge-output, and autonomous and controlled jump delay maps (Δ_a and Δ_c, respectively) may be added as above. (2) The model includes the "unified" model posed by Branicky, Borkar, and Mitter (**BBM**; [14]) and thus several other previously posed hybrid systems models [3, 4, 19, 36, 43, 46]. It also includes systems with impulse effect [5] and hybrid automata [23]. (3) In particular, our unified BBM model is, briefly, a **controlled hybrid system**, with the form $[\mathbb{Z}_+, [\{\mathbb{R}^{d_i}\}_{i=0}^\infty, \mathbb{R}_+, \{f_i\}_{i=0}^\infty, U], \mathbf{A}, \mathbf{V}, \mathbf{G}, \mathbf{C}, \mathbf{F}]$. The **admissible control actions** available in this model are the continuous controls $u \in U_q$, exercised in each constituent regime; the **discrete controls** $v \in V_q$, exercised at the autonomous jump times (i.e., on hitting set A); and the **intervention times**

---

[1]Here, we may take the view that the system evolves on the state space $\mathbb{R}^* \times Q$, where $\mathbb{R}^*$ denotes the set of finite, but variable-length real-valued vectors. For example, Q may be the set of labels of a computer program and $x \in \mathbb{R}^*$ the values of all currently-allocated variables. This then includes Smale's tame machines [6].

[2]Think of modeling the closure time of a discretely-controlled hydraulic valve or trade mechanism imperfections in economic markets.

and **destinations** of controlled jumps (when the state is in C). Control results for this model are given in [14, 17].

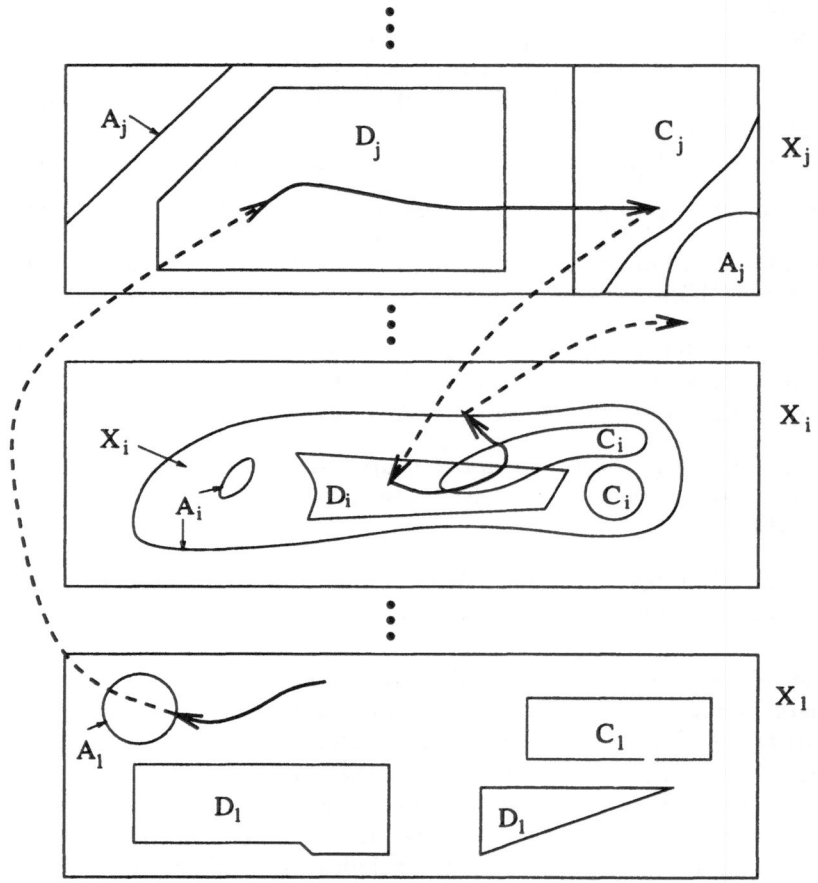

Figure 3: Example dynamics of CGHDS.

4   Paradigms for Hybrid Systems

I see four basic paradigms for the study of hybrid systems: aggregation, continuation, automatization, and systemization. The first two approaches deal with the different sides—analog and digital—of hybrid systems. They attempt to suppress the hybrid nature of the system by converting it into a purely discrete or purely continuous one, respectively. The last two approaches are more general and potentially more powerful. Under them, a hybrid system is seen directly as an interacting set of automata or dynamical systems; they complement the input-output and state-space paradigms, respectively, of both control theory and computer science. More specifically, the approaches are:

Aggregation. That is, suppress the continuous dynamics so that the hybrid system is a finite automaton or discrete-event dynamical system (DEDS). This is the approach most often taken in the literature, e.g., [3]. The drawback of this approach is three-fold.

- *Nondeterminism*: one usually obtains nondeterministic automata [3, 27].

- *Nonexistence*, i.e., even if clever constructions are used, no finite automaton may exist that captures the combined behavior [24].

- *Partition Problem.* It appears a conceptually deep problem to determine when there exist partitions of just a continuous system such that its dynamics is captured by a meaningful finite automaton. "Meaningful," since we note that every system is homomorphic to one with a single equilibrium point [41]. The answer thus depends on the dynamical behavior one is interested in capturing and the questions one is asking.

The aggregation program has been fully carried out so far only under strong assumptions on the hybrid system [1, 24].

Continuation. The complement of aggregation, that is, suppress the discrete dynamics so that the hybrid system becomes a differential equation. This original idea of Prof. Sanjoy Mitter and myself is to convert hybrid models into purely continuous ones—modeled by differential equations—using differential equations that simulate finite automata. In this familiar, unified realm one could answer questions of stability, controllability, and observability, converting them back to the original model by taking a "singular limit." For instance, one would like tools that allow one to conclude the following: if a "sufficiently close" continuation of a system is stable, then the original system is stable. Such a program is possible in light of the existence of simple continuations of finite automata [13, 18] and pushdown automata and Turing machines [13]. The drawback of this approach is three-fold.

- *Arbitrariness*, i.e., how one accomplishes the continuation is largely arbitrary. For example, to interpolate or "simulate" the step-by-step behavior of a finite automaton Brockett used his double-bracket equations [19] and the author used stable linear equations [7, 13]. In certain cases this freedom is an advantage. However, care must be taken to insure that the dynamics used does not introduce spurious behavior (like unwanted equilibria) or that it itself is not hard to analyze or predict.

- *Hiding Complexity.* One cannot generally get rid of the underlying discrete dynamics, i.e., the complexity is merely hidden in the "right-hand side" of the continuation differential equations [9].

- *Artificiality.* It can lead to a possibly unnatural analytical loop of going from discrete to continuous and back to discrete. Cf. Chen's recent results in stochastic approximation vis-à-vis Kushner's [21].

The combination of these points has been borne out by some experience: it can be easier to examine the mixed discrete-continuous system. Cf. my switched aircraft controller analysis [10] and Megretsky's relay system analysis [32].

Automatization. Treat the constituent systems as a network of interacting automata [36, p. 325]. The focus is on the input-output or language behavior. This language view has been largely taken in the computer science literature in extending the dynamical behavior of finite automata incrementally toward full hybrid systems (see [1, 25] for background).

Automatization was pioneered in full generality by Nerode and Kohn [36]. The viewpoint is that systems, whether analog or digital, are automata. As long as there is compatibility between output and input alphabets, links between automata can be established. However, there is still the notion of "reconciling different time scales" [36, p. 325]. For instance, a finite automaton receives symbols in abstract time, whereas a differential equation receives inputs in "real time." This reconciliation can take place by either of the following: forcing synchronization at regular sampling instants [36, p. 333], or synchronizing the *digital* automaton to advance at event times when its input symbols change [36]. For hybrid systems of interest, the latter mechanism appears more useful. It has been used in many hybrid systems models, e.g., [3, 19]. For a fruitful example of this approach see [23].

Systemization. Treat the constituent systems as interacting dynamical systems [41]. The focus is on the state-space. This state-space view has been taken most profitably in the work of Witsenhausen [46] and Tavernini [43].

Systemization is developed in full generality in the thesis [12]. The viewpoint is that systems, whether analog or digital, are dynamical systems. As long as there is compatibility at switching times when the behavior of a system changes in response to a logical decision or event occurrence, links between these dynamical systems can be established. Again, there is still the notion of reconciling dynamical systems with different time scales (i.e., transition semigroups). For instance, a finite automaton abstractly evolves on the positive integers (or on the free monoid generated by its input alphabet), whereas a differential equation evolves on the reals. This reconciliation can take place by either or both of the following: sequentially synchronizing the dynamical systems at event times when their states enter prescribed sets, or forcing uniform semigroup structure via "timing maps." Both approaches are introduced here, but the concentration is on the former.

Systemization is established in my formulation of hybrid dynamical systems above. It has been used in examining complexity and simulation capabilities of hybrid systems [13], analyzing their stability [10, 11], and in establishing the first comprehensive state-space paradigm for the control of hybrid systems [14].

5 Thoughts on Hybrid Systems

Modeling. There are a myriad of hybrid systems models (see [12] and the collections [25, 2] for references). Of less prevalence are results proven about these models. I mention a few here. Tavernini [43], for instance, has given a model plus assumptions and proven properties of trajectories and simulations of trajectories of hybrid systems. In early work, Witsenhausen [46] gave a model and derived some optimal control results. Deshpande [23] took an automata approach to modeling and derived results for viable control of their hybrid systems. Above, and in [14, 12] I have a given a model which (1) captures the phenomena associated with hybrid systems, (2) subsumes many of the models presented in the literature, (3) is amenable to the posing and solution of optimal control problems (see the discussion below).

One needs to explore the plethora of modeling choices available in hybrid systems. Since hybrid systems include dynamical systems as a subset, subclasses which permit efficient simulation, analysis, and verification should be explored. I believe that such a program is indeed being carried out by the computer scientists. Control theorists should do the same in their field in examining the hybrid control of hybrid systems.

Analysis. In general, hybrid systems analysis is theoretically intractable. The reason is that even simple hybrid systems possess the power of universal computation [1, 12]. Thus, any purported general solution of reachability or stability would lead to the impossible situation of a solution to the halting problem. There are, however, at least two ways to proceed. The first is to limit the hybrid systems one wants to consider. The second is to use general conditions for, say, stability of hybrid systems as *sufficient* conditions for design, i.e., design controls so that the system is easy to analyze.

The first approach has been taken by the computer scientists (as mentioned above). The second approach has had success in control theory. For example, even though it is in general, notoriously hard to find Lyapunov functions to prove stability. Yet, most stability proofs in adaptive control treat Lyapunov stability as a sufficient condition and design stable adaptive controllers by first choosing a Lyapunov function and then controls that enforce its decay.

In the case of hybrid systems, the analysis tools I have developed [10, 11] in general deal not only with the behavior of the constituent systems, but also impose conditions on the behavior of the system at switching times. Thus, control design arising from the constraints of [11] is a topic of further research.

There are also theoretical issues to be explored. Some examples include the stability of systems with multiple equilibrium points, the stability of switched systems, relations between fixed-point theory and Lyapunov stability, and the stability and dynamics of ordinary differential equations driven by Markov chains whose transition probabilities are a function of the continuous state. The latter may provide a link to the large literature on jump systems.

Another important topic of further research is to incorporate developed analysis tools into software engineering tools. This will allow application of these tools to complicated examples in a timely manner.

Control. Specific open theoretical issues were discussed in [14]. Another has to do with the robustness of our hybrid controls with respect to state. Here, the transversality assumptions made in [14] should combine with Tavernini's result on continuity with respect initial condition to yield continuity of control laws on an open dense set.

Algorithms for synthesizing optimal hybrid controllers have also been given [12, 17], along with some academic examples. An important area of current research is to develop good computational schemes to compute near-optimal controls in realistic cases. Analysis of rates of convergence of discretized algorithms should be explored. Later, the development of software tools to design such controllers automatically will become an important area of research.

Design. Finally, from modeling, through analysis and control, we come to design of complex, hybrid systems. Here, some of the interaction between levels is under our jurisdiction. What would we do with such freedom, coupled with our new-found analysis and control techniques? For example, we might design a flexible manufacturing system that not only allows quick changes between different product lines, but allows manufacture of new ones with relative ease.

Consider the so-called reflex controllers of [37, 44], which constitute a dynamically consistent interface between low-level servo control and higher-level planning algorithms that ensures obstacle avoidance. Thus as a step in the direction of hierarchical, hybrid design, the reflex control concept is an example of how to incorporate a new control module to allow rapid, dynamically transparent design of higher-level programs. Further, there are some structures for the control programs used in research aircraft [28] that may lend themselves to such an approach. In each case, these designs incorporate structures which allow engineers to separate the continuous and logical worlds.

These controllers provide the inspiration, our analysis and control results the foundation, and our steps toward efficient algorithms the impetus, for setting a course toward design of hybrid systems. Ultimately, it is hoped they will lead to truly intelligent engineering systems.

References

[1] R. Alur *et al.*. Hybrid automata: An algorithmic approach to the specification and verification of hybrid systems. In Grossman et al. [25], pp. 209–229.

[2] P. Anstaklis *et al.*, eds. *Hybrid Systems II*. Springer, New York, 1995.

[3] P. J. Antsaklis, J. A. Stiver, and M. D. Lemmon. Hybrid system modeling and autonomous control systems. In Grossman et al. [25], pp. 366–392.

[4] A. Back, J. Guckenheimer, and M. Myers. A dynamical simulation facility for hybrid systems. In Grossman et al. [25], pp. 255–267.

[5] D. Bainov and P. Simeonov. *Systems with Impulse Effect*. Ellis Horwood, 1989.

[6] L. Blum, M. Shub, and S. Smale. On a theory of computation and complexity over the real numbers. *Bulletin American Math. Society*, 21(1):1–46, July 1989.

[7] M. S. Branicky. Equivalence of analog and digital computation. In *Workshop on Continuous Algorithms and Complexity*, Barcelona, October 1993. Abstract.

[8] M. S. Branicky. Topology of hybrid systems. In *Proc. IEEE Conf. Decision and Control*, pp. 2309–2314, San Antonio, December 1993.

[9] M. S. Branicky. Analog computation with continuous ODEs. In *Proc. IEEE Workshop Physics and Computation*, pp. 265–274, Dallas, November 1994.

[10] M. S. Branicky. Analyzing continuous switching systems: Theory and examples. In *Proc. American Control Conf.*, pp. 3110–3114, Baltimore, June 1994.

[11] M. S. Branicky. Stability of switched and hybrid systems. In *Proc. IEEE Conf. Decision and Control*, pp. 3498–3503, Lake Buena Vista, December 1994.

[12] M. S. Branicky. *Studies in Hybrid Systems: Modeling, Analysis, and Control*. PhD thesis, Massachusetts Institute of Technology, EECS Dept., June 1995.

[13] M. S. Branicky. Universal computation and other capabilities of hybrid and continuous dynamical systems. *Theoretical Comp. Science*, 138(1):67–100, 1995.

[14] M. S. Branicky, V. Borkar, and S. K. Mitter. A unified framework for hybrid control. Technical Report LIDS-P-2239, Lab. for Information and Decision Systems, Massachusetts Institute of Technology, April 1994. Also see: [15, 16].

[15] M. S. Branicky, V. S. Borkar, and S. K. Mitter. A unified framework for hybrid control. In G. Cohen and J.-P. Quadrat, eds., *Proc. 11th INRIA Intl. Conf. Analysis and Optimization of Systems*, vol. 199 of *Lecture Notes Control and Info. Sciences*, pp. 352–358, New York, 1994. Springer. Extended Abstract.

[16] M. S. Branicky, V. S. Borkar, and S. K. Mitter. A unified framework for hybrid control. In *Proc. IEEE Conf. Decision and Control*, pp. 4228–4234, Lake Buena Vista, December 1994.

[17] M. S. Branicky and S. K. Mitter. Algorithms for optimal hybrid control. In *Proc. IEEE Conf. Decision and Control*, pp. 2661–2666, New Orleans, December 1995.

[18] R. W. Brockett. Smooth dynamical systems which realize arithmetical and logical operations. In H. Nijmeijer and J. M. Schumacher, eds., *Three Decades of Mathematical Systems Theory*, pp. 19–30. Springer, Berlin, 1989.

[19] R. W. Brockett. Hybrid models for motion control systems. In H. L. Trentelman and J. C. Willems, eds., *Essays in Control*, pp. 29–53. Birkhäuser, Boston, 1993.

[20] C. Chase, J. Serrano, and P. J. Ramadge. Periodicity and chaos from switched flow systems. *IEEE Trans. Automatic Control*, 38(1):70–83, 1993.

[21] H.-F. Chen. Stochastic approximation and its new applications. In *Proc. Hong Kong Intl. Workshop New Directions of Control and Manuf.*, pp. 2–12, 1994.

[22] D. F. Delchamps. Stabilizing a linear system with quantized state feedback. *IEEE Trans. Automatic Control*, 35(8):916–924, August 1990.

[23] A. Deshpande. *Control of Hybrid Systems*. PhD thesis, U. Cal. Berkeley, 1994.

[24] S. Gennaro, C. Horn, S. Kulkarni, P. Ramadge. Reduction of timed hybrid systems. In *Proc. IEEE Conf. Decision and Cont.*, pp. 4215–4220, December 1994.

[25] R. L. Grossman *et al.*, eds. *Hybrid Systems*. Springer, New York, 1993.

[26] D. Harel. Statecharts: A visual formalism for complex systems. *Science of Computer Programming*, 8:231–274, 1987.

[27] C. S. Hsu. *Cell-to-Cell Mapping*. Springer, New York, 1987.

[28] C. S. Hynes. Flight control systems that can be formally verified. Technical report, NASA Ames Research Center, 1993. Extended abstract.

[29] R. Koplon and E. Sontag. Sign-linear systems as cascades of automata and continuous variable systems. In *Proc. IEEE Conf. Decision and Control*, pp. 2290–2291, San Antonio, December 1993.

[30] S. R. Kulkarni and P. J. Ramadge. On the existence and memory requirements of convergent on-line decision rules. In *Proc. IEEE Conf. Decision and Control*, pp. 3022–3027, New Orleans, December 1995.

[31] G. A. Lafferriere. Discontinuous stabilizing feedback using partially defined Lyapunov functions. In *Proc. IEEE Conf. Decision and Control*, pp. 3487–3491, Lake Buena Vista, December 1994.

[32] A. Megretsky. Robustness analysis of a relay system. Lab. for Information and Decision Systems Seminar, Massachusetts Institute of Technology, April 1995.

[33] G. Meyer. Design of flight vehicle management systems. In *Proc. IEEE Conf. Decision and Control*, Lake Buena Vista, December 1994. Plenary Lecture.

[34] A. S. Morse. Supervisory control of families of linear set-point controllers—Part 1: Exact matching, Part 2: Robustness. Preprints, March 1993, November 1994.

[35] A. S. Morse. Control using logic-based switching. In A. Isidori, ed., *Trends in Control*, pp. 69–113. Springer, Berlin, 1995.

[36] A. Nerode and W. Kohn. Models for hybrid systems: Automata, topologies, stability. In Grossman et al. [25], pp. 317–356.

[37] W. S. Newman and M. S. Branicky. Experiments in reflex control for industrial manipulators. In *Proc. IEEE Intl. Conf. on Robotics and Automation*, May 1990.

[38] P. Peleties and R. DeCarlo. Asymptotic stability of m-switched systems using Lyapunov-like fncs. *Proc. Amer. Cont. Conf.*, pp. 1679–1684, Boston, June 1991.

[39] P. J. Ramadge. On the periodicity of symbolic observations of piecewise smooth discrete-time systems. *IEEE Trans. Automatic Control*, 35(7):807–813, 1990.

[40] L. J. Serrano. *The Effects of Time Sampling and Quantization in the Discrete Control of Continuous Systems*. PhD thesis, Princeton, EE Dept., October 1990.

[41] K. S. Sibirsky. *Intro. to Topological Dynamics*. Noordhoff, Netherlands, 1975.

[42] E. D. Sontag. *Mathematical Control Theory*. Springer, New York, 1990.

[43] L. Tavernini. Differential automata and their discrete simulators. *Nonlinear Analysis, Theory, Methods, and Applications*, 11(6):665–683, 1987.

[44] T. S. Wickman, M. S. Branicky, and W. S. Newman. Reflex control for robot system preservation, reliability, and autonomy. *Computers and Electrical Engineering*, 20(5):391–407, 1994. Special Issue on Fault Tolerance and Robotics.

[45] M. A. Wicks, P. Peleties, and R. A. DeCarlo. Construction of piecewise Lyapunov functions for stabilizing switched systems. In *Proc. IEEE Conf. Decision and Control*, pp. 3492–3497, Lake Buena Vista, December 1994.

[46] H. S. Witsenhausen. A class of hybrid-state continuous-time dynamic systems. *IEEE Trans. Automatic Control*, 11(2):161–167, 1966.

Complexity of Hybrid System Models

John Guckenheimer
Center for Applied Mathematics
Cornell University, Ithaca, NY 14853

This is a "position paper," not an account of accomplished research. Its intent is to be provocative and ask basic questions about the design of control systems from a computational viewpoint that emphasizes the problems associated with model simulation rather than the physical engineering foundations of models. In previous work [1], we have proposed a definition of "hybrid dynamical systems" that is quite general and allows the representation of discrete and continuous dynamical systems on manifolds, piecewise smooth dynamical systems, switching control systems, singularly perturbed dynamical systems, etc. We recall this description and adopt it as the basis for our discussion.

The problem domain is a disjoint union of open, connected subsets of $\mathbf{R}^n$, called charts. Each chart has associated with it a vector field. Inside each chart is a *patch*, an open subset with closure contained inside the chart. The patch boundaries are assumed piecewise smooth. The evolution of the system is implemented as a sequence of trajectory segments where the endpoint of one segment is connected to the initial point of the next by a transformation that may be trivial. Even if there is no change of continuous state, states of a system have also a discrete part, and *switches* that change the discrete part of the system state do occur. Time is divided into contiguous periods separated by instances where a state transformation is applied at times referred to as *events*.

We have implemented a problem solving environment for the simulation of hybrid systems [1] and I have used it to design a controller for a double pendulum on a cart that travels along a one dimensional track [2]. This controller has the property that the domain of stability for the inverted pendulum is much larger than the stability domain of other controllers that have been described. We have also constructed a theory that classifies the types of dynamical behavior that can be expected from "generic" hybrid systems with two dimensional charts and patches [3]. We rely upon the theory of iterations of one dimensional mappings and the classical theory of structural stability for planar vector fields. Despite these achievements, I have been dissatisfied with the prospects that control using hybrid systems will become a significant new direction that impacts control design in industrial settings. This talk is an attempt to explain my poorly formulated misgivings and raise questions about how one can surmount the obstacles that I think currently impede progress.

The information required to describe a hybrid system model is significantly greater than the information required to define a single smooth vector field on R^n. This is largely a practical matter, and it is far from trivial when one faces the task of designing controllers. Indeed, the use of a class of models for studying devices or physical phenomena depends upon both the importance of the models and the human effort that is required to study them. This effort is related both to model construction and to model analysis. I lay out issues from the perspective of a "dynamicist" rather than a "control theorist." Emphasis will be placed upon "nonlinearity" in the form of system behavior that is not present in strictly linear models. With a bit of chutzpah, I justify this attitude on the basis that when I listen to representatives of industry describe problems in control, I usually hear them talk about problems of design, sometimes "control configured design."

Determining the dynamics of nonlinear systems via analytic methods is a hard problem. There are no general techniques that lead to solutions, so investigation of many systems requires numerical computation. The "design" problem can be thought of as the selection of parameter values from a high dimensional parameter space that meet the problem specifications. When these specifications involve the dynamics of the system (e.g., find a stable limit cycle, ensure a minimal size to the basin of attraction for an attractor, etc.), qualitative analysis requires significant computation. Moreover, there is little guidance from theory about how one should gain insight from the problem formulation into its dynamics. Intuition and insight can come from study of particular classes of examples, but there are few situations in which this insight can be systematically summarized and taught to others. To some degree, the study of dynamics and control in an engineering context consists of such enterprises. Nonetheless, we frequently fool ourselves into believing that we know more than we do. Given the vast increases in computational capability that have occurred over the past two decades, it is reasonable to reexamine carefully how computation can become a larger part of the design process.

Tools for the investigation of dynamical systems have been developed, but I think that these are still primitive compared to what is possible on small computers today. The computational challenges are largely ones that involve algorithms rather than raw computing speed for highly detailed models. Much of the focus in the analysis of dynamical systems involves the computations of structures that have some type of mathematical singularity. Even for small problems, care must be exercised to formulate algorithms that compute these objects robustly. This difficulty is compounded by the fact that features that are important in the qualitative theory often appear on very small scales. Sorting out these phenomena from the artifacts of numerical integration is a challenge. The development of appropriate tools for investigating dynamical systems in the context of applications has been a focus of my research for several years. Progress in developing better algorithms and developing "packages" with more efficient user interfaces is being made, but this is a slow and tedious business as are most enterprises that involve numerical computation. The goal of these efforts is the development of environments that automatically compute and present the salient features of dynamical systems involving a small number of parameters.

Computers have become indispensable tools for solving two different kinds of "large" problems. On the one hand, there are problems in fluid dynamics where the description of problems with simple geometries is relatively compact, but one seeks to compute highly resolved spatial structures. The codes for such problems become large and complicated, yet conducting a simulation requires only moderate effort with "standard" codes. On the other hand, VLSI circuits may have millions of elements. Their design requires computer packages that provide a highly efficient environment and specialized languages for the description of the designs. The complexity of hybrid system models grows rapidly with the number of components unless the architecture of the models is constrained in ways that allow compact descriptions. One of the principle questions that I wish to raise is the question as to whether there are classes of models that are sufficiently general to satisfy design objectives, but sufficiently constrained that the design process becomes feasible with modest effort. Factored into the question is the issue as to whether one can develop sufficient insight into the model behavior from its description. In the case of VLSI design, this is clearly possible in regimes that do not strain physical limitations on the sizes or spacing of circuit elements.

My intuition is that different types of control problems will require different strategies and perhaps significantly different design tools. The tools that are so useful for VLSI design are not suitable to answer questions about transient stability of power systems, even though both deal with electrical circuits. Let me end my remarks with a proposed example that attempts to generalize the philosophy behind my work on stabilization of a double pendulum [2]. The problem I discuss is one of stabilization of an equilibrium that is difficult to stabilize with linear feedback control. In the case of the pendulum, the basin of attraction for the closed loop system is long and thin.

The strategy I propose for solving the stabilization problem is a conservative one based in part upon intuition about how animals solve such problems. The idea is to surround the desired basin of attraction for the equilibrium with a piecewise smooth boundary. To each component of the boundary, we look for a simple control law so that the closed loop vector field points into the basin of attraction from that component of the boundary. While this strategy is hardly "optimal" in terms of a cost function, it has the property that desired properties of the control law can be verified without having to integrate the vector field. Simply put, the system has a region of "comfort." When one crosses the boundary of the comfort region, the system is triggered to be alert and take corrective action. Note that one of the features of such a strategy that distinguishes it from piecewise smooth systems is that hysteresis is allowed. The discrete states of the system change on boundaries that are part of the controller description, but there is no requirement that the regions associated with different control laws should be disjoint.

Let me end with the question as to whether the type of control laws I have described can be readily designed to give satisfactory solutions to the problem of preventing catastrophic failure of many systems. Computationally, the task of verifying transversality of a vector field across a surface reduces to showing that a scalar function (dot product of the gradient of a function defining the boundary component with the vector field) does not change sign inside the boundary component. This is a problem that has been extensively studied using techniques of integral arithmetic, especially when the boundary components are parameterized by "rectangles" in a suitable coordinate system. In any case, this strategy avoids details of the dynamical behavior inside the basin of attraction and merely concentrates on whether we remain close to our objective. As I showed in the case of the pendulum, such a strategy can be easily augmented with more sophisticated procedures like linear feedback control where those strategies work well.

Acknowledgments: This research was partially supported by the Department of Energy, Office of Naval Research and the National Science Foundation.

References

[1] A. Back, J. Guckenheimer, and M. Myers, A Dynamical Simulation Facility for Hybrid Systems, in Springer Lecture Notes in Comp. Sci., 736, 255-267,1993.

[2] J. Guckenheimer, A Robust Hybrid Stabilization Strategy for Equilibria, IEEE Trans. Automatic Control, 40, 321-326, 1995.

[3] J. Guckenheimer and S. Johnson, Planar hybrid systems, in Springer Lecture Notes in Comp. Sci., 999, 1995.

Embedded System Design and Hybrid Systems

A. Sangiovanni-Vincentelli
Department of Electrical Engineering and Computer Sciences
University of California at Berkeley

Abstract

Reactive real-time embedded systems are pervasive in the electronics system industry. The design of embedded controllers, i.e., of embedded systems that perform control actions on physical systems, is addressed. Hybrid and heterogeneous systems are used as mathematical models to define a design methodology that could shorten considerably the time from the conception of the system to its implementation.

1 Introduction

In this paper, we focus our attention on the design of embedded systems. Embedded systems cover a broad range of applications, from microwave ovens and watches to telecommunication network management and control functions. *Embedded systems* are electronic components of a physical system such as a vehicle, a household appliance, or a communications system, that typically

- monitor variables of the physical system such as temperature, pressure, traffic, chemical composition, position, or speed via sensor measurements;

- process this information making use of one or more mathematical models of the physical system;

- output signals, such as controls for power circuitry to open or close valves, to ignite sparks, to open or close switches, that influence the behavior of the physical system to control its function and to optimize its performance.

These embedded systems are integrated onto the physical system itself and hidden from the user. The most challenging embedded systems, also called reactive real-time systems, are the ones that have to control the behavior of physical systems with tight timing constraints. In the remainder of this paper, we call reactive real-time embedded systems *embedded controllers*.

The design of these systems consists of first specifying their functionality and the performance indices used to measure the quality of the design. Then a set of tasks, including control strategies, implementing the functionality of the system is derived. The implementation of such systems can vary from a full hardware configuration, (all the tasks to be performed by the embedded system are translated into a suitable set of customized integrated circuits), to a full software implementation, (all the tasks are implemented as software routines run on a programmable component such as a micro-processor or a Digital Signal Processor (DSP)) or, most often, to a mix of hardware and software configuration, where most of the "intelligence" is implemented in software and involves discrete variables, while a limited amount of signal management and signal transformations from the sensors to the micro-controller and from the micro-controller to the actuators may be implemented in hardware.

In general, the design process is fairly well understood at the implementation stage while it is far from being fully satisfactory at the specification and task determination level. For example, specifications are almost always given in informal, imprecise terms, often using natural languages. For example, in automotive applications, *drivability* is one of the most important criteria to value the quality of an engine control system. Drivability is a qualitative criterion related to the *feel* a driver has of the behavior of the engine. Hence, it is very difficult to analyze the correctness of the design by evaluating its conformance to the specifications. Control tasks are often very simple qualitative heuristic rules. These rules can be thought of as derived from a simplified model of the system to be controlled. The performance of the control actions though have to be assessed on a detailed model of the system or on a prototype implementation.

Once the tasks that the controller has to perform have been defined, the design problem consists of finding a hardware or a software implementation. The implementation has to satisfy constraints on cost, power, and timing that restrict the range of possible choices. If we had as many processing elements as tasks then the design would be relatively straightforward. However, the resources to be used are limited and have to be shared among different tasks. Next, the choice of processing units and the scheduling of tasks onto these units is the core of the design problem. The decisions on which processing units to use and on the scheduling strategy have to be validated. This is far from being easy even though at this stage of the design we are essentially concerned with the discrete domain.

Tasks are often classified according to the kind of computation they involve in implementation. Finite State Machines (FSMs), Data Flow (DF) graphs, Discrete Event Systems (DES) are model of computations commonly used. They are all characterized by different mathematical properties and by different representation power. A unified model which includes different models of computations and can be efficiently manipulated has not been developed yet and there are reasons to believe that such a model will never be developed [11]. However, when mixing different models of computation, it is difficult to assess the properties of the overall system. The situation is further complicated by the need of modeling interconnected and hierarchical systems where models of computation mix freely. Mixed discrete systems are often called *heterogeneous systems*. Only recently the study of the formal properties of heterogeneous systems has caught the attention of researchers.

The goal of this paper is to propose a design methodology for embedded controllers and to present several open problems that, if solved, should greatly help the design of complex systems. In particular, the following issues will be explored:

- how to model the determination of the control strategies as a hybrid system problem;

- how to define and represent the finite component of the hybrid system;

- how to validate the design of an embedded controller.

In the conclusions, I will present some thoughts about the *implementation* of embedded controllers and the associated synthesis problems. This has been a very active research area for the past 5 years. We are now at a stage (see [25, 8, 9, 31]) where we can synthesize and verify almost automatically complex embedded systems with performance and dimensions equal to or better than manual designs for "control-dominated" application with integrated design environments such as POLIS developed as a collaboration among University of California at Berkeley, Magneti Marelli, the second largest electronic subsystems provider to the automotive industry in Europe, and Cadence, the leading company in design automation of electronic systems with the support of Daimler-Benz, Hitachi and SGS-Thomson.

The point of views presented here reflect my past experience in the design of and design tools for complex digital integrated systems and in the industry dealing with automotive embedded controllers.

2 Embedded Controller Design

It has been argued that VLSI design can deal with millions of components efficiently while control strategy design has difficulty even with systems of medium complexity [18]. In this section, the key points that make the design of VLSI circuits possible are pointed out and an approach to embedded controller design is presented that leverages some similarities with the VLSI case.

2.1 VLSI Design

In the case of VLSI devices such as micro-processors, the design process has been fairly successful because of:

- *Abstraction mechanisms*, made possible by the use of digital circuits, that allow to hide details related to the physical implementation of the circuits, and to "see" the design as a set of Boolean functions. For example, were we to consider a micro-processor as complex as the Pentium at the differential equation level (where all signals in the device are represented as continuous waveforms), 10 years of compute time of the most powerful computer would not have sufficed to verify a fraction of its behavior with detailed circuit simulators such as SPICE [1].

- *Hardware description languages*, that, although still not theoretically sound, offer a mostly precise and unambiguous way of specifying the functionality of the design;

- *Validation tools* such as simulation, emulation and formal verification that are capable of handling very large designs in affordable, albeit sometimes very long, compute time;

- *Synthesis tools* that allow to map in a nearly optimal fashion Boolean functions into gates available in libraries offered by IC manufacturers.

To make embedded controller design more efficient, *we argue that a similar set of paradigms and tools must be developed.* In particular, I believe that abstraction and validation are the keys to complexity management in design [37].

2.2   Abstraction in Embedded Controller Design

Embedded controllers are characterized by cost, performance and size constraints. In important application domains such as automotive electronics, embedded controllers consist of

- micro-controllers that are relatively incapable of carrying out complex calculations without violating real-time constraints;

- sensors that give approximate measurements, often just detecting whether a signal is above or below a threshold;

- actuators that often have a finite number of settings, sometimes just two (on/off);

Because of its limited compute power, the controller must issue controls based on simple models of the system to be controlled (the plant) that may be effective only in limited ranges of operation. In addition, often the control signals and the output variables of the plant as observed by the sensors can take only a finite set of values. In these cases, *the controller acts as a mapping from a finite set to a finite set.* A big advantage of considering the problem in a finite setting, is that the number of alternatives to consider is limited and that exhaustive analysis is, at least in principle, possible.

This abstraction is clearly similar to the Boolean abstraction in VLSI design. However, in VLSI design this abstraction is validated by a design methodology that exploits the basic principle of operation of the transistor and that has been brought to a high level of sophistication. In the practice of embedded system design the finite approximation is still applied in a heuristic fashion.

---

[1]SPICE is a circuit simulator developed at Berkeley in the late 1970s where the circuit is represented with basic components such as transistors, resistors and capacitors and the circuit variables are computed by solving a system of mixed algebraic-differential equations.

Discrete-variable mathematical objects such as Finite State Machines (FSMs), Discrete Event Dynamical Systems (DEDS), synchronous languages and automata together with the associated specification languages and manipulation tools (e.g., StateCharts and StateMate [19, 15], Signal and Lustre [3]) have been used rather extensively by industrial control and automotive control practitioners (e.g. [4, 2, 38, 14]).

Note that in most applications, while the embedded controller and the control strategies can be seen in a discrete setting, the "real" plants have almost always continuous dynamics. Hence, to design and to validate embedded systems correctly, we have to "interface" the discrete and the continuous world. The concepts developed in the field of hybrid system [2] analysis and design are a step towards dealing with the discrete abstraction in a more formal setting.

The view of an embedded controller as a discrete system connected to a continuous plant is used in several papers such as, among others, the ones by Nerode and Kohn (e.g., [23]) and by Caines(e.g., [12]). However, a more general form of hybrid systems, as also proposed by Caines, corresponds to a form of hierarchical control, where the discrete system interacts with a set of subsystems which can be discrete themselves or continuous consisting of a continuous plant controlled by a continuous controller. The structure with a discrete controller connected to a continuous system subject to "internal" feedback can be used to represent switching controllers in the sense of Morse [27]. If the subsystems are discrete, they can represent another level of the hierarchy where each subsystem is connected to another set of subsystems of the same form.

We now consider an embedded controller representation as a "hybrid" system. The representation is willingly generic since it is intended to cover a range of applications. Nevertheless it is useful to expose problems and to describe approaches to design.

The discrete controller, in its abstracted (sometimes referred to "high level") representation, is a finite state system where:

- States correspond to different "internal views" (or models) of the plant and/or to different control actions. For example, *switching control* as defined by Morse [27] involves a supervisor, a finite state system, that decides which controller in a finite set of linear candidates to apply based on the region of operation of the plant. A state of the supervisor in this case corresponds to the linear controller chosen. In gear control in automatic transmission, the controller selects among a finite number of gears the one to apply according to a set of operating conditions of the vehicle. In this case, a state corresponds to the selected gear.

- The transition structure of the finite state system is (at least in part) determined by the dynamics of the plant and/or the lower-level controller (which may be continuous). A transition to a new state is "guarded" by a condition on the trajectory of the plant or of the plant as controlled

---

[2]The term hybrid system is used here in its broadest contest as a system described by a mixed set of discrete and continuous equations

by the lower level controller. Whenever the trajectory triggers an input to the controller indicating that a new region of operation is reached, the transition to a new state is enabled and a new control is applied that depends on the reached state. In this framework the regions of operations may be defined either by the structure of the controller, e.g., the choices available to the supervisor in switching control or by the structure of the plant, e.g., the transmission system of an automobile whose dynamics change discontinuously according to the position of the clutch.

The properties of the behavior of the closed loop system depend upon the dynamics of the plant, upon the discretization of the observed variables and controls and upon the control laws applied by the controller. Many results on hybrid systems are concerned with the *analysis of the behavior and properties of a given continuous-discrete structure*, e.g. [29, 1, 27, 12]. While analysis is essential to assess the quality of the design, I believe that emphasis should be now placed on the embedded controller *design* problem that is concerned with *deriving the discrete abstraction so that a given set of properties holds.*

3   A Design Methodology for Embedded Controllers

A methodology for the design of embedded controllers has to reflect the characteristics of the building blocks available to the designer. The compute power of most micro-controllers is limited, hence the approximations of the plant and the calculations needed to determine the control to apply, have to be "inexpensive". Furthermore the control signals must take into account the simple structure of the actuators available for most of the applications. In addition, the measurements have to be fast and simple.

A design methodology along the lines presented in the previous sections consists of the following steps:

- Define the control goal, e.g., desired behavior, stability, robustness;

- Define the control constraints, e.g., time to reach the desired behavior, cost of the architecture, types and number of control signals and of measurements available;

- Determine the decomposition into regions of the domain of the variables defining the behavior of the plant.

- Determine how to switch from one region to another, e.g., how to map the state of one approximation when a boundary of a region is reached into the "initial" state of the new region and how to apply the new control signals.

- Analyze the partition with respect to the goals and constraints.

- Implement the control strategy in software and/or hardware components chosen in terms of characteristics that include cost, computing power, and reliability.

This design methodology is implicitly followed in several applications, heuristic criteria are widely used and little attention is paid to guarantee that the approximations used are consistent with the control objectives. In addition, little has been achieved in terms of how to switch from one of the region to another so that the desired properties hold globally.

Two steps are of particular importance to shorten development time and to guarantee that the specifications are met:

- the *decomposition into regions* that provides the bases of what we call the finite abstraction;

- the *validation methods* that are closely linked to the finite abstraction.

4 Decomposition into Regions

The design problem examined here is *how to define the regions of operations of the plant and how to determine a control law based on an approximated model of the plant in each region.* [3]

The following considerations should be kept in mind when determining the set of regions:

- the boundaries of the regions should be described *compactly* in terms of variables that can be measured with the set of available sensors;

- the behavior of the plant *with respect to the control goals* should be approximated with a "representative" model characterized by:

    - compact mathematical description,

    - the existence of simple, but with proven properties, control laws that achieve the desired behavior when applied to the approximate model.

For example, a representative model can be a linearized version of the plant for which compact representations and control laws with guaranteed properties are available, or even a model of the form $\dot{x} = c$, c a constant; the approximations used in each region do not have to be all the same, in the sense that an approximation in a region can be linear and of the form $\dot{x} = c$ in another;

- the number of regions is minimized, since the parameters that are used to represent these regions may take a fairly large amount of memory, a rather expensive item for most of the applications.

---

[3] A proposal in this direction has been formulated in [36] for the design of automotive embedded controllers.

Remarks

- The selection strategy described above involves trading off different aspects: for example, simplicity in the mathematical representation of the region boundaries and of the approximations within a region often is competing with the minimization of the number of regions. Because of the complexity of these trade-offs, it is difficult to find general strategies for region determination: the abstractions involved are likely to be highly problem dependent.

- In the selection of the regions of operation and of the approximate models, an approach is suggested that goes beyond standard approximation theory: the approximations do not have to be accurate with respect to the actual trajectory of the plant, they must be such that the control signals applied by the controller make the actual plant behave as desired. In spirit, this strategy is close to the mechanisms used in formal verification to abstract away details of the behavior of the system that are inconsequential with respect to the property being tested [37].

- The decomposition need not be a partition: sometimes it is more effective to decompose the domain into overlapping regions For example, in [17] the inverse pendulum problem is solved by decomposing the state space of the pendulum into partially overlapping regions where the dynamics of the pendulum are approximated by $\dot{x} = c$ where c is a constant that depends on the region. Note that the abstraction and the procedure followed to obtain the control strategy guarantee that the solution is achieved with strong stability properties.

- The partition may be explicitly defined *a priori* or may be implicitly defined by a set of rules followed by the controller (in this case we have a form of "adaptive" control).

I believe it is interesting to look at some well-known numerical analysis techniques as examples of decomposition and approximation strategies.

4.1   Nonlinear Algebraic Equations

Consider an iterative algorithm for the solution of $f(x) = 0$, where $x \in R^n$ and $f : R^n \to R^n$, continuously differentiable and with non singular Jacobian at the solution $\hat{x}$. Consider the plant to be described by f and the design goal to determine a control strategy to bring the plant from an initial point x^0 to $\hat{x}$. Most of the numerical techniques are iterative and obtain successive approximations to $\hat{x}$, $x^1, ..., x^k, ...$ At a generic x^k, the "control" determines a direction to follow in the search for $\hat{x}$ and a rule that stops the search along that direction i.e., that tells when x^{k+1} is reached and it is time to compute a new direction. The sequence of iterates defines the trajectory of the plant, the controller acts in discrete fashion and the states of the discrete system are implicitly defined by the algorithm. For example, the Newton-Raphson algorithm consists of the following iteration:

$$J(x^k)(x^{k+1} - x^k) = f(x^k), k = 0, ..., \tag{1}$$

where $J(x^{k+1})$ is the Jacobian of f at the $k + 1$st iteration. Then the search direction is given by the Jacobian, the rule by which we decide to recompute a new direction and that determines the iterate is given by Eq. 1. The controller acts by modeling the plant as a linear system and "solving" the linear system to determine the search direction and the new iterate.

Other popular techniques such as finite differences differ from Newton-Raphson in the way the search direction is computed to simplify the computational complexity of the algorithm.

A different class of techniques proposed when solving VLSI circuit equations is based on the approximation of the nonlinear equations describing the circuit components with a piecewise linear map. Then the set of nonlinear equations describing the circuit is also piecewise linear but the boundaries of the different linear regions are implicitly defined by the boundaries of the linear regions of the circuit components. The solution process consists of moving along the direction determined by the combination of the linear equations of the devices until the first boundary of one of the component is hit. At this point the search is stopped and a new direction is computed. This class has significant computational advantages since no derivative needs to be computed and the new search direction can be found by using rank-one updates. Note that this method is based on a pre-computed approximation, while the Newton-Raphson method and its derivatives can be seen as an "adaptive" approximation mechanism.

4.2 Ordinary Differential Equations

Consider a set of ordinary differential equations of the form:

$$\dot{x} = f(x), \ x(t_0) = x_0. \tag{2}$$

Popular integration methods are the linear multi-step methods including the Backward Euler (BE) method. BE computes the solution approximating 2 as follows:

$$x_{k+1} - x_k = h_k f(x_{k+1}), \ k = 0, 1, ... \tag{3}$$

where h_k is the time-step at step k. Eq. 3 is solved by Newton-Raphson or similar techniques. The more sophisticated implementations of linear multi-step methods use an adaptime time-step control method that determines the time-step by estimating how accurate the approximation of Eq. 3 is. After x_{k+1} has been computed, an error evaluation is carried out and, if the error with respect to the "real" solution is too large with respect to a "reference error" determined by the user, the time-step is shrunk and the current x_{k+1} rejected. In this case the analogy to a control problem is even closer. The controller determines the sequences of time-steps and evaluates the plant response with respect to the reference signal (error) before issuing a new control. Once more the controller actions are discrete and they are computed by the controller using an approximation of the real plant. Note that in the interval of time $t_{k+1} - t_k (= h_k)$ the plant is characterized by $\dot{x} = constant$, the approximation that is used in [17] in the different regions of operation of the plant.

In numerical integration, the criterion used to switch from one "plant" approximation to another is the accuracy with which the "real" solution is approximated by the sequence $x_0, x_1, ..., x_k, ...$ In control, I believe that the criterion to switch should be how closely the desired behavior of the "real" plant is matched. For example, the desired behavior may be related only to whether an attractor is found and not to the actual trajectories.

An example of this kind of criteria comes from VLSI Transistor-level circuit simulation. The Newton-Raphson method is known to have certain convergence problems when the initial guess is fairly distant from the solution. In VLSI design, DC (direct current) solutions of the circuit are needed to study the behavior of the circuit as well as to compute the initial conditions for the calculation of transient responses. SPICE solves the DC equations of the circuit with Newton-Raphson techniques and, sometimes, has difficulty converging due to the difficulty of guessing where the solution lies.

One of the most used techniques to improve convergence of the DC solution is to convert the DC problem into a "pseudo"-transient problem, where the initial values of the circuit variables are all known (for example zero) and the desired DC solution is the equilibrium point of the pseudo-transient problem. If the equations are continuous and the time step is chosen sufficiently small, x_{k+1} is close to x_k and the Newton-Raphson iteration has better chances to converge using x_k as initial guess for x_{k+1}. In this case, the accuracy with which the trajectory is computed is of no interest. What is important is to reach the equilibrium point with the minimum amount of computation. The time-step is then controlled by the number of iterations needed to compute x_{k+1} and not by the error criterion.

5   Validation

An important consideration in the design of embedded controllers is related to system validation: it is common knowledge that validation has to be performed with much more attention than in the past. In fact, most embedded systems have strong safety requirements, human lives may be exposed. While in the past the somewhat limited complexity of the designs and the relatively long design cycles allowed the use of efficient *ad hoc* validation techniques, such as prototyping and breadboarding, high complexity and time-to-market pressure are placing a great burden on designers to validate their designs early and with techniques that allow little space for errors.

In this context, it is necessary to develop a validation methodology based on a combination of:

- *A priori guaranteed performance* in view of the approximations and of the structure of the plant representation.

- *Formal verification* (see [37]), a technique that constructs a proof that the given strategy achieves all (or some) of the desired goals within the set of constraints considering the particular instance of the plant to be controlled [4].

- *Simulation* of the given embedded controller and the plant.

- *Emulation*.

All the approaches to validation are based on a representation at various levels of abstraction of the embedded controller and of the plant. Discrete representations play an essential role. Before reviewing briefly the various validation techniques the discrete representation issue will be addressed.

5.1   Discrete Representations: Model of Computation

This section reviews some of the mathematical representations (*Model of Computations* MoC) for finite systems following the approach of [25]. The most used classification subdivides finite representations into *control-dominated* and *data-flow*. Control models are based on Finite State Machines (FSMs) and privilege the aspect of timely reaction to an input event. Data Flow (DF) models represent model of computations where data values and their efficient manipulation are of importance. The basic FSM models are not Turing complete and hence their representation power is limited, while basic DF models are Turing complete and have strong representation power. On the other hand, many problems on DF models are undecidable, while for FSM problems there are procedures that yield solutions or decide that no solutions exists in finite time. Several extensions have been proposed for FSMs (Extended Finite State Machines or EFSMs) that make them able to deal with data but that are Turing complete and hence suffer of the same computational limitation as DF models. On the other hand, DF models have been restricted (Synchronous DF or SDF) to eliminate undecidability. Describing the computation involved in the discrete domain in DF or FSM format biases the implementation. For example, software DF computations are best implemented on DSPs while FSMs are better implemented on micro-controllers. Similar distinctions can be made when implementing the tasks in hardware.

Even if FSMs and DF models are finite, the number of states needed to describe the model may be very large. A way of beating the curse of representation complexity is to use hierarchy and decomposition [37]. Hierarchy and decomposition are commonly used by designers when describing complex discrete systems. In this case, states at the highest level of the hierarchy represent themselves FSMs that operate at lower levels. In addition, a behavior may be represented as the interconnection of one or more concurrent FSMs that communicate with a particular protocol. For example, in the automotive industry, it is by now fairly common to use hierarchical communicating

---

[4]Note that it may be impossible to prove that a property holds with respect to all plant models with a given structure but it may be possible to prove that a particular instance of the model satisfies the desired properties

FSMs for embedded controllers with a language such as StateCharts [15]. Several versions of hierarchical interconnected FSMs have been proposed, mostly regarding different communication mechanisms. The more general the communication mechanism, the more difficult is to prove general properties but the more adherent to reality is the model [25].

We will now examine in more details the aspects of finite representations making use of the notion of *hierarchical process network* as a framework for the analysis.

A process network is a structural model that consists of a set of *nodes* representing *processes* and/or *sub-networks*, and a set of *edges* representing *communication links*. Its semantics is determined by: a node model and a communication model.

Communication among processes involves *writing* and *reading* information among partners. Sometimes just the act of reading or writing is used for synchronization purposes, and the transmitted information is ignored. We distinguish between two basic mechanisms, depending on whether a single piece of information (written once) can be read once or many times:

1. *Destructive read*, if the information written by one partner can be read *at most once* by the other partner. In this case, we further distinguish four cases, depending on whether the write and read operations are *blocking* or not. A write operation is blocking if it cannot be completed if the current or previous information would be lost (e.g., because a buffer is full or because a partner is not ready to perform an unbuffered communication). A read operation is blocking if it cannot be completed until some data has been or is being written by a partner.

2. *Non-destructive read*, if the information written by one partner can be read *many times* by the other partner. The simplest form is a variable shared among partners.

Communication can be:

1. *Synchronous* if the write and read operations *must* occur simultaneously, and

2. *Asynchronous*, otherwise. In this case there can be a finite or an infinite number of *buffer locations*, where the information is stored between the instant in which it is written and the instant in which it is read.

Note that non-destructive read is generally asynchronous, and implies a buffer with a single cell (the shared variable).

The standard distinctions between *uni-directional* and *bi-directional* communication, as well as between *point-to-point* and *broadcast*, are also useful for classification purposes.

Each node of a network can be:

1. A process ("leaf" node), which reads data from incoming edges, computes a function based on those data and writes information along outgoing edges.

2. An instance of a network. The instantiation mechanism can be *static* if the total number of nodes in a given hierarchy is fixed, or *dynamic* if it is variable.

Dynamic instantiation may increase the expressive power of the model, while static instantiation is only a convenient mechanism to reduce the size of the representation. We can further distinguish two cases of dynamic instantiation:

(a) recursive instantiation, in which a node can instantiate a network that includes the node itself,

(b) non-recursive instantiation.

In both cases, an appropriate mechanism to dynamically update the communication structure must also be provided. Generally this occurs via a formal/actual edge substitution mechanism. This means that a network of processes has some input and output edges which are connected with processes only at one end. The instantiating node has exactly the same number and type of input and output edges, and corresponding edges are connected together during the instantiation procedure.

We can now proceed to classify various models described in the literature along these lines.

5.1.1 Finite State Machines

Classical Finite State Machines A Finite State Machine (FSM) is a process whose input/output function can be computed by a finite automaton. ([21]). The edges of the automaton are labeled with input/output data pairs. A network of FSMs uses broadcast *synchronous non-blocking* communication among the leaf nodes.

Extended Finite State Machines A network of *extended synchronous FSMs* is similar to a network of FSMs, but also allows non-destructive synchronous communication of *integer values*. The (no longer finite) state of each EFSM is the Cartesian product of a finite-valued variable and a set of integer variables. The next state and output functions of the nodes are usually described, for notational convenience, using relational, arithmetic and Boolean operators. The descriptive power of EFSMs is equivalent to that of Turing machines. (by reduction of counter machines [21]).

Sometimes in the literature, FSMs which can perform arithmetic and relational tests on integer *subrange* variables are referred to as "extended FSMs". In this case, though, the set of possible states of the system is finite. Hence this "extension" does not augment the power of FSMs beyond regular languages. The only advantage is that the next state and output functions can be written more compactly by using arithmetic and relational operators.

Note that hybrid systems of the kind addressed in this paper have structure similar to EFSMs: the next state and output functions of the high-level discrete representation of the controller depend on the evolution of dynamical systems described by a set of ordinary differential equations.

It has been pointed out in other papers in this book (e.g. [18, 4]) that discrete representations suffer from an explosion in complexity with respect to continuous representations. It is indeed true if the representations are thought of as a set of states and transitions explicitly described. However, significant advances have been made in representing and manipulating FSMs *implicitly* defined [7, 35]. These implicit representations are based on Binary Decision Diagrams, a data structure used to describe Boolean functions [5] and on characteristic functions of the transition relation of the FSM. [5] "Small" implicit descriptions are known to exist for VERY large FSMs, i.e., FSMs with 10^{20} states and more. The notion of small representation has to be taken in the context of these modern techniques for FSM manipulations. I believe that an in-depth study has still to be undertaken to classify implicit discrete representations arising in hybrid systems.

5.1.2   FSM-based languages

Esterel   Esterel [3] belongs to a family of synchronous languages which includes Lustre and Signal. The *synchronous hypothesis*, common to all these languages, states that time is a sequence of instants, between which nothing interesting occurs. In each instant, some events occur in the environment, and a reaction is computed *instantly* by the modelled system. This means that computation and internal communication take no time. This hypothesis is very convenient, because it allows modeling the complete system as a single FSM. This has the advantage that the behavior is totally predictable, because there is no problem of synchronizing or interleaving concurrent processes. One problem is that this hypothesis entails the implicit or explicit computation of a fairly large FSM, which may be problematic for large specifications with lots of concurrency.

The Esterel language is very simple and includes constructs for hierarchy, pre-emption, concurrency and sequencing. Hierarchy is handled via procedure calls and module instantiations, like in normal programming languages. Pre-emption consists of two basic exception raising and exception handling constructs, one which allows the module to terminate its computation for the current instant, and one that does not. Concurrency is specified by using a parallel composition construct (otherwise the module behavior is sequential but instantaneous). To avoid exponential blow up, the current Esterel compiler preserves concurrency, as much as possible, all the way to the implementation level. Communication, as in many models for reactive systems, is based on synchronous non-blocking read and non-blocking write (possibly with associated

---

[5] A transition relation is a Boolean relation defined on $S \times S \times I$, where S is the set of states of the FSM and I is the set of inputs. Given the bits of the encoding of a state and the inputs at that states, it computes the bits of the encoding of the states that can be reached applying the inputs applied to the given state. A characteristic function of the transition relation is a Boolean function $\Xi : S \times S \times I \to \{0,1\}$ that is 1 for all states and inputs that are related by the transition relation.

values).

Like all FSM based control-dominated models, data manipulation cannot be done very naturally. Also, having a synchronous model makes it hard to specify components of a system that operates at different rates. Hence Esterel by itself can only be used for process level modeling while the system level modeling of asynchronous communicating processes should be done using another formalism. The perfect synchrony hypothesis simplifies the design process, but also forces the timing constraints to be specified outside Esterel. The rigor and simplicity of specifying the behavior of individual modules makes the description of the entire system, including constraints, somewhat more difficult.

StateCharts StateCharts [15] is a graphical specification language that extends classical Finite State Machines by allowing hierarchical decomposition, timing specifications, concurrency, synchronization, and subroutines. Hierarchical decomposition is accomplished by clustering states through the use of AND or OR operations. AND clustering models concurrency (it corresponds to the classical FSM product), while OR clustering models sequencing (the FSM can be in only one of several OR states at a time). Transitions in and out of states or hierarchical states can be specified without restriction. The emphasis of this hierarchical composition is on condensing information. Timing is specified by allowing the use of linear inequalities as timing constraints on states at any hierarchical level. Concurrent FSMs can be synchronized through transition edges either explicitly with conditions, or implicitly by going through hierarchical states. An implicit stack allows a state to "call" a sequence of states and then "return" to the calling state.

The underlying mathematical model for StateCharts is an extended synchronous FSM. Hence, it has the same problem as ESTEREL with data-flow specification and system representation in a co-design framework[6]. In addition, the communication protocol implies that communications are synchronous, while in several practical applications, this is a serious restriction. he semantic of transitions that involve several levels of the hierarchy at once is not always clear and leaves ambiguities that make life difficult when trying to prove properties of the description. Further, if formal verification techniques are to be used, we are often forced to flatten and aggregate the representation resulting in very large FSMs even using implicit techniques to perform the flattening.

Despite these limitations, StateCharts have a natural implementation as software programs, and substantial research has been done on synthesizing StateCharts as hardware components. StateCharts can be a very good visual language for specifying the synchronous components in an asynchronous system.

---

[6] A descendent of the original StateCharts, called Statemate, utilizes StateCharts for control, and Activity Charts (a graphical hierarchical data-flow specification) for data.

5.1.3   Control/data flow-based models

We next turn our attention to process modeling with Control/Data Flow Graphs
(CDFGs). While FSM-based specification naturally separates hardware from
software at the task or process level, the use of control/data flow graphs facilitates partitioning at the operation level. The former is usually called coarse-
grain partitioning and the latter is usually called fine-grain partitioning. The
next two approaches are both geared toward automatic fine-grain partitioning.

Flow Graphs   A hierarchical Flow Graph model is used for all partitioning,
scheduling and synthesis manipulations in the Vulcan system [16]. The model
consists of nodes and edges, where nodes denote operations and edges denote
dependencies. Operations can be conditional operations, computational operations, wait, and loop. Conditional operations allow data-dependent paths
in the graph model. Computational operations includes logical, arithmetic,
and relational operations. Both non-blocking and blocking read operations,
called "receive" and "wait" respectively, are available. The blocking read can
result in a non-deterministic delay. Data-dependent loops can also have a non-
deterministic delay.

5.1.4   Petri nets and data-flow networks

Petri nets   A Petri net ([32], [28]) is a flat hierarchy. Nodes (usually called
"transitions") "fire" by reading from each input and writing to each output.
Communication is *asynchronous*, with *infinite buffers* (usually called "places"),
with blocking read and non-blocking write. In the pure Petri net model no
value is transferred by communications, the only significant information being
the possible transition firing sequences.

Data flow networks   A data-flow network ([22], [13], [33]) is similar to
a Petri net, but each communication can transfer a value (e.g., integer or
Boolean), and buffers have FIFO semantics. Little or no restriction is posed
on the function computed by each leaf node in response to a set of communications on its input edges, apart from terminating in finite time. Note that
due to the blocking read communication, a node cannot test an input buffer
for emptiness. Nonetheless nodes can decide from which input(s) and to which
output(s) communications will be performed. [7]. The main result concerning
data flow networks, which is directly connected with the blocking read communication mechanism, is their *determinacy*. This means that the *sequence*
of output values at each node does not depend on the order in which nodes
are selected to execute computations and communications, as long as the order
does not cause deadlock (e.g., by scheduling a process when one of the buffers
it will read from is empty) or starvation (e.g., by never scheduling a process
with non-empty input buffers).

---

[7]Networks in which this choice is not possible are called *synchronous* in [6].

5.2 Formal Verification

Formal verification aims at proving properties of the design that are true independent of input choice, while simulation is as good as the test inputs that are fed to the simulator. The design and the properties to be verified have to be specified in formal terms, using some of the models described in the previous section. Unfortunately, formal verification today requires skills in formulating properties and apply design exploration techniques that are beyond the reach of most designers. Hence, in "bleeding" edge applications where errors can be extremely expensive (think of the cost of recalling millions of Pentiun chips that exhibited a bug in the floating point unit) designers are teamed up with formal verification experts. In the case of embedded controllers the description of the behavior of the plant with differential equations makes formal verification a real theoretical challenge even beyond the normal challenge that this technique has to face in the digital circuit design case [1]. *In passim* I would like to point out that these methods are not a substitute for a well formed theory. If indeed some general properties of a control algorithm can be proven, it is a waste of time and effort to try to use the expensive procedures needed for formal verification. In some sense, formal verification is a brute force approach and should be used only when necessary and possible. I believe that the finite abstraction is key for the successful application of formal verification to the validation of embedded controllers. In fact, the approaches that can be reasonable made automatic such as *model matching* and *language containment*(see e.g. [24, 10, 26, 35]) are based on the computation of all reachable states, that is feasible almost exclusively for finite systems or for other systems whose abstracted behavior is equivalent to finite systems. For example, the concept of *timed automata* was introduced to represent the time behavior of a system with an FSM in [1].

5.3 Simulation

Simulation has been the work-horse of design verification in modern engineering. Efficient simulation is the result of a trade-off between simulation speed and accuracy. For example, when simulating digital VLSI circuit it is often an over-kill to represent the components of the design at the transistor level. Such a representation would require very large compute time since it involves the numerical solution of a system of ordinary differential equations with a number of variables that is proportional to the number of circuit components. If the system to be simulated is finite (or can be approximated by a finite model e.g., a digital circuit) and the model of computation allows to abstract away timing considerations, then simulation can be as fast as the actual implementation [31] or close to it [37]. Simulation in the design of embedded controllers involves the simulation of the controller, of the approximated model of the plant the controller uses to determine the control law and of an accurate model of the plant. While the controller and the approximated model can be simulated very fast, the accurate model of the plant often involves differential equations. Hence simulation requires the co-ordinated solution of a mixed set of equations. Mixed-mode simulation has been the object of much research in the VLSI community for many years. The main difficulty is how to synchronize fast execution of the discrete model with the much slower execution of the continuous model. To maximize speed, decoupling of the two sets has been often considered and event-driven scheduling of the two types of execution used. Care has to be exercised in choosing the execution sequence to maintain given accuracy requirements [30]. In automotive applications [34], the plant is as complex as a combustion engine. Models of different levels of accuracy have been devised and evaluated with tools such as Matrixx or Matlab at lower levels of accuracy or Saber at higher levels of accuracy. In both cases simulation can not replace experimentation to validate the design.

5.4   Emulation

Recently emulation has been used to shorten the development cycle and to catch design errors earlier in the design. In this case, a board containing the actual micro-processor that will be used in the application surrounded by interface circuitry is connected with actuators and sensors which are in turn hooked to a "box" containing DSPs that process the discretized set of equations describing the engine. Validation can now be performed allowing much more accurate and faster assessment of system behavior.

6   Concluding Remarks

Issues about the high-level design of embedded controllers have been presented with particular emphasis on discrete abstractions. A design methodology and a set of models of computation have been presented. After the high-level design is completed the hardware and software components of the embedded system must be implemented. Hardware-software co-design is a fairly new branch of design technology that aims at developing a set of algorithms and tools to shorten the design time and make designs more robust.

The inputs to the problem are a specification, a set of resources and possibly a mapping onto an architecture. The objective is to realize the specification with the minimum cost.

Area and code size must be balanced against performance, which often dominates due to the real-time characteristics of many embedded systems. Cost considerations generally suggest the use of software running on off-the-shelf processors, whenever possible. This choice, among other things, allows one to separate the software from the hardware synthesis process, relying on some form of pre-designed or customized interfacing mechanism.

Performance and flexibility can be improved simultaneously using re-programmable hardware, like Field Programmable Gate Arrays. FPGAs can be re-programmed either off-line, as is commonly done to replace embedded software by changing a ROM, or on-line, to speed up the currently executed algorithm.

The hardware synthesis task for ASICs used in embedded systems (whether they are implemented on FPGAs or not) is generally performed according to the classical high-level and logic synthesis methods.

The software synthesis task for embedded systems, on the other hand, is a relatively new problem. Traditionally, software synthesis has been regarded with suspicion, mainly due to the excessive claims made in earlier times. In fact, the problem is much more constrained in the case of embedded systems than in the case of general-purpose computing. For example, embedded software cannot use unconstrained dynamic memory allocation nor virtual memory. This is due to the physical constraints (the absence of a swapping device), to real-time constraints, and to the need to partition the specification between software and hardware. For some highly critical applications even the use of a stack may be forbidden, and everything must be dealt with by polling and static variables. Algorithms also tend to be simpler, with a clear division into cooperating tasks, each solving one specific problem from digital filtering to control and decision-making. In particular, the problem of translating cooperating Finite State Machines into a software implementation has been successfully solved in a number of ways.

Software synthesis methods that have been proposed in the literature can be classified according to the following general lines:

1. the specification formalism, which may be more or less similar to a programming language,

2. the specification- and the implementation-level interfacing mechanisms,

3. the scheduling method.

Some sort of scheduling is required by almost all software synthesis methods (except by [3], which resolves all concurrency at compilation time) to sequence the execution of a set of originally concurrent tasks. Concurrent tasks are an excellent specification mechanism, but cannot be implemented as such on a standard CPU. The scheduling problem (see, e.g., [20] for a review) amounts to finding a linear execution order for the elementary operations composing the tasks, so that all the timing constraints are satisfied.

I strongly believe that the high-level design problem including the selection of the control laws should be carefully coupled with the design implementation to obtain a final result that truly optimizes designers criteria.

References

[1] R. Alur and T.A. Henzinger. Logics and models of real time: A survey. In J.W. de Bakker, C. Huizing, W.P. de Roever, and G. Rozenberg, editors, *Real-Time: Theory in Practice. REX Workshop Proceedings*, 1992.

[2] M. Abate and M. Scarnera. Personal Communication, Magneti-Marelli Engine Control Division, 1995.

[3] G. Berry, P. Couronné, and G. Gonthier. The synchronous approach to reactive and real-time systems. *IEEE Proceedings*, 79, September 1991.

[4] K. Butts, I. Kolmanovsky, N. Sivashankar, and J. Sun. Hybrid systems in automotive control applications. In S. Morse, editor, *this book*. Springer-Verlag, 1996.

[5] R. Bryant. Graph-based algorithms for boolean function manipulation. *IEEE Transactions on Computers*, C-35(8):677–691, August 1986.

[6] J. T. Buck. *Scheduling Dynamic Dataflow Graphs with Bounded Memory Using the Token Flow Model*. PhD thesis, U.C. Berkeley, 1993. UCB/ERL Memo M93/69.

[7] O. Coudert, C. Berthet, and J. C. Madre. Verification of Sequential Machines Using Boolean Functional Vectors. In *IMEC-IFIP Int'l Workshop on Applied Formal Methods for Correct VLSI Design*, pages 111–128, November 1989.

[8] M. Chiodo, D. Engels, P. Giusto, H. Hsieh, A. Jurecska, L. Lavagno, K. Suzuki, and A. Sangiovanni-Vincentelli. A case study in computer-aided codesign of embedded controllers. *Design Automation for Embedded Systems*, 1(1-2), January 1996.

[9] M. Chiodo, P. Giusto, H. Hsieh, A. Jurecska, L. Lavagno, and A. Sangiovanni-Vincentelli. Hardware/software codesign of embedded systems. *IEEE Micro*, 14(4):26–36, August 1994.

[10] E. Clarke, O. Grumberg, and D. Long. Model checking and abstraction. *ACM Transactions on Programming Languages and Systems*, 16(5):1512–1542, September 1994.

[11] W-T. Chang, A. Kalavade, and E. Lee. Effective heterogeneous design and co-simulation. In G. De Micheli, editor, *Nato Advanced Study Institute*. Kluwer Academic Publisher, 1996.

[12] P.E. Caines and Y-J Wei. Hierarchical hybrid control systems. In S. Morse, editor, *this book*. Springer-Verlag, 1996.

[13] J. B. Dennis. First version data flow procedure language. Technical Report MAC TM61, Massachusetts Institute of Technology, May 1975.

[14] A. Damiano and P. Giusto. Specification by Volvo, Personal Communication, Magneti-Marelli Automotive Electronics Division, 1995.

[15] D. Drusinski and D. Har'el. Using statecharts for hardware description and synthesis. *IEEE Transactions on Computer-Aided Design*, 8(7), July 1989.

[16] R. K. Gupta, C. N. Coelho Jr., and G. De Micheli. Program implementation schemes for hardware-software systems. *IEEE Computer*, pages 48–55, January 1994.

[17] J. Guckenheimer. A robust hybrid stabilization strategy for equilibria. In *IEEE Transactions on Automatic Control*, volume AC-40, pages 742–754, February 1995.

[18] J. Guckenheimer. Complexity of hybrid system models. In S. Morse, editor, *this book*. Springer-Verlag, 1996.

[19] D. Har'el, H. Lachover, A. Naamad, A. Pnueli, et al. STATEMATE: a working environment for the development of complex reactive systems. *IEEE Transactions on Software Engineering*, 16(4), April 1990.

[20] W.A. Halang and A.D. Stoyenko. *Constructing predictable real time systems*. Kluwer Academic Publishers, 1991.

[21] J. E. Hopcroft and J. D. Ullman. *Introduction to Automata Theory, Languages and Computation*. Addison-Wesley, Reading, Mass., 1979.

[22] G. Kahn. The semantics of a simple language for parallel programming. In *Proceedings of IFIP Congress*, August 1974.

[23] W. Kohn, A. Nerode, J. Remmel, and X. Ge. Multiple agent hybrid control: Carrier manifolds and chattering approximations to optimal control. In *Proceedings of 33rd IEEE Conference on Decision and Control*, pages 4221–4227, December 1994.

[24] R. P. Kurshan. *Automata-Theoretic Verification of Coordinating Processes*. Princeton University Press, 1994.

[25] L. Lavagno, A. Sangiovanni-Vincentelli, and H. Hsieh. Models and algorithms for embedded system synthesis and validation. In G. De Micheli, editor, *Nato Advanced Study Institute*. Kluwer Academic Publisher, 1996.

[26] K. McMillan. *Symbolic model checking*. Kluwer Academic, 1993.

[27] A. S. Morse. Control using logic-based switching. In A. Isidori, editor, *Trends in Control*, pages 69–114. Springer-Verlag, 1995.

[28] T. Murata. Petri Nets: Properties, analysis and applications. *Proceedings of the IEEE*, pages 541–580, April 1989.

[29] J. McManis and P. Varaiya. Suspension automata: a decidable class of hybrid automata. In *Proceedings of the Sixth Workshop on Computer-Aided Verification*, pages 105–117, 1994.

[30] A.R. Newton. Logic and mixed-mode simulation. In P. Antonietti, H. DeMan, and D.O. Pederson, editors, *Nato Advanced Study Institute*. Kluwer Academic Publisher, 1991.

[31] C. Passerone, M. Chiodo, W. Gosti, L. Lavagno, and A. Sangiovanni-Vincentelli. Evaluation of trade-offs in the design of embedded systems via co-simulation. Technical Report UCB/ERL M96/12, U.C. Berkeley, 1996.

[32] C. A. Petri. *Kommunikation mit Automaten*. PhD thesis, Bonn, Institut für Instrumentelle Mathematik, 1962. (technical report Schriften des IIM Nr. 3).

[33] P. Panangaden and E.W. Stark. Computations, residuals, and the power of indeterminacy. In *Proc. ICALP'88*, pages 439–454, Berlin, 1988. Springer-Verlag. Lecture Notes in Computer Science 317.

[34] C. Rossi. Personal Communication, Magneti-Marelli Engine Control Division, 1995.

[35] T. Shiple, A. Aziz, F. Balarin, S. Cheng, R. Hojati, T. Kam, S. Krishnan, V. Singhal, H. Wang, R. Brayton, and A. Sangiovanni-Vincentelli. Formal design verification of digital systems. In *Proceedings of TECHCON*, 1993.

[36] A. L. Sangiovanni-Vincentelli. Project ades: Architectures and design of advanced electronic systems. Technical report, Research Proposal for a Consortium Involving Cadence Design Systems, Consiglio Nazionale delle Ricerche, Magneti-Marelli and SGS-Thomson, February 1994.

[37] A.L. Sangiovanni-Vincentelli, R. McGeer, and A. Saldanha. Verification of elecronic systems. In *Proceedings of 1996 Design Automation Conference*, June 1996.

[38] R. vanHaxleden. Personal Communication, Daimler Benz Research, Berlin, 1995.

Hierarchical Hybrid Control Systems*

Peter E. Caines[†]and Yuan-Jun Wei
Department of Electrical Engineering
and
Centre for Intelligent Machines
McGill University
3480 University St, Montreal, Quebec, Canada H3A 2A7
e-mail: peterc@cim.mcgill.edu

Abstract

The notion of dynamical consistency ([19],[7]) is extended to hybrid systems so as to define the set of dynamically consistent hybrid partition machines associated with a continuous system S. Following the formulation of the notions of between-block and in-block controllable hybrid partition machines, the lattice $HIBC(S)$ of hybrid in-block controllable partition machines is defined and investigated. A sufficient condition for the existence of a non-trivial lattice $HIBC(S)$ is given and two examples are presented.

1 Introduction

An important class of hybrid systems is that characterised by a continuous system in a feedback configuration with a discrete system; there is now an extensive literature on this subject from which we cite [1]-[6],[8],[10]-[15],[17] and [18]. These works all employ some form of decomposition of the state spaces of the continuous subsystems. The dynamics of these subsystems necessarily depend upon the joint discrete and continuous state components by virtue of the feedback configuration; similarly, the discrete subsystem dynamics depend upon the joint system state. In some cases, these system configurations may be viewed as facilitating the effective design of discrete controllers, and that is one of the objectives of the hierarchical hybrid system theory presented in this paper. If successful, the advantage of such an approach would evidently be the enhanced realizability of controllers by computer systems.

A common approach to hybrid control (see e.g.,[14],[5]) is to decompose the state space into a countable set of blocks and represent each block by a symbol

---

*Work supported by NSERC grant A1329 and the NSERC-NCE-IRIS II project.
[†] Also affiliated with the Canadian Institute for Advanced Research.

that plays the role of a state of the discrete system as viewed by the controller. We shall call a control law which requires the application of a single sequence of control actions on every state in such a block a block invariant control law.

A general model employed by Branicky, Borkar and Mitter in [3] subsumes the models of [1],[4],[13]-[15] and [17]. It permits what are called autonomous and controlled jumps of the joint system state. This occurs when the continuous component hits the boundary manifold of certain obstacle-like sets defined in the currently active copy of the continuous state space (each copy of which is associated with a discrete state component). Apart from the controlled jumps, the state is subject to piecewise constant real vector valued controls. A variety of optimal control problems are solved for the systems in this class.

A. Nerode and W. Kohn introduced the notion of a small topology [13],[2], which is the information space of the declarative controller. Furthermore, in [11] and [12], they convert an optimal control problem (with ϵ-tolerance) into a linear programming problem based upon the discrete topology. In this framework, the state of a discrete event system (DES) is defined as a segment of a trajectory generated by a fixed control function within a pre-decided time interval, and a discrete control action is interpreted as a transition from one control function to the next.

Motivated by Lyapunov stability theory, M. Glaum [10] related algebraic properties of the vector fields associated with a control system to topological obstructions to achievable closed-loop dynamics produced as solutions to the stabilization problem. This approach necessarily involves controls which are not block invariant since the stabilizing controls are state dependent.

In the present paper, we approach the problem of state quantization for the design of discrete controllers from a hierarchical control point of view. We view a hybrid system as one whose continuous subsystems are supervised via their feedback relationships with a discrete controller and for which each of the continuous subsystems is itself (internally) subject to feedback control. This notion of hierarchical control is realized via the idea of dynamical consistency (see [19],[7]) wherein an abstract transition from one partition block of states to a second one is defined whenever every state in the first block can be driven directly into the second block. In general, such control actions are not block invariant.

2   Hybrid DC Partition Machines

Consider the differential system

$$\mathcal{S}: \quad \dot{x} = f(x, u), \qquad x \in \Re^n, u \in \Re^m, u(\cdot) \in \mathcal{U}, f : D \times \Re^m \subset \Re^{n+m} \to \Re^n, \quad (1)$$

where we assume that $\mathcal{U}$ is the set of all bounded piecewise $C^q(\Re^1)$ $(q \geq 1)$ functions of time with limits from the right (i.e. functions which are q times continuously differentiable, except possibly at a finite number of points in any compact interval, and which are bounded and have limits from the right). We further assume that $f \in C^1(\Re^{n+m})$ and that for each $u \in \mathcal{U}$, $f(x, u(t))$ satisfies a global Lipschitz condition on D, uniformly in $t \in \Re^1$. D is assumed to be a closed set and to have a non-empty, path-connected, interior. We shall refer to these conditions on f and $\mathcal{U}$ as the *standard (control) ODE conditions*. Subject to these conditions, a unique solution $x(\cdot) : \Re^1 \to \Re^n$ exists for any initial condition $x^0 \in in(D)$ and control function $u \in \mathcal{U}$ until a possible escape time from D (which does not exist in case D is $\Re^n$).

Definition 2.1 [18],[6] A *finite* (C^r) *partition* of the state space $D \subset \Re^n$ of (1) is a pairwise disjoint collection of subsets $\pi = \{X_1, X_2, \cdots, X_p\}$ such that each X_i is open, path-connected, and is such that

$$D = \bigcup_{i=1}^{p}(X_i \cup \partial X_i),$$

and, further, is such that the boundary of every block X_i is a finite union of $n - 1$ dimensional connected C^r manifolds. $\Pi(D)$ shall denote the set of all (C^r) partitions of D. A partition $\pi \in \Pi(D)$ is said to be *non-trivial* if $|\pi| > 1$.

□

The reader may note that the state space partition used here is simpler in its definition than those appearing in the regular synthesis theory of optimal control (see e.g.[9] and [16]).

The DC condition for a pair of partition blocks of a continuous system is defined in two steps. The first states that each interior state of the first block can be locally driven to some state on an appropriate boundary; the second states that each such boundary state can be driven into the second block.

Definition 2.2 [18],[6] For $\pi \in \Pi(D)$, $\langle X_i, X_j \rangle \in \pi \times \pi$ is said to be a *dynamically consistent* (DC) pair (with respect to S) if and only if:

either $i = j$,

or, if $i \neq j$, for all x in X_i, there exists $u_x(\cdot) \in \mathcal{U}$, defined upon $[0, T_x], 0 < T_x < \infty$, and there exists $y \in \partial X_i \cap \partial X_j$, such that:

    (i) $\forall t \in [0, T_x), \phi(t, x, u_x) \in X_i$, and $\lim_{t \to T_x^-} \phi(t, x, u_x) = y$;

and for the state y in (i) there exists $u_y \in \mathcal{U}$ defined on $[0, T_y), 0 < T_y < \infty$, such that

    (ii) $\forall t \in (0, T_y), \phi(t, y, u_y) \in X_j$;

where $\phi(\cdot, \cdot, \cdot)$ in (i) and (ii) are the integral curves of the vector field $f(\cdot, \cdot)$ with respect to the control functions $u_x, u_y \in \mathcal{U}$ and the initial conditions x, y, respectively.

□

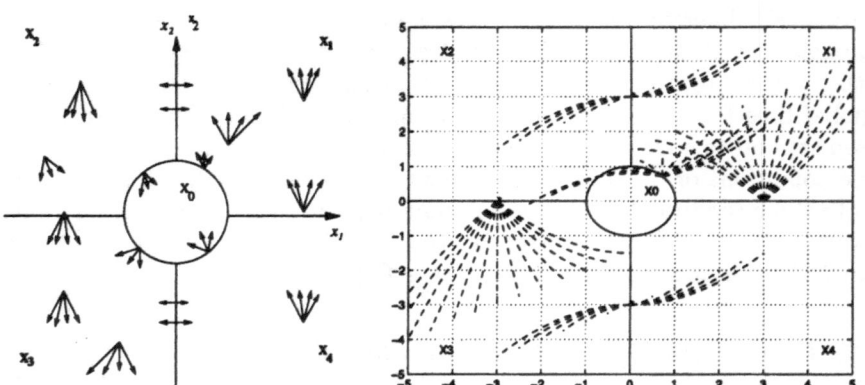

Figure 1: Left: The vector fields. Right: Trajectories w.r.t. constant controls $u \epsilon [-3, 3]$

In the example below, we take D to be the space $\Re^2$.

Example 1

$$\begin{bmatrix} \dot{x}_1 \\ \dot{x}_2 \end{bmatrix} = \begin{bmatrix} 0 & 0 \\ 1 & 0 \end{bmatrix} \begin{bmatrix} x_1 \\ x_2 \end{bmatrix} + \begin{bmatrix} 1 \\ 0 \end{bmatrix} u \qquad (2)$$

Consider the partition $\pi_1 = \{X_i; 0 \leq i \leq 4\}$ displayed in Figure 1.

It is clear from the above figures that $\langle X_0, X_1 \rangle$ is a DC pair. However $\langle X_1, X_0 \rangle$ is not DC, since, for example, any trajectory starting from $(0.5, 2.0) \epsilon$ X_1 must first enter X_2 in order to enter X_0.

Definition 2.3 [18],[6] Given $\pi \epsilon \Pi(D)$, the *hybrid DC partition machine*

$$\mathcal{M}^{\pi} = \langle X^{\pi} = \{\overline{X}_1, \cdots, \overline{X}_p\}, U = \{\overline{U}_i^j; 1 \leq i, j \leq p\}, \Phi^{\pi} \rangle,$$

based upon the system $\mathcal{S}$, is the finite state machine defined by $\Phi^{\pi}(\overline{X}_i, \overline{U}_i^j) =$ $\overline{X}_j$, for all i, j, $1 \leq i, j \leq |\pi|$, if and only if $\langle X_i, X_j \rangle$ is DC. $\Pi(\mathcal{S})$ shall denote all hybrid partition machines of $\mathcal{S}$. $\mathcal{M}^{\pi}$ is called *non-trivial* if π is non-trivial.

□

Definition 2.4 A Partial Ordering of Partition Machines [7],[6]
Let $\mathcal{M}^{\pi_1}, \mathcal{M}^{\pi_2} \epsilon \Pi(\mathcal{S})$, we say that $\mathcal{M}^{\pi_2}$ is *weaker* than $\mathcal{M}^{\pi_1}$, denoted $\mathcal{M}^{\pi_1} \preceq$ $\mathcal{M}^{\pi_2}$, if, for every $X_i \epsilon \pi_1 = \{X_1, \cdots, X_n\}$, there exists $Y_j \epsilon \pi_2 = \{Y_1, \cdots, Y_m\}$ such that $X_i \subset Y_j$.

□

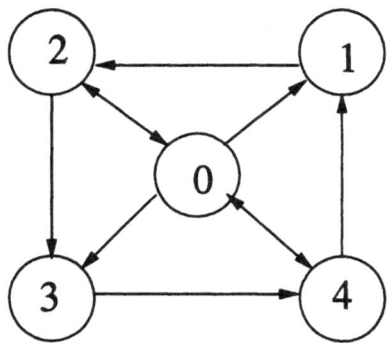

Figure 2: The partition machine corresponding to π_1

Definition 2.5 Hybrid Between-Block and In-block Controllability
[18],[6] A hybrid partition machine $\mathcal{M}^\pi$ is called *hybrid between-block control-lable* (HBBC) if any block state $\overline{X}_i$ can be reached from any other block state $\overline{X}_j$ by applying a finite number of block transitions. $\mathcal{M}^\pi$ is called *hybrid in-block controllable* (HIBC) if for every $X_i \, \epsilon \, \pi$, and for all $x, y \, \epsilon \, X_i$, the following holds:

$$\exists u(\cdot) \, \epsilon \, \mathcal{U}, \exists T, 0 \leq T < \infty, \, (\forall t, 0 \leq t \leq T, \phi(t, x, u) \, \epsilon \, X_i) \wedge \phi(T, x, u) = y. \quad (3)$$

In other words, each block $X_i \, \epsilon \, \pi$ is *controllable for the system* $\mathcal{S}$. We shall use $HBBC(\mathcal{S})$ (respectively, $HIBC(\mathcal{S})$) to denote the set of all HBBC (respectively, HIBC) machines of $\mathcal{S}$. □

The following theorem is an immediate consequence of Definition 2.5; it indicates how a state to state control problem is decomposed into a pair of high and low level control problems.

Theorem 2.1 [18],[6] *An HIBC machine $\mathcal{M}^\pi$ is HBBC if and only if any pair of states $x, y \, \epsilon \, in(D) \setminus (\cup_{i=1}^{|\pi|} \partial X_i)$ are mutually accessible with respect to $in(D)$.* □

3 The Lattices of Hybrid In-block Controllable Machines

The partially ordered set $HIBC(\mathcal{S})$ is the principal object of interest in our work and the following theorems, proved in [18], form the foundation of our investigation.

Theorem 3.1 [18],[6]
Hybrid in-block controllability is closed under chain union, i.e. $\mathcal{M}^{\pi_1}, \mathcal{M}^{\pi_2} \, \epsilon \, HIBC(\mathcal{S})$ implies $\mathcal{M}^{\pi_1 \cup^c \pi_1} \, \epsilon \, HIBC(\mathcal{S})$. □

Theorem 3.2 Existence of HIBC Lattices [18],[6]
If for every pair $\mathcal{M}^{\pi_1}, \mathcal{M}^{\pi_2} \in HIBC(\mathcal{S})$, *the set*

$$LOW(\mathcal{M}^{\pi_1}, \mathcal{M}^{\pi_2})\underline{\Delta}\{\mathcal{M}^{\pi}; \mathcal{M}^{\pi} \in HIBC(\mathcal{S}), \mathcal{M}^{\pi} \preceq \mathcal{M}^{\pi_1}, \mathcal{M}^{\pi} \preceq \mathcal{M}^{\pi_2}\},$$

is not empty, then the greatest lower bound partition machine $\mathcal{M}^{\pi_1 \sqcap \pi_2}$ *exists and is given by the machine defined on the following partition:*

$$\pi_1 \sqcap \pi_2 = \bigcup^c \{\pi; \pi \in LOW(\mathcal{M}^{\pi_1}, \mathcal{M}^{\pi_2})\}.$$

The tuple $\langle HIBC(\mathcal{S}), \cup^c, \sqcap, \preceq \rangle$ *forms a lattice, called the hybrid in-block controllable (HIBC) lattice of* $\mathcal{S}$; *further, if* $\mathcal{S}$ *is controllable with respect to* $in(D)$, *then* $HIBC(\mathcal{S})$ *has top element* $\mathcal{M}^{\pi_{tr}}$.

□

4   Lattices of Annular Partitions for a Double Integrator System

Consider the system in Example 1. Let $N > 1$ be an integer and

$$D = \{(x, y); x^2 + y^2 \leq N\}.$$

Let

$$A_{i,j} = \{(x, y); i < x^2 + y^2 < j\}, \qquad 1 \leq i < j \leq N.$$

Then for any list of integers $1 = i_1 < i_2 < \cdots < i_n = N$, the following set forms an annulus partition of D:

$$\pi = A_0 \cup \{A_{i_k, i_{k+1}}; 1 \leq k \leq n - 1\},$$

where $A_0 = \{(x, y); x^2 + y^2 < 1\}$. Let $Ann(D)$ denote the set of all such (integer) annular partitions of D.

Lemma 4.1 *For* $A_{i,j}, A_{k,l} \in Ann(D)$, *if* $A_{i,j} \cap A_{k,l} \neq \emptyset$, *then*

$$A_{i,j} \cup A_{k,l} = A_{min(i,k),max(j,l)}.$$
$$A_{i,j} \cap A_{k,l} = A_{max(i,k),min(j,l)}.$$

□

Hence $Ann(D)$ is closed under both $\cup^c$ and $\cap$. It is easy to prove the following lemma which is illustrated in Figure 3 and Figure 4.

Lemma 4.2 *For* $1 \leq i, j \leq N$, $1 < k \leq N$,
(i) A_0 *and any* $A_{i,j} \in Ann(D)$ *are IBC.*
(ii) Both $\langle A_{i,j}, A_{j,k} \rangle$ *and* $\langle A_{j,k}, A_{i,j} \rangle$ *are DC.*
(iii) Both $\langle A_0, A_{1,k} \rangle$ *and* $\langle A_{1,k}, A_0 \rangle$ *are DC.*

□

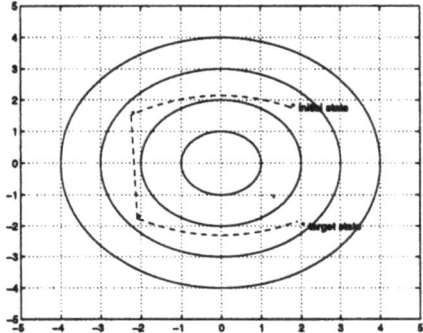

Figure 3: A trajectory from (1.76,1.76) to (2.0,-2.0) under control sequence $u_1 = -4, \Delta_1 = 1, u_2 = 0.1, \Delta_2 = 1.5, u_3 = 4, \Delta_3 = 1$.

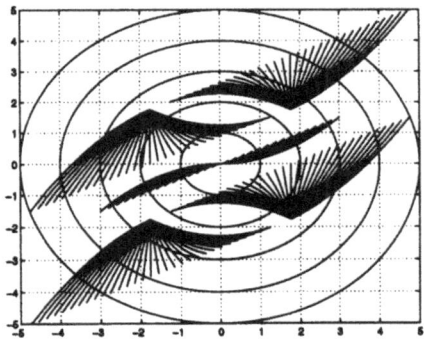

Figure 4: Trajectories with respect to constant controls $u \in [-3, 3]$.

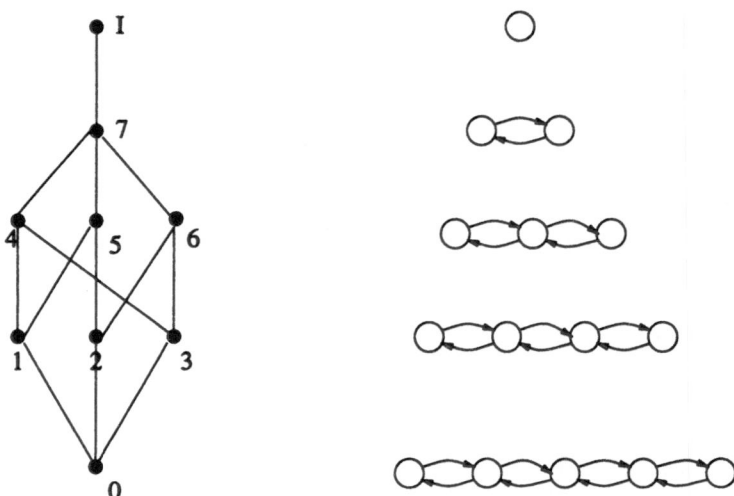

Figure 5: HIBC lattice for the circular annular partitions $Ann(D)$ together with the corresponding partition machines (with all self-loops omitted).

Hence we have

Theorem 4.1 $\langle Ann(D), \cup^c, \cap, \subset \rangle$ *forms a lattice, with bottom element*

$$\pi_{id} = \{A_0, A_{i,i+1}; 1 \leq i \leq N-1\},$$

and top element D, which is identical to the associated lattice of hybrid in-block controllable partition machines.

$\square$

We list here the annular lattice for $N = 5$.

$$\pi_{id} = \{A_0, A_{i,i+1}; 1 \leq i \leq 4\}, \ \pi_1 = \{A_0, A_{1,3}, A_{3,4}, A_{4,5}\},$$

$$\pi_2 = \{A_0, A_{1,2}, A_{2,4}, A_{4,5}\}, \ \pi_3 = \{A_0, A_{1,2}, A_{2,3}, A_{3,5}\},$$

$$\pi_4 = \{A_0, A_{1,3}, A_{3,5}\}, \ \pi_5 = \{A_0, A_{1,4}, A_{4,5}\},$$

$$\pi_6 = \{A_0, A_{1,2}, A_{2,5}\}, \ \pi_7 = \{A_0, A_{1,5}\}.$$

The corresponding lattice is displayed in Figure 5 below.

Acknowledgements The authors gratefully acknowledge conversations concerning this work with Hector Sussmann and Ekaterina Lemch.

References

[1] A. Back, J. Guckenheimer, and M. Meyers. A dynamical simulation facility for hybrid systems. In A. Ravn R. Grossman, A. Nerode and H.Rischel, editors, *Hybrid Systems, LNCS vol 736*. Springer Verlag, 1993.

[2] M.S. Branicky. Topology of hybrid systems. In *Proceedings of the 32nd IEEE Conference on Decision and Control*, pages 2309–2314, St.Antonio, Texas, 1993.

[3] M.S. Branicky, V. S. Borkar, and S.K. Mitter. A unified framework for hybrid control. In *Proceedings of the 33rd IEEE Conference on Decision and Control*, pages 4228–4234, Lake Buena Vista, FL, 1994.

[4] R. Brockett. Language driven hybrid systems. In *Proceedings of the 33rd IEEE Conference on Decision and Control*, pages 4210–4214, Lake Buena Vista, FL, 1994.

[5] P. E. Caines and R. Ortega. The semi-lattice of piecewise constant controls for non-linear systems: A possible foundation for fuzzy control. In *Proceedings, IFAC-NOLCOS 95, Tahoe, CA*, June 1995.

[6] P.E. Caines and Y-J. Wei. On dynamically consistent hybrid systems. In W. Kohn and A. Nerode, editors, *Proceedings of the 1994 Cornell University Workshop on Hybrid Systems & Autonomous Control*, Lecture Notes in Computer Science, NYC, 1995. Springer Verlag.

[7] P.E. Caines and Y-J. Wei. The hierarchical lattices of a finite machine. *Systems and Control Letters*, pages 257–263, July, 1995.

[8] D. F. Delchamps. Stabilizing a linear system with quantized state feedback. *IEEE Trans. on Automatic Control*, 35(8):916–924, August 1990.

[9] W.H. Fleming and R. W. Rishel. *Deterministic and Stochastic Optimal Control*. Springer Verlag, New York, 1976.

[10] M. Glaum. Smooth partitions, transversality and graphs. Master's thesis, Department of Mathematics, The University of British Columbia, Vancouver, Canada, 1993.

[11] W. Kohn, J. James, A. Nerode, and N. DeClaris. A hybrid systems approach to integration of medical models. In *Proceedings of the 33nd IEEE Conference on Decision and Control*, pages 4247–4252, Lake Buena Vista, FL, 1994.

[12] W. Kohn, A. Nerode, J. Remmel, and X. Ge. Multiple agent hybrid control: Carrier manifolds and chattering approximations to optimal control. In *Proceedings of the 33nd IEEE Conference on Decision and Control*, pages 4221–4227, Lake Buena Vista, FL, 1994.

[13] A. Nerode and W. Kohn. Models for hybrid systems: Automata, topologies, stability. Technical report, Cornell University, Ithaca, NY, 1992.

[14] J. A. Stiver and P.J. Antsaklis. On the controllability of hybrid control systems. In *Proceedings of the 32nd IEEE Conference on Decision and Control*, pages 294–299, San Antonio, Texas, 1993.

[15] J. A. Stiver, P.J. Antsaklis, and M.D. Lemmon. Digital control from a hybrid perspective. In *Proceedings of the 33rd IEEE Conference on Decision and Control*, pages 4241–4246, Lake Buena Vista, FL, 1994.

[16] H. J. Sussmann. Regular synthesis for time-optimal control of single-input real analytic systems in the plane. *SIAM. J. Control and Opt.*, 25(5):1145–1162, September, 1987.

[17] L. Tavernini. Differential automata and their discrete simulators. *Nonlinear Analysis: Theory, Methods and Applications*, 11(6), 1987.

[18] Y-J Wei. *Logic Control: Markovian Fragments, Hierarchy and Hybrid Systems*. PhD thesis, Department of Electrical Engineering, McGill University, Montreal, Canada, October 1995.

[19] Y-J. Wei and P.E. Caines. Lattice structures and hierarchical COCOLOG for finite machines. In *Proceedings of the 33rd IEEE Conference in Decision and Control*, pages 3125–3130, Lake Buena Vista,FL, December 1994.

The Analysis of Piecewise Linear Dynamical Systems

N.B.O.L. Pettit
Control Systems Centre
UMIST, P.O.Box 88
Manchester, M60 1QD, UK

Abstract

The motivation for developing an analysis tool for piecewise linear
(**PL**) systems arose from the desire to model some automotive control
systems in a coherent way. These systems appear to have some common properties, in that the system nonlinearities are often represented
as piecewise linear functions (e.g. saturation, backlash and relays) and
the controllers use logic that can also be treated as piecewise linear functions. This type of system model appears to capture a wide class of
complex systems. This observation led to a project at the Control Systems Centre to analyse piecewise linear systems. As a consequence a
concept was developed that relies on exploiting the computer as an analysis tool and this idea was tested on the example of a simplified model
of an Anti-Lock Braking System (ABS) controlled with a rule-base.

This paper outlines the work done at the Control Systems Centre,
UMIST, to develop a basic analysis concept and gives a summary of the
efforts to extend the tool to handle as wide a class of systems as possible.
The principal conclusions of the ABS example study are also given, as
they provide a good insight into the potential of this approach.

1 Introduction

Peter Wellstead initiated a study at the Control Systems Centre, UMIST to
look at specific automotive control systems and see if traditional analysis and
control ideas could be applied. The systems chosen were the traction control
and Anti-Lock Braking systems used in cars. These systems were selected since
the controllers developed by the Industry were based on extensive testing and
simulation coupled with engineering intuition to obtain an acceptable rule-
based controller. There was (and still is) a desire by the companies (in our
case, Lucas Automotive Ltd) to see if more a rigorous analysis of these systems
was possible. The main reason was to obtain better methods of assessing the
safety of these systems, rather than control, since the failure of such a system
could lead to expensive litigation.

The study [1] found that traditional analysis techniques were not suitable for these systems; the main problem being the presence of discontinuous actions in the dynamics from either the controller or system nonlinearities. This led to characterizing the systems using piecewise linear models. These models have the advantage that both the logic in the controllers and nonlinearities in the system model (such as saturations, hysteresis and backlash) appear as piecewise linear functions, with the system dynamics described by standard integration elements as with linear systems. This gives rise to a specific geometrical interpretation of the systems in the state space, as illustrated by figure 1.

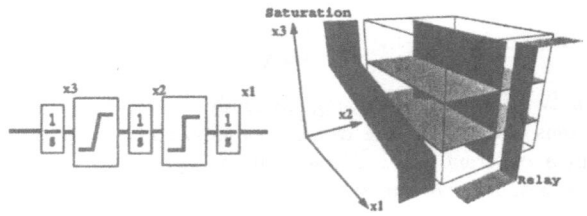

Figure 1: A geometric interpretation of a piecewise linear system in state space.

Here the model description adopted causes a partitioning of the state space into cells. These cells have distinctive properties in that the dynamics within each cell are described by a linear dynamic equation and the cell boundary forms a convex polytope. The boundaries are in effect switches between different linear systems, the switches arising from the breakpoints in the piecewise linear functions of the model. This natural partitioning of the state space opens the way for analysing the local behaviour of each cell and then aggregating the information to obtain global information.

2   A Computational Analysis Approach

The idea for the analysis is derived from considering the phase portrait. The phase portrait offers a powerful graphical technique that presents global dynamic behaviour for linear, piecewise linear and continuous nonlinear model descriptions in a form that allows easy interpretation of system behaviour. However it is essentially restricted to models with two states only (or perhaps 3 states with todays graphics). The aim of the analysis developed for PL systems is to preserve the graphical nature of presenting the dynamics of the model, but allow more general models to be analysed.

The basic procedure used is illustrated in figures 2 and 3. The first step is to take each cell or polytope generated from the model and map the movement of the trajectories between the polytope boundaries as dictated by the linear system defined for that polytope's interior. The trajectory tangents will have two principle behaviours approaching the interior boundaries of the cell: either they will point out indicating an 'exit' from the cell; or they point in, indicating an 'entrance' to the cell. The cell boundaries can thus be partitioned in to **Xface**'s and **Nface**'s, indicating exit and entrance sections of the cell boundary respectively. The set of points defining the Nface can be considered as a set of initial conditions for a bundle of trajectories. These can then be tracked through the cell to find which Xfaces (if any) the bundle will leave the cell via. If each Nface and Xface is represented as a node, then any dynamics that move from an Nface to a particular Xface can be represented as a directed connection between the nodes. This is illustrated in figure 2.

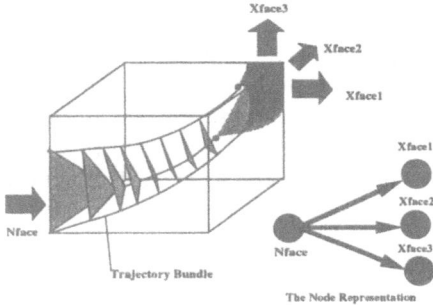

Figure 2: Identifying a connected graph for a piecewise linear system cell.

The local graphs generated by the analysis can then be linked up to form a global graph of the model. Each breakpoint or logic rule will appear in the graph as a set of unique nodes, with the connections indicating possible dynamic behaviours around the various switches. This idea of a global graph is illustrated in figure 3.

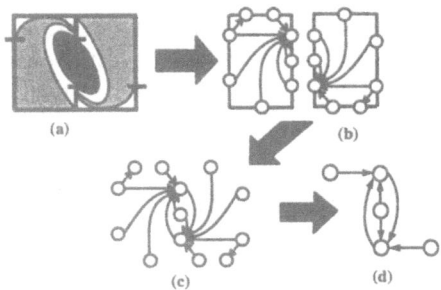

Figure 3: Forming a graph of a piecewise linear system. (a) The state space model, (b) The local graphs, (c) The global graph, (d) Simplifying the graph.

The stages of the analysis are described in [2] and [3] with a comprehensive description being given in [4]. This paper will outline the main stages of the analysis and the principle techniques used.

2.1   Representing the PL system

The question of representing PL models is one of interpretation. How do I represent my model data to be of greatest use to the analysis? Some form of implicit model description is essential if realistic analysis is to be achieved. This is because the systems to be analysed are complex by nature, and can easily contain hundreds or even thousands of cells in their state space realization. Each cell is adjacent to a unique subset of the other cells and will contain its own linear dynamic equation together with its boundary description. All this information must be readily and automatically available for analysis without a crippling memory overhead. This question of representing PL systems has been looked at by researchers in the field of circuit theory, showing that implicit descriptions of these system can be developed [5], [6]. The representation scheme developed here allows for SIMULINK^{TM} to act as a model interface. It is based around describing the model as the set of equations in (1).

$$
\begin{aligned}
n &= R^i m + k^i \\
m &= Fx + Gn + ku \\
\dot{x} &= Jx + Kn + lu
\end{aligned}
\tag{1}
$$

Here n is the output vector of all piecewise linear functions, m is the input, R^i is a gain matrix and k^i is an offest vector. The i index is a vector that dictates which segment of each of the PL functions is active. In other words it acts as a pointer to the cells that are generated by the model. u is just a constant reference signal. This representation is useful since for any chosen i, the cell represented by i can be determined in terms of its bounding hyperplanes and linear dynamic equation associated with the cell interior. Also, all the system matrices and vectors in the equations in (1) can be identified directly from a SIMULINK^{TM} model. The details of this idea can be found in [7] and [4]. Currently the representation is restricted to simple forms of PL functions such as saturation, relays and quantization, but it should be possible to extend the range of functions that can be handled. This restriction is purely an artifact of the representation, not the analysis.

2.2   Analysing the PL model

Given a local cell, its boundaries and internal linear dynamic equation, the first step is to partition the boundaries into Nfaces and Xfaces. This is a relatively straightforward operation. Given a simple interior linear equation and one boundary of the cell:

$$
\dot{x} = Ax + b \text{ and } \gamma x = c
\tag{2}
$$

the dynamics will touch the boundary at a tangent when:

$$
\gamma \dot{x} = 0 \Rightarrow \gamma Ax = -\gamma b
\tag{3}
$$

Equation (3) gives a new hyperplane equation. If this hyperplane intersects the part of the hyperplane $\gamma x = c$ that makes up the cell, then that boundary is divided into an Xface and Nface along that intersection. This process is repeated for all boundaries.

The second stage is to trace the connections between Nfaces and Xfaces. This is a more complex numerical problem relying on using an efficient algorithm for solving the problem. The principle is straightforward. Both Nfaces and Xfaces will be convex polytopes by construction, and thus can be defined by their vertices. Given the vertices are treated as a set of initial conditions, then the evolution of the trajectories from the vertices will form a set of points after some fixed time. Due to the linear dynamics, this set of points will form another convex polytope. This convex polytope will contain all the points from the initial Nface projected by the fixed time. Thus by tracking the vertices, the evolution of the Nface can be traced through the cell (see [2]). The Nface can also be defined by an equality constraint and a set of inequality constraints. These constraints can also be projected using the linear equation of the cell and the Nface tracked via its inequality constraints. The choice of approach dictates what algorithms are used. This in turn places restrictions on the complexity and dimension of system that can be analysed. Some of these issues are discussed in [8]. The most effective technique at present appears to be as follows. Given an Nface and Xface described in terms of their constraining hyperplanes:

$$
\begin{aligned}
N : A_N x &\leq b_N \\
X : A_X x &\leq b_X
\end{aligned}
\tag{4}
$$

where N is a projected Nface and X is the Xface that is being checked for a connection.

The connection problem is formulated as an l_∞ problem of the form:

$$
\min_\infty \|x_1 - x_2\|_\infty \text{ subject to } Mx - v \leq 0
$$
$$
M = \begin{bmatrix} A_N & 0 \\ 0 & A_X \end{bmatrix}, v = \begin{bmatrix} b_N \\ b_X \end{bmatrix}
\tag{5}
$$

This can then be formulated as a linear program that calculates the l_∞ norm (see [9]) which in this case will be the l_∞ distance between the Nface and Xface.

By determining a 'time bound' between which a connection between the Xface and evolved Nface must occur and then numerically searching this bound using the connection algorithm outlined in eqns 4 and 5 above, the connections between boundaries can be built up for each cell.

2.3   Generating the graphs

The connections found for each cell will form directed arcs within a graph, with each Nface and Xface forming nodes in the graph. Thus the output of the analysis will be in the form of a connection matrix. This graph can then be analysed via standard graph tools. Patterns such as cycles take on meaning in terms of the global dynamics within the system. Since sets of nodes can relate to particular logic rules, the possible impact of a particular rule can be traced through the graph. Some of these ideas are outlined in the paper [10]. Our work in this area has so far been to develop a MATLAB^{TM} interface that allows interactive manipulation of a graph. Certain pattern combinations tend to indicate types of dynamic behaviour, such as a spiral pattern linked with a cycle has appeared for both a stable limit cycle in three states as well as a chaotic system, where the chaos was a scroll behaviour 'embedded' in the graph as a cycle. Thus the graph, coupled with knowledge from the model and analysis, such as what nodes represent singularities and which nodes are derived from particular system logic, allows global dynamic behaviours to be found. It is an open question as to what can be inferred from the graph. Is there sufficient information to define a form of controllability from the graph? Can the conservatism of the graph be quantified? i.e. how much information has been sacrificed to construct the graph? and can the graph be used directly in control design?

The mapping of the system information to a graph offers an attractive technique of dealing with the complexity present in the original model. Given effective tools for interpreting graph patterns in terms of system behaviour, large graphs can be generated and then analysed effectively.

3   Discussion on Logic Switching using the ABS example

The example of an Anti-Lock Braking system (ABS) for a car provides a good illustration of some of the issues in logic switching. The aim of the ABS is to improve the effectiveness of vehicle braking by maintaining the tyre braking torque at or near its maximum value. The principle used is simple. The key factor is the tyre adhesion to the road as braking torque is applied. This is usually given as the torque curve shown in figure 4.

The tyre adhesion is at its highest between wheel slip values marked A and B on figure 4. If wheel slip increases beyond B, the wheel 'locks', tyre adhesion decreases and more importantly the driver looses the ability to steer the vehicle, i.e. the system is considered unstable. The aim of the ABS controller is to keep wheel slip between A and B in the figure. A basic rule controller that can do this is:

```
if wheel slip > B then brake pressure = 0
if wheel slip < A then brake pressure = maximum
```

although in principle other control techniques for nonlinear system would do a better job.

The problem with this approach is that the tyre adhesion curve is not fixed. It will change constantly due to normal variations found in the road surface. Whats more sudden 'discrete event' changes will occur if the wheels move over different surface types, for instance an icy road will cause the curve to flatten and lose its distinct peak altogether. This means it is not possible to select

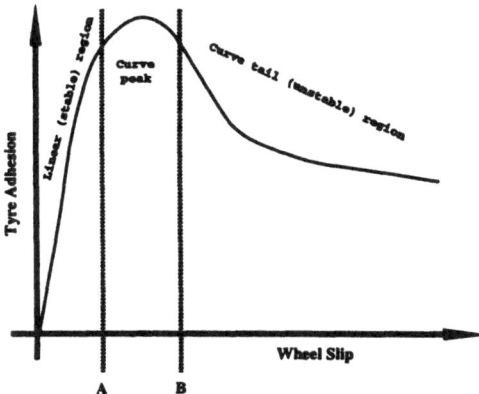

Figure 4: The Tyre Adhesion curve for brake control

meaningful threshold values such as A and B. A more sensible controller would be:

```
if wheel slip increasing and brake force increasing
then brake pressure = maximum
if wheel slip increasing and brake force decreasing
then brake pressure = 0
```

This is an incomplete rule set, but the idea is that the rules can track the position along the tyre torque curve and act accordingly. This still has problems in icy conditions, but more importantly neither brake force nor wheel slip can be measured. Instead the pressure of the brake hydraulics can be measured at some point before the four ABS valves. This coupled with the history of each of the the hydraulic valves allows a rough estimate of braking force at each wheel to be inferred. Similarly wheel slip cannot be measured. Typically the wheel speed of all four wheels is measured and used to infer wheel slip and road velocity. Thus at best the controller can only infer a very approximate measure of position along the tyre torque curve.

This poor measurement data can, however, be coupled with prior system knowledge. For example, if a car wheel register's a deceleration greater than that equivalent to gravity, then the wheel must be approaching the unstable region of the tyre adhesion curve since mechanics show car deceleration can't exceed this limit. In practice the deceleration limit is closer to 0.8 g.

Additional restrictions are imposed by the necessity of a discrete valve controlling brake pressure. Generally three discrete output signals are used: (a) increase brake pressure at a set rate; (b) decrease brake pressure at a set rate; and (c) hold brake pressure at current value.

The practical difficulties of the ABS system have been presented because there is a strong sense in which these problems dictate the use of a switched logic controller. The controller cannot know what the tyre adhesion curve will be at any time, so scheduling for different behaviours is not possible. The measurements are more suited to qualitative assessments of the wheel behaviour, e.g. fast deceleration of the front wheels coupled with a large wheel speed differential between the front and rear wheels imply the wheels are entering the unstable region. Thus the rule-based controller was adopted and is able to achieve what has now been refined to be excellent control of the system.

However, the process is still ad-hoc and tuning the controllers for each new car takes months. These difficulties were illustrated when an very simple ABS model was developed and a rule-based controller designed. Details of the model can be found in [4] and [1]. In outline, a three state model was developed that allowed the three actuator actions on brake pressure described above. Four principle tyre torque curves were used, one representing icy conditions, two in the 'normal' conditions range and one for 'sticky tarmac' conditions. A rule-based controller of ten rules was constructed. These rules required 9 variables to be tuned. Four were threshold values for the acceleration of the wheel slip as these dictated which portion of the tyre torque curve the controller had reached. Two were emergency limits and three related to the choice of rate at which brake pressure was to be increased or decreased, according to which setting was selected on the actuator.

The initial control design was carried out via simulation. Something like 1000 simulations were used to obtain values for the rule variables that appeared to give relatively good control. The model was then analysed using the concept outlined in section 2. Something like 50 cells existed for each tyre adhesion curve. The analysis found that the rules were forcing the control to spend the majority of its cycle in the unstable region of the curve. In the case of ice, the emergency rule, indicating a dangerous length of time spent in the unstable curve region, was always activated.

A correction to the scheduled values was carried out based on the problems highlighted by the analysis. This correction appeared sufficient in simulation. A repeated analysis revealed that the correction had forced an unwanted sliding behaviour to occur that, if entered, would keep the wheel in the unstable curve region for most of the braking time. The simulation missed the condition because it did not occur in normal operation, but would be triggered in an unusual situation such as hitting an oil patch. This problem was an artifact of the controller logic interacting with the system in an unpredictable manner.

A second redesign obtained an acceptable control behaviour for all four tyre adhesion conditions.

This example illustrates two features. One is that the analysis idea developed here for piecewise linear systems can provide insight into a logic system that interacts with a dynamical system. More generally and perhaps more importantly, it illustrates the difficulties of applying a qualitative 'intuitive' control design such as the rule base. When such controllers interact with complex dynamic systems, the effects can be counter-intuitive and certainly unpredictable. The start of the example suggested that the ABS system had so many difficulties that a nonlinear switching control seemed the only practicable solution. This may be true, but unless solid theory and design procedures are established it will always be a very costly control option in terms of manpower, cost of simulation, prototyping and testing and the danger of unpredictable interactions between the logic and dynamics can't be eliminated.

4 Any general observations?

The project started with the example of the Anti-Lock brake. The system had severe nonlinearities coupled with a controller that was governed by logic rules. This combination led to investigating methods to analyse hybrid systems. Our difficulty was that the example was of a relatively complex system. Most tools available for hybrid systems were not sufficiently developed to handle such an example. Our solution was to construct an analysis technique that exploited the ability to simulate hybrid systems. This kind of intelligent simulation gave a good compromise between the flaws of straight simulation and rigor of exact analysis. This compromise mix of simulation concepts with analysis offers a powerful way to deal with large scale hybrid systems.

Acknowledgements
The following people have made invaluable contributions to the project. The project has been headed by Professor Peter Wellstead. The work on the Anti-Lock Brake was carried out by Russell Wilson-Jones. I would also like to thank Vincent Besson for his contribution to representing piecewise linear systems using SIMULINK and Theodore Manavis for developing the MATLAB graph manipulation tool. I am also indebted to Jim Farrelly for his comments on measuring techniques for ABS systems. The project has been funded by the EPSRC with contributions from Lucas Automotive Advanced Engineering Centre.

References

[1] R. Wilson-Jones. *A generalised phase portrait for piecewise linear systems.* PhD thesis, Control Systems Centre, UMIST, P.O.Box 88, Manchester, M60 1QD, UK. 1993.

[2] N.B.O.L. Pettit and P.E. Wellstead. A graphical analysis method for piecewise linear systems. In *Proc. of 33rd Conference on Decision and Control*, pages 1122-1127, Florida, USA, December, 1995. IEEE.

[3] N.B.O.L. Pettit and P.E. Wellstead. Analyzing Piecewise Linear Dynamical Systems, *IEEE Control Systems Magazine*, to appear in October issue, 1995.

[4] N.B.O.L. Pettit. *Analysis of piecewise linear dynamical systems*, volume 3 of *UMIST Control Systems Centre Series*. Research Studies Press, John Wiley and Sons, Taunton, UK, 1995.

[5] W.M.G. van Bokhoven. *Piecewise linear modeling and analysis*, PhD thesis, Eindhoven University of Technology, Eindhoven, The Netherlands, 1981.

[6] L.O. Chua. Section-wise piecewise linear functions: canonical representation, properties and application. *Proceedings of the IEEE*, 65:915-929, 1977.

[7] V. Besson, N.B.O.L. Pettit and P.E. Wellstead. Representing piecewise linear systems for analysis and simulation, In *Proc. 3rd Conference on Control Applications*, pages 1815-1820, Glasgow, UK, August, 1994. IEEE.

[8] N.B.O.L. Pettit and P.E. Wellstead. Designing a computation environment for analysing piecewise linear systems. In *Proc. of the Nonlinear Control Systems Design Symposium (NOLCOS '95)*, pages 947-952, Tahoe City, USA, June, 1995. IFAC.

[9] I. Barrodale and C. Phillips. ALGORITHM 495 - Solution of an Overdetermined System of Linear equations in the Chebyshev Norm, *ACM Trans. Mathematical Software*, 1:264-270, 1975.

[10] N.B.O.L. Pettit, T.Manavis and P.E.Wellstead. Using graph theory to visualise piecewise linear systems, In *Proc. 3rd European Control Conference*, pages 1631-1636, Rome, Italy, September, 1995.

Hybrid Controller Design for Multi-Agent Systems *

John Lygeros, Datta N. Godbole and Shankar Sastry

Intelligent Machines and Robotics Laboratory
University of California, Berkeley
Berkeley, CA 94720
lygeros, godbole, sastry@eecs.berkeley.edu

Abstract

A design and verification methodology for hybrid dynamical systems, based on optimal control and game theory, is presented. The hybrid design is seen as a game between two players. One is the disturbances that enter the dynamics. The disturbances can encode the actions of other agents (in a multi-agent setting), the actions of high level controllers or the usual unmodeled environmental disturbances. The second player is the control, which is to be chosen by the designer. The two players compete over a cost function that encodes the properties that the closed loop hybrid system needs to satisfy (e.g. safety). The control "wins" the game if it can keep the system "safe" for any allowable disturbance. The solution to the game theory problem provides the designer with continuous controllers as well as sets of safe states where the control "wins" the game. The sets of safe sets can be used to construct an interface that guarantees the safe operation of the combined hybrid system.

The motivating example for our work is Automated Highway Systems. We show how to cast the lower level, multi-agent vehicle following problem in the game theoretic setting and give an algorithm that can produce a safe design.

1 Introduction

The control of large scale dynamical systems is one of the major challenges facing control engineers today. Recent technological advances, such as faster computers, cheaper and more reliable sensors and the integration of control considerations in the product design and manufacturing process, have made

---

*Research supported by the Army Research Office under grant DAAH 04-95-1-0588 and the PATH program, Institute of Transportation Studies, University of California, Berkeley, under MOU-135 and MOU-238.

it possible to extend the practical applications of control to systems that were impossible to deal with in the past. Large scale systems are common in applications such as chemical process control, power generation and distribution, highway and air traffic control, etc. Of particular interest are multi-agent, scarce resource problems, where a large number of agents, equipped with communication and control capabilities have to make efficient use of a scarce resource. For example, in highway systems, the vehicles can be viewed as agents competing for the highway (which plays the role of the resource) while in air traffic management systems the aircraft compete for air space and runway space. The size and complexity of these systems makes it difficult to approach them using tools from the classical, continuous control literature. To manage the system complexity designers are often forced to use hybrid control schemes, i.e. designs that make use of both discrete and continuous controllers.

Design and verification of hybrid controllers has been an active research area recently. This paper will attempt to present a rather general technique for designing hybrid controllers for complex systems. The main emphasis will be on multiagent systems where many agents are trying to make optimum use of a congested, common resource. Typically in systems like these, the optimum policy for each agent does not coincide with the "common good". Therefore, compromises need to be made. To achieve the common optimum we should ideally have a centralized control scheme that computes the global optimum and commands the agents accordingly. A solution like this may be prohibitively complex, expensive and unreliable. Therefore, a control scheme based on semi-autonomous agent operation is proposed. In this case each agent is trying to optimize its own usage of the resource and coordinates with neighboring agents in case there is a conflict of objectives. Such an architecture is naturally suited for hybrid designs. *At the continuous level each agent chooses it's own optimal strategy, while discrete coordination is used to resolve conflicts.* This is the class of hybrid systems that we will be most interested in.

Background

A common approach to the design of hybrid controllers involves independently coming up with a reasonable design for both the discrete and continuous parts. The combined hybrid controller is then put together by means of interfaces, and verification is carried out to ensure that it satisfies certain properties.

This approach has been motivated by the success of verification techniques for finite state machines and the fact that many hybrid designs already operating in practice need to be verified, without having to redesign them from scratch. Verification algorithms for finite state machines have been in use for years and efficient programs exist to implement them (COSPAN [1], HSIS [2], STATEMATE [3], etc.). They have proved very successful in discrete problems such as communication protocols [4] and software algorithms. The push towards stronger verification techniques has been in the direction of extending the standard finite state machine results to incorporate progressively more complicated continuous dynamics. Theoretical results have established conditions under which verification problems for systems with clocks can be solved computationally [5, 6], and algorithms have been developed to implement the verification process (for example timed COSPAN [7] and KRONOS [8]). Verification of timed systems has proved useful in applications such as digital circuit verification [9] and real-time software [10]. Recently the theory has been extended to systems where the dynamics can be modeled by rectangular differential inclusions [11, 6].

Progress in the direction of automatic verification has been impeded for two fundamental reasons, **undecidability** and **computational complexity**. To guarantee that an automatic verification algorithm will terminate in finite number of steps with an answer, the system needs to satisfy very stringent technical requirements [12]. On the other hand, relatively simple hybrid systems lead to very large numbers of discrete states when looked at from the point of view of automatic verification. Even though efficient algorithms exist, the problem may still be prohibitively large for current computers [10].

A different approach has been to design the hybrid controller so that performance is a-priori guaranteed [13, 14, 15]. This eases the requirements on verification somewhat as a large part of the complexity can be absorbed by careful design. The techniques presented in this paper fit in with this way of thinking.

Design Approach

We start by modeling the systems dynamics at the continuous level. Two factors affect the system evolution at this level. The first is the **control**, that the designer has to determine. The second is the **disturbances** that enter the system, over which we assume no control. We will distinguish two classes of disturbances:

- **Class 1:** Exogenous signals, such as unmodeled forces and torques in mechanical systems, sensor noise, etc and unmodeled dynamics.

- **Class 2:** The actions of other agents, in a multiagent setting.

Disturbances of Class 1 are standard in classical control theory. Class 2 will be the most interesting one from the point of view of hybrid control. Recall that at this stage we are merely modeling the plant, therefore we assume no cooperation between the agents. As a result, each agent views the actions of its neighbors as uncontrollable disturbances.

In the continuous domain specifications about the closed loop system can be encoded in terms of cost functions. Acceptable performance can be encoded by means of thresholds on the final cost. Our objective is to derive a continuous design for the control inputs that guarantees performance despite the disturbances. If it turns out that the disturbance is such that the specifications can not be met for any controller the design fails. The only way to salvage the situation is to somehow limit the disturbance. For disturbances of Class 2 this may be possible by means of communication and coordination between the agents. The objective then is to come up with a discrete design that limits the disturbance so that a continuous design is feasible.

Summarizing, our approach to hybrid controller design consists of determining continuous control laws and conditions under which they satisfy the closed loop requirements. Then a discrete design is constructed to ensure that these conditions are satisfied. This process eliminates the need for automatic verification as the hybrid closed loop system is by design guaranteed to satisfy the specifications.

In this paper we will limit our attention to the design of continuous control laws and interfaces between these laws and the discrete world. The design of the continuous laws will be optimal with respect to the closed loop system requirements. An ideal tool for this kind of set up is game theory. In the game theoretic framework the control and the disturbances are viewed as adversaries in a game. The control seeks to improve system performance while the disturbance seeks to make it worse. Games like these do not necessarily have winners and losers. If we set a threshold on the cost function to distinguish acceptable from unacceptable performance we can say that the control wins the game if the requirement is satisfied for any allowable disturbance, while the disturbance wins otherwise.

Game theoretic ideas have already been applied in this context to problems with disturbances of Class 1 and quadratic cost functions. The resulting controllers are the so called H_∞ or L_2 optimal controllers (see for example [16]). We will try to extend these ideas to the multiagent, hybrid setting.

Limitations

The main limitation of the application of gaming and optimal control ideas to hybrid design is that the resulting problems are usually very hard to solve analytically. As a result efficient numerical algorithms are needed to apply these techniques to more complicated problems. In addition extensive designer input is needed during the controller design, verification or abstraction process.

This approach is best suited to address questions of "reachability". Problems like these can usually be cast as pursuit evasion problems in the gaming framework. Extensions to other important discrete questions such as fairness, liveness and language containment requirements should also be possible, however more work is needed in this direction.

2   Game Theoretic Framework

As our starting point we use a rather general state space model for the continuous plant. Let $x(t) \in \Re^n$ represent the state, $u(t) \in \Re^m$ represent the input and $d(t) \in \Re^p$ represent any disturbance that affects the dynamics, at time t.[1] The plant dynamics will be described by a differential equation and the value

---

[1]The design methodology also extends to systems where the plant dynamics themselves are hybrid (e.g. hopping robots).

of the state at a given time, say t_0:

$$\dot{x}(t) = f(x(t), u(t), d(t), t) \tag{1}$$
$$x(t_0) = x^0 \tag{2}$$

The behavior of the system at time t is assumed to be monitored through a set of outputs $y(t) \in \Re^q$. Their value depends on a map:

$$y(t) = h(x(t), u(t), d(t), t) \tag{3}$$

Physical considerations (such as actuator saturation, etc.) impose certain restrictions on the system evolution. We assume that these restrictions are encoded in terms of constraints on the state, inputs and disturbances.[2]

$$x() \in \mathcal{X} \subset PC^1(\Re, \Re^n) \tag{4}$$
$$u() \in \mathcal{U} \subset PC(\Re, \Re^m) \tag{5}$$
$$d() \in \mathcal{D} \subset PC(\Re, \Re^p) \tag{6}$$

Of particular interest is the case where the constraints can be encoded by requiring that the state, input and disturbance lie in a certain set for all times, that is for all $t \in \Re$:

$$x(t) \in \mathbf{X} \subset \Re^n \tag{7}$$
$$u(t) \in \mathbf{U} \subset \Re^m \tag{8}$$
$$d(t) \in \mathbf{D} \subset \Re^p \tag{9}$$

We assume that the differential equation (1) has unique solutions.

Both inputs and disturbances are exogenous functions of time. The difference is that the disturbances are assumed to be beyond the control of the designer and can take arbitrary values within the constraint set $\mathcal{D}$. As discussed in the introduction, the disturbances can represent a number of factors: unmodeled dynamics or environmental inputs, the actions of other agents or the actions of higher layers (in a multi-layered, hierarchical design). The designer's objective is to use the inputs to regulate the outputs, despite the actions of the disturbances. We will assume that the whole state is available for feedback. Extensions to the case of output feedback should also be possible.

In our framework for hierarchical control the desired properties, that the system trajectory needs to satisfy, will be specified by the discrete layers. The continuous layer will be responsible for the regulation, i.e. the selection of inputs u to try to achieve these properties. The interface provides communication between the two layers, providing the lower layer with the desired properties (in continuous terms) and the higher layer with feedback on whether the requirements can be met. Here we will primarily be concerned with the design of the continuous layer and the interface from continuous to discrete.

---

[2] $PC(\cdot, \cdot)$ denotes the set of piecewise continuous functions whereas $PC^1(\cdot, \cdot)$ represents the set of piecewise differentiable functions.

2.1 Discrete Layer

In this piece of work we will not deal at all with the design of the discrete layer. In terms of the continuous layer the outcome of the discrete design will be represented by:

- A sequence of desired way points x_i^d that should be tracked (i is an integer).

- A set of design specifications (J_i, C_i), where J_i are cost functions:

$$J_i : \Re^n \times PC \times PC \longrightarrow \Re \qquad (10)$$

$i = 1, \ldots, N$ that encode desired properties and $C_i, i = 1, \ldots, N$ are thresholds that specify acceptable limits on the cost functions. An acceptable trajectory is one for which $J_i(x^0, u, d) \leq C_i$ for all $i = 1, \ldots, N$

We assume that the cost functions are ordered in the order of decreasing importance. Qualitatively, the most important cost functions encode things such as safety, while the least important ones encode performance aspects such as resource utilization. The design should be such that the most important constraints are not violated in favor of the less important ones, in other words the design should lead to $J_i \leq C_i$ whenever possible, even if this means that $J_j > C_j$ for some $j > i$. It should be noted that at this stage we make no assumption about the high level design, or even the language used at the higher level. We are merely looking at the higher level from the point of view of the continuous plant, i.e. after the interface. Questions about how the high level objectives, which are usually given linguistically, get parsed to way points, cost functions and thresholds are not addressed.

2.2 Continuous Layer

We now present a technique for systematically constructing controllers which carry out the objectives set by the discrete layer and are optimal with respect to the given cost functions.

At the first stage we treat the design process as a two player, zero sum dynamic game with cost J_1. One player, the control u, is trying to minimize the cost, while the other, the disturbance d, is trying to maximize it. Assume that the game has a saddle point solution, i.e. there exist input and disturbance trajectories, u_1^* and d_1^* such that:

$$
\begin{aligned}
J_1^*(x^0) &= \max_{d \in \mathcal{D}} \min_{u \in \mathcal{U}} J_1(x^0, u, d) \\
&= \min_{u \in \mathcal{U}} \max_{d \in \mathcal{D}} J_1(x^0, u, d) \\
&= J_1(x^0, u_1^*, d_1^*)
\end{aligned}
$$

Consider the set:

$$V_1 = \{x \in \Re^n | J_1^*(x) \leq C_1\}$$

This is the set of all initial conditions for which there exists a control such that the objective on J_1 is satisfied for the worst possible allowable disturbance (and hence for any allowable disturbance).

u_1^* can now be used as a control law. It will guarantee that J_1 is minimized for the worst possible disturbance. Moreover if the initial state is in V_1 it will also guarantee that the performance requirement on J_1 is satisfied. u_1^* however does not take into account the requirements on the remaining J_i's. To include them in the design let:

$$\mathcal{U}_1(x^0) = \{u \in \mathcal{U}|J_1(x^0, u, d_1^*) \leq C_1\} \tag{11}$$

Clearly:

$$\mathcal{U}_1(x^0) \begin{cases} = \emptyset & \text{for } x^0 \notin V_1 \\ \neq \emptyset & \text{for } x^0 \in V_1, \text{ as } u_1^* \in \mathcal{U}_1(x^0) \end{cases}$$

The set $\mathcal{U}_1(x^0)$ is the subset of admissible controls which guarantee that the requirements on J_1 are satisfied, whenever possible. Within this class of controls we would like to select the one that minimizes the cost function J_2. Again we pose the problem as a zero sum dynamic game between control and disturbance. Assume that a saddle solution exists, i.e. there exist u_2^* and d_2^* such that:

$$\begin{aligned} J_2^*(x^0) &= \max_{d \in \mathcal{D}} \min_{u \in \mathcal{U}_1(x^0)} J_2(x^0, u, d) \\ &= \min_{u \in \mathcal{U}_1(x^0)} \max_{d \in \mathcal{D}} J_2(x^0, u, d) \\ &= J_2(x^0, u_2^*, d_2^*) \end{aligned}$$

Consider the set:

$$V_2 = \{x \in \Re^n | J_2^*(x) \leq C_2\}$$

As the minimax problem only makes sense when $\mathcal{U}_1(x^0) \neq \emptyset$ we assume that $V_2 \subset V_1$. V_2 represents the set of initial conditions for which there exists a control such that for any allowable disturbance the requirements on both J_1 and J_2 are satisfied. To introduce the remaining cost functions to the design we again define:

$$\mathcal{U}_2(x^0) = \{u \in \mathcal{U}_1(x^0)|J_2(x^0, u, d_2^*) \leq C_2\} \tag{12}$$

i.e. the subset of admissible controls that satisfy the requirements on both J_1 and J_2 for any disturbance.

The process can be repeated for the remaining cost functions. At the $i + 1^{\text{st}}$ step we are given a set of admissible controls $\mathcal{U}_i(x^0)$ and a set of initial conditions V_i such that for all $x^0 \in V_i$ there exists $u_i^* \in \mathcal{U}_i(x^0)$ such that for all $d \in \mathcal{D}$, $J_j(x^0, u_i^*, d) \leq C_j$, where $j = 1, \ldots, i$. Assume the two player, zero sum dynamic game for J_{i+1} has a saddle solution, u_{i+1}^*, d_{i+1}^*:

$$\begin{aligned} J_{i+1}^*(x^0) &= \max_{d \in \mathcal{D}} \min_{u \in \mathcal{U}_i(x^0)} J_{i+1}(x^0, u, d) \\ &= \min_{u \in \mathcal{U}_i(x^0)} \max_{d \in \mathcal{D}} J_{i+1}(x^0, u, d) \\ &= J_{i+1}(x^0, u_{i+1}^*, d_{i+1}^*) \end{aligned}$$

Define:

$$V_{i+1} = \{x \in \Re^n | J_{i+1}^*(x) \le C_{i+1}\}$$

and:

$$\mathcal{U}_{i+1}(x^0) = \{u \in \mathcal{U}_i(x^0) | J_{i+1}(x^0, u, d_{i+1}^*) \le C_{i+1}\} \tag{13}$$

The process can be repeated until the last cost function. The result is a control law u_N^* and a set of initial conditions $V_N = V$ such that for all $x^0 \in V_N$ and for all $d \in \mathcal{D}$ $J_j(x^0, u_N^*, d) \le C_j$ where $j = 1, \ldots, N$. The controller can be extended to values of the state in the complement of V using the following switching scheme:

$$u^*(x) = \begin{cases} u_N^*(x) & x \in V \\ u_{N-1}^*(x) & x \in V_{N-1} \setminus V \\ \ldots & \ldots \\ u_1^*(x) & x \in \Re^n \setminus V_2 \end{cases} \tag{14}$$

The algorithm presented above is sound in theory but can run into technical difficulties when applied in practice:

1. There is no guarantee that the dynamic games will have a saddle solution.

2. There is no straight-forward way of computing $\mathcal{U}_i(x^0)$

3. There is no guarantee that the sets V_i (and consequently $\mathcal{U}_i(x^0)$) will be non-empty.

2.3 Interface

The sets V are such that for all initial conditions in them all requirements on system performance are guaranteed. These sets impose conditions that the discrete switching scheme needs to satisfy. The discrete layer should not issue a new command (way point) if the current state does not lie in the set V for the associated controller. Essentially these sets offer a way of consistently abstracting performance properties of the continuous layer.

It should be noted that, by construction, the sets V_i are nested. Therefore there is a possibility that an initial condition lies in some V_i but not in V. This implies that certain requirements on the system performance (e.g. safety) are satisfied, while others (e.g. efficient resource utilization) are not. This allows the discrete design some more freedom. It may, for example, choose to issue a new command if it is dictated by safety, even though it violates the requirements of efficiency. This construction provides a convenient way of modeling gradual performance degradation, where lower priority performance requirements are abandoned in favor of higher priority ones.

In the next section, we illustrate this design methodology by an example.

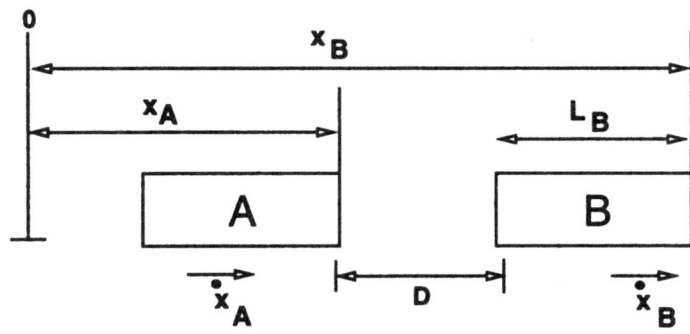

Figure 1: Vehicle Following

3 Automated Vehicle Following Example

In this section we investigate how the approach developed in Section 2 can be useful in applications. We consider the example of vehicle following on an automated highway. To simplify the notation, time dependency of the states inputs and outputs will be suppressed unless explicitly stated.

3.1 Problem Statement

Consider two vehicles (labeled A and B) moving along a single lane highway (Figure 1). Assume that the vehicles have lengths L_A and L_B and let x_A and x_B denote their positions with respect to a fixed reference on the road. Assume that vehicle B is leading, i.e. $x_B > x_A > 0$. The problem we are interested in is the vehicle following problem: we assume no control over vehicle B and try to control vehicle A.

Leading Vehicle: The dynamics of the leading vehicle will be abstracted by a second order ordinary differential equation:

$$\ddot{x}_B = d$$

with $d(t) \in [d_{min}, d_{max}] \subset \Re$ and $-\infty < d_{min} \leq 0 \leq d_{max} < \infty$. The values of d_{min} and d_{max} are dictated by the road and vehicle conditions. We will use $d_{min} = -5ms^{-2}$ and $d_{max} = 3ms^{-2}$. This abstraction is justified by the laws of motion and the assumed sensor arrangement to be discussed shortly.

To make the problem more realistic, a restriction on the speed of vehicle B is imposed:

$$\dot{x}_B \in [v_{min}, v_{max}]$$

For highway operation it is assumed that vehicles will not be allowed to go backwards, therefore $v_{min} = 0$ will be used. v_{max} is imposed by engine limitations. One objective of the controllers we design will be fuel efficiency. As a consequence the engine will not have to be pushed to its limits for maximum speed and therefore v_{max} will not feature in the calculations. Therefore, we will assume $v_{max} = \infty$ to simplify the analysis.

Trailing Vehicle: The dynamics of the trailing vehicle will be approximated by a third order ordinary differential equation:

$$\dddot{x}_A = b_A(\dot{x}_A, \ddot{x}_A) + a_A(\dot{x}_A)v_A$$

a_A and b_A are complicated nonlinear functions of the state with $a_A(\dot{x}_A) \neq 0$. The first two derivatives on x_A arise from the laws of motion. The third has to do with the operation of the engine, which can be modeled by the throttle angle acting through an integrator on some nonlinear dynamics (involving engine time constants, aerodynamic drag, etc.). For our purposes the details of the nonlinear functions b_A and a_A are not important. Following the designs of [17], we will assume that feedback linearization has already been carried out, i.e.:

$$v_A(t) = \frac{-b_A(\dot{x}_A, \ddot{x}_A) + u}{a_A(\dot{x}_A)}$$

$$\dddot{x}_A = u$$

We will design controllers for the resulting linear dynamics.

As for the leading vehicle we assume that the dynamics are constrained by the engine, tire and road conditions. More specifically it is required that:

$$\dot{x}_A \in [v_{min}, v_{max}] = [0, \infty)ms^{-1}$$
$$\ddot{x}_A \in [a_{min}, a_{max}] \approx [d_{min}, d_{max}]$$
$$u(t) \in [j_{min}, j_{max}]$$

The design will use $a_{min} = -5ms^{-2}, a_{max} = 3ms^{-2}, j_{min} = -50ms^{-3}$ and $j_{max} = 50ms^{-3}$.

Sensors: Based on current technology, it is assumed that vehicle A is equipped with sensors that can measure its own velocity and acceleration, as well as relative position and velocity with respect to vehicle B. We will assume that the acceleration of vehicle B can not be measured and is not communicated to vehicle A (as it is, for example, for the platoon following scenario of [18]).

The vehicles are also equipped with communication devices. It is assumed that they will only be used to exchange discrete messages for coordinating the agent operation. In case the vehicle is equipped with multiple, redundant sensors or communication capabilities it will be assumed that the necessary sensor fusion has been carried out before hand and the designer has access to a unique and accurate measurement.

Specifications: The objective is to design a controller for vehicle A. The requirements we impose on the design are safety, passenger comfort and efficiency. It is assumed that safety takes precedence over the other two requirements. Comfort and efficiency will be treated as equally important and a compromise between them will be sought. Safety will be assumed to mean no collision between vehicles A and B, at any time[3]. For passenger comfort it is required that the input u is kept "as small as possible". In the transportation literature, $\dddot{x}_A \leq 2.5ms^{-3}$ is often quoted as the limit for a comfortable ride. Finally for efficiency it is required that all maneuvers (for example convergence of the vehicle spacing to a desired value) take place "as quickly as possible". All the above statements are quantified in the next section.

3.2 Game Theoretic Formulation

Both the design specifications and the system dynamics are independent of the absolute vehicle position. To completely remove the absolute position from the problem we introduce a new variable to measure the spacing between vehicles A and B:

$$D = x_B - x_A - L_B$$

All pertinent information can now be encoded by the state vector:

$$x = \begin{bmatrix} \dot{x}_A \\ \ddot{x}_A \\ D \\ \dot{D} \end{bmatrix} = \begin{bmatrix} x_1 \\ x_2 \\ x_3 \\ x_4 \end{bmatrix} \in \Re^4$$

¿From Section 3.1 the dynamics are:

$$\dot{x} = \begin{bmatrix} 0 & 1 & 0 & 0 \\ 0 & 0 & 0 & 0 \\ 0 & 0 & 0 & 1 \\ 0 & -1 & 0 & 0 \end{bmatrix} x + \begin{bmatrix} 0 \\ 1 \\ 0 \\ 0 \end{bmatrix} u + \begin{bmatrix} 0 \\ 0 \\ 0 \\ 1 \end{bmatrix} d$$

$$= Ax + Bu + Dd$$

$$x(0) = x^0$$

For the vehicle following problem we are interested in regulating the spacing and relative velocity to a desired fixed point. This requirement can be encoded in terms of two outputs:

$$y = \begin{bmatrix} D \\ \dot{D} \end{bmatrix} = \begin{bmatrix} 0 & 0 & 1 & 0 \\ 0 & 0 & 0 & 1 \end{bmatrix} x$$

$$= Cx$$

Note that the assumed sensor arrangement can provide full state measurements.

To complete the picture we also need to encode all the constraints in the new coordinates. From the discussion in Section 3.1 it is required that, for all

---

[3]Different definitions of safety, for example no high relative velocity collisions can also be accommodated in this framework

t:

$$x(t) \in X \quad = \quad \left\{ x \in \Re^4 | x_1 \in [v_{min}, v_{max}], x_2 \in [a_{min}, a_{max}], x_4 + x_1 \in [v_{min}, v_{max}] \right\}$$
$$u(t) \in U \quad = \quad [j_{min}, j_{max}]$$
$$d(t) \in D \quad = \quad [d_{min}, d_{max}]$$

The analysis can be greatly simplified if the input constraints are modified somewhat to guarantee that the state constraints are satisfied. In particular we will assume that:

$$u(t) \quad \in \quad \begin{cases} [j_{min}, j_{max}] & \text{if } x_2 \in (a_{min}, a_{max}) \\ [0, j_{max}] & \text{if } x_2 = a_{min} \\ [j_{min}, 0] & \text{if } x_2 = a_{max} \end{cases} \qquad (15)$$

$$d(t) \quad \in \quad \begin{cases} [d_{min}, d_{max}] & \text{if } x_4 + x_1 \in (v_{min}, v_{max}) \\ [0, d_{max}] & \text{if } x_4 + x_1 = v_{min} \\ [d_{min}, 0] & \text{if } x_4 + x_1 = v_{max} \end{cases} \qquad (16)$$

This will ensure that the constraints on x_2 and $x_1 + x_4$ will never be violated if $x^0 \in X$.

The only state constraint that we have to worry about is $x_1 \in [v_{min}, v_{max}]$. For the reasons discussed above we will not be concerned with the upper bound too much. We can work around the lower bound by assuming that $\ddot{x}_A = x_2$ becomes zero when $x_1 = v_{min}$ is reached (recall that when x_1 reaches v_{min}, $x_2 \le 0$).

The two players are the unmeasured acceleration, d, of vehicle B and the controller, u, of vehicle A. They compete in three fronts over the cost functions:

1. **Safety:**

$$J_1(x^0, u, d) = -\inf_{t \ge 0} x_3(t) \qquad (17)$$

A safe maneuver is one where:

$$J_1(x^0, u, d) \le C_1 = 0 \ m$$

Allowing $J_1 = 0$ meters makes the limiting case (where the vehicles just touch with zero relative velocity) acceptable.

2. **Comfort:**

$$J_2(x^0, u, d) = \sup_{t \ge 0} |u(t)| \qquad (18)$$

A comfortable maneuver is one where:

$$J_2(x^0, u, d) \le C_2 = 2.5 ms^{-3}$$

3. **Efficiency:**

$$J_3(x^0, u, d) = \int_0^{\infty} (y(\tau) - y_d)^T P(y(\tau) - y_d) d\tau \qquad (19)$$

where y_d is the desired fixed point for a given maneuver and P is positive definite.

The solution to the system equations can be obtained using the variation of constants formula [19]:

$$x(t) = e^{At}x^0 + \int_0^t e^{A(t-\tau)}Bu(\tau)d\tau + \int_0^t e^{A(t-\tau)}Dd(\tau)d\tau$$

By the fact that A is nilpotent ($A^3 = 0$) we obtain:

$$x(t) = \begin{bmatrix} x_1^0 + tx_2^0 \\ x_2^0 \\ -t^2x_2^0/2 + x_3^0 + tx_4^0 \\ -tx_2^0 + x_4^0 \end{bmatrix} + \int_0^t \begin{bmatrix} t-\tau \\ 1 \\ -(t-\tau)^2/2 \\ -t+\tau \end{bmatrix} u(\tau)d\tau + \int_0^t \begin{bmatrix} 0 \\ 0 \\ t-\tau \\ 1 \end{bmatrix} d(\tau)_{\cdot}$$

(20)

This equation is valid, under the assumed constraints (15, 16), as long as $x_1(t) \in [v_{min}, v_{max}]$.

3.3 Design for Safety

As for the train gate problem, we will start by guessing a saddle solution. Let:

$$T_1 = \frac{a_{min} - x_2^0}{j_{min}} \qquad T_2 = -\frac{x_1^0 + x_4^0}{d_{min}}$$

Consider the candidate saddle strategy:

$$u^*(t) = \begin{cases} j_{min} & \text{if } t \le T_1 \\ 0 & \text{if } t > T_1 \end{cases} \qquad (21)$$

$$d^*(t) = \begin{cases} d_{min} & \text{if } t \le T_2 \\ 0 & \text{if } t > T_2 \end{cases} \qquad (22)$$

In other words, both A and B try to come to a stop as quickly as possible, under the given constraints.

For the candidate saddle solution $x(t)$ can be written explicitly:

- If $0 \le t \le \min\{T_1, T_2\}$:

$$x(t) = e^{At}x^0 + \begin{bmatrix} t^2/2 \\ t \\ -t^3/6 \\ -t^2/2 \end{bmatrix} j_{min} + \begin{bmatrix} 0 \\ 0 \\ t^2/2 \\ t \end{bmatrix} d_{min}$$

- If $T_1 \le t \le T_2$:

$$x(t) = e^{At}x^0 + \begin{bmatrix} tT_1 - T_1^2/2 \\ T_1 \\ -T_1^3/6 + tT_1(T_1-t)/2 \\ -tT_1 + T_1^2/2 \end{bmatrix} j_{min} + \begin{bmatrix} 0 \\ 0 \\ t^2/2 \\ t \end{bmatrix} d_{min}$$

- If $T_2 \leq t \leq T_1$:

$$x(t) = e^{At}x^0 + \begin{bmatrix} t^2/2 \\ t \\ -t^3/6 \\ -t^2/2 \end{bmatrix} j_{min} + \begin{bmatrix} 0 \\ 0 \\ tT_2 - T_2^2/2 \\ T_2 \end{bmatrix} d_{min}$$

- If $t \geq \max\{T_1, T_2\}$:

$$x(t) = e^{At}x^0 + \begin{bmatrix} tT_1 - T_1^2/2 \\ T_1 \\ -T_1^3/6 + tT_1(T_1 - t)/2 \\ -tT_1 + T_1^2/2 \end{bmatrix} j_{min} + \begin{bmatrix} 0 \\ 0 \\ tT_2 - T_2^2/2 \\ T_2 \end{bmatrix} d_{min}$$

Let T_3 be the stopping time for vehicle A. If $0 \leq T_3 \leq T_1$:

$$T_3^2 j_{min}/2 + T_3 x_2^0 + x_1^0 = 0$$

$$\Rightarrow \quad T_3 = \frac{-x_2^0 - \sqrt{(x_2^0)^2 - 2j_{min}x_1^0}}{j_{min}}$$

Note that the second solution would lead to $T_3 < 0$ (assuming that $j_{min} < 0$ and $x_1^0 \geq 0$). If $T_1 \leq T_3$:

$$(T_1 T_3 - T_1^2)j_{min} + T_3 x_2^0 + x_1^0 = 0$$

$$\Rightarrow \quad T_3 = \frac{(a_{min} - x_2^0)^2 - 2j_{min}x_1^0}{2j_{min}a_{min}}$$

¿From the discussion at the end of Section 3.2 the times of interest are now restricted to the interval $t \in [0, T_3]$. A simple calculation shows that:

Lemma 1 *If $x^0 \in X$ then $x(t) \in X$ for all $t \in [0, T_3]$.*

To calculate $J_1(x^0, u^*, d^*)$ recall that $x_3(t)$ is a differentiable function of time, with derivative $x_4(t)$, defined on a compact interval $[0, T_3]$. Therefore:

Lemma 2 *There exist $\hat{T} \in [0, T_3]$ such that:*

$$J_1(x^0, u^*, d^*) = -x_3(\hat{T})$$

Moreover:

$$\hat{T} \in \{0, T_3, \{T \in (0, T_3)|x_4(T) = 0\}\}$$

The calculations for analytically determining $J_1^*(x^0) = J_1(x^0, u^*, d^*)$ from this lemma are rather messy. However if the set of times T where $x_4(T) = 0$ is finite we can easily carry out the calculation numerically. After a few steps of algebra we can determine that:

- For $0 \leq T \leq \min\{T_1, T_2\}$:

$$T = \frac{d_{min} - x_2^0 \pm \sqrt{(d_{min} - x_2^0)^2 + 2j_{min}x_4^0}}{j_{min}}$$

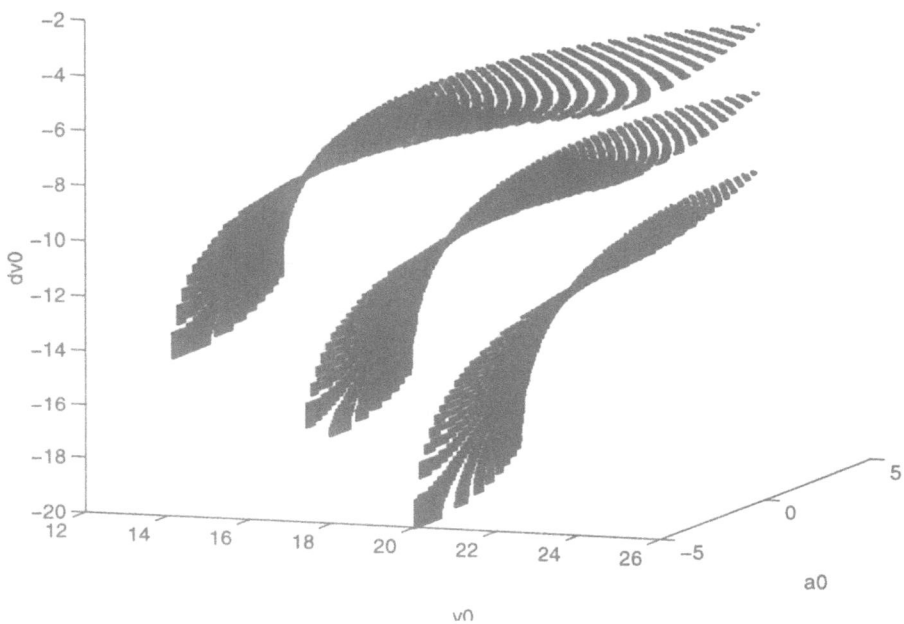

Figure 2: Safe set of initial conditions for $x_3^0 = 15, 20$ and 25 meters

- If $T_1 \le T \le T_2$:

$$T = \frac{x_4^0 + T_1^2 j_{min}/2}{a_{min} - d_{min}}$$

- If $T_2 \le T \le T_1$:

$$T = \frac{-x_2^0 \pm \sqrt{(x_2^0)^2 + 2j_{min}(T_2 d_{min} + x_4^0)}}{j_{min}}$$

- If $T \ge \max\{T_1, T_2\}$:

$$T = \frac{x_4^0 + T_1^2 j_{min}/2 + T_2 d_{min}}{a_{min}}$$

A computer program was written to calculate $J_1^*(x^0)$ for various values of x^0 and the constants given in Section 3.1. It should be noted that x_3^0 only enters the calculations as a constant offset on $x_3(t)$. Figure 2 shows the surface where $J_0^*(x^0) = 0$ for some values of x_3^0. Any initial condition on or above these surfaces will not lead to a collision under the strategy (u^*, d^*).

To complete the safety calculation we need to show:

Lemma 3 (u^*, d^*) *is globally a saddle solution for cost* $J_1(x^0, u, d)$.

Proof: For (u^*, d^*) to be a saddle point we need to show that a unilateral change in strategy leaves the player who decided to change worse off. Let $x^*(t)$ denote the state at time t under the inputs (u^*, d^*).

$$x_3^*(t) = \begin{bmatrix} 0 & -t^2/2 & 1 & t \end{bmatrix} x^0 - \int_0^t \frac{(t-\tau)^2}{2} u^*(\tau) d\tau + \int_0^t \frac{t-\tau}{2} d^*(\tau) d\tau$$

Fist fix u^* and allow d to vary. Let $x(t)$ denote the state at time t under the inputs (u^*, d). Then:

$$x_4^*(t) = \begin{bmatrix} 0 & -t & 0 & 1 \end{bmatrix} x^0 - \int_0^t (t-\tau) u^*(\tau) d\tau + \int_0^t d^*(\tau) d\tau$$

$$x_4(t) = \begin{bmatrix} 0 & -t & 0 & 1 \end{bmatrix} x^0 - \int_0^t (t-\tau) u^*(\tau) d\tau + \int_0^t d(\tau) d\tau$$

$$\Rightarrow x_4(t) - x_4^*(t) = \int_0^t (d(\tau) - d^*(\tau)) d\tau$$

We need to distinguish two cases:

1. $t \leq T_2$. The bounds on d imply that:

$$d(\tau) \geq d^*(\tau) \quad \Rightarrow \quad \int_0^t (d(\tau) - d^*(\tau)) d\tau \geq 0$$

$$\Rightarrow \quad x_4(t) - x_4^*(t) \geq 0$$

2. $t \geq T_2$. Then, by definition of T_2, $\int_0^t d^*(\tau) d\tau = -(x_1^0 + x_4^0)$. The state constraints imply that $x_1(t) + x_4(t) \geq 0$ (vehicle B does not go backwards), therefore $\int_0^t d(\tau) d\tau \geq -(x_1^0 + x_4^0)$. Subtracting, $\int_0^t (d(\tau) - d^*(\tau)) d\tau \geq 0$.

In both cases, we are able to conclude that $\dot{x}_3(t) = x_4(t) \geq x_4^*(t) = \dot{x}_3^*(t)$. Integrating this inequality from 0 to t and using the fact that $x_3^*(0) = x_3(0) = x_3^0$:

$$x_3(t) \geq x_3^*(t) \text{ for all } t$$

$$\Rightarrow -\inf_{t \geq 0} x_3(t) \leq -\inf_{t \geq 0} x_3^*(t)$$

$$\Rightarrow J_1(x^0, u^*, d) \leq J_1(x^0, u^*, d^*) \tag{23}$$

Now fix $d = d^*$ and let u vary. Let $x(t)$ denote the state at time t under the inputs (u, d^*). Then:

$$x_4(t) = \begin{bmatrix} 0 & -t & 0 & 1 \end{bmatrix} x^0 - \int_0^t (t-\tau) u(\tau) d\tau + \int_0^t d^*(\tau) d\tau$$

$$\Rightarrow x_4(t) - x_4^*(t) = \int_0^t (t-\tau)(u^*(\tau) - u(\tau)) d\tau$$

As above we need to distinguish two cases:

1. $t \leq T_1$. The bounds on u imply that:

$$u(\tau) \geq u^*(\tau) \quad \Rightarrow \quad \int_0^t (t - \tau)(u^*(\tau) - u(\tau))d\tau \leq 0$$
$$\Rightarrow \quad x_4(t) - x_4^*(t) \leq 0$$

2. $t \geq T_1$. Recall that $d^*(t)$ is piecewise constant, with a discontinuity at T_2 (which may be either greater or less than T_1). Therefore $x_4(t)$ and $x_4^*(t)$ are piecewise differentiable, with derivatives $-x_2(t)$ and $-x_2^*(t)$ respectively. By definition of T_1, $x_2^*(t) = a_{min} \leq x_2(t)$ in the interval of interest. Therefore, as $x_4(0) = x_4^*(0) = x_4^0$, $x_4^*(t) \geq x_4(t)$.

In both cases $\dot{x}_3(t) = x_4(t) \leq x_4^*(t) = \dot{x}_3^*(t)$. Using the fact that $x_3^*(0) = x_3(0) = x_3^0$ and integrating:

$$x_3(t) \quad \leq \quad x_3^*(t) \text{ for all } t$$
$$\Rightarrow -\inf_{t \geq 0} x_3(t) \quad \geq \quad -\inf_{t \geq 0} x_3^*(t)$$
$$\Rightarrow J_1(x^0, u, d^*) \quad \geq \quad J_1(x^0, u^*, d^*) \tag{24}$$

Combining inequalities 23 and 24:

$$J_1(x^0, u^*, d) \leq J_1^*(x^0) \leq J_1(x^0, u, d^*) \tag{25}$$

for all d and u. By definition, (u^*, d^*) is a saddle solution. $\square$

Lemma 3 implies that the surfaces of Figure 2 are 2D slices of the 3D boundary of the safe set $V \subset X$ for various values of x_3^0.

3.4 Completing the Design

The analysis of the previous section provides a design that will satisfy the requirements for safety. By construction any initial condition in the safe set (on or above the corresponding surface of Figure 2) will not lead to a collision, if the control of equation (21) is applied. In the interior of the safe set, where safety is guaranteed, the other performance aspects (passenger comfort and efficiency) also become important. In order to complete the continuous design we now need to come up with a controller that optimizes with respect to these performance aspects in the interior of the safe set.

The safe set also provides guidelines for the discrete design. The discrete level should be such that initial conditions outside the safe set are never encountered. Such initial conditions may be caused for example in a multilane highway, when vehicle B changes lane in front of vehicle A. The safe set provides requirements on the vehicle states for a safe lane change to take place. It is up to the discrete layer to guarantee that these requirements are satisfied.

4   Concluding Remarks

Hierarchical, hybrid system design and verification has attracted significant attention over the last few years. In this paper we presented an approach to the design of hybrid controllers for complex systems. Our approach fits nicely with a multiagent scenario. We proposed a control architecture based on semi-autonomous agent operation. Our scheme naturally leads to a hierarchical, hybrid design, with continuous controllers trying to optimize each agents resource utilization at a lower level and discrete controllers resolving interagent conflicts at a higher level. An algorithm was presented to produce the continuous controllers, as well as abstractions of their performance in terms of the discrete level. The algorithm makes use of ideas from game theory, treating the design process as a two player, zero sum game between the controller of an agent and the disturbance generated by the actions of other agents. The resulting abstractions can be though of as guidelines for the design of the discrete layer. The implication is that if the resulting continuous controllers are used and the discrete controller satisfies the guidelines, the closed loop hybrid system is, by design, guaranteed to exhibit the desired behavior.

Our approach was illustrated by means of the vehicle following problem. We also discussed how similar ideas can be used to solve problems of verification and discrete abstraction generation.

Future work in this area involves deriving conditions under which the continuous problems are solvable (saddle and optimal solutions exist, etc.), as well as the techniques for designing the higher (discrete) layer. We hope to be able to extend older results in game theory [20] and optimal control [21] in order to solve the problems in the continuous domain and also to develop algorithms for determining the sets of guaranteed performance, V_i.

The problem of designing the discrete layer also involves the choice of descriptive language at the discrete layer. The current approach only provides switching guidelines which can be used by the designer to come up with a discrete design. An interesting issue that needs to be addressed in the discrete setting is what happens if the abstractions indicate that conflicting objectives (e.g. safety and capacity) can not be met. As discussed in the introduction, the problem may still be solvable in this case by using inter-agent coordination. The role of coordination is to cut down on the set of allowable disturbances generated by other agent actions and hence bias the game in the controllers favor. The questions of how are the higher levels to be designed to achieve sufficient coordination and where is the line between feasible and infeasible performance requirements are very interesting and still require a lot of work to be answered.

References

[1] Z. Har'El and R. Kurshan, *Cospan User's Guide.* AT&T Bell Laboratories, 1987.

[2] Adnan Aziz, et al., "HSIS: a BDD-based environment for formal verification," in *ACM/IEEE International Conference on CAD*, 1994.

[3] M. Heymann, "Hierarchical decomposition of hybrid systems." (preprint), 1994.

[4] A. Hsu, F. Eskafi, S. Sachs, and P. Varaiya, "Protocol design for an automated highway system," *Discrete Event Dynamic Systems*, vol. 2, no. 1, pp. 183–206, 1994.

[5] R. Alur, C. Courcoubetis, and D. Dill, "Model checking for real-time systems," *Logic in Computer Science*, pp. 414–425, 1990.

[6] R. Alur, C. Courcoubetis, T. A. Henzinger, and P. H. Ho, "Hybrid automaton: An algorithmic approach to the specification and verification of hybrid systems," in *Hybrid System* (R. L. Grossman, A. Nerode, A. P. Ravn, and H. Rischel, eds.), no. 736 in LNCS, pp. 209–229, New York: Springer Verlag, 1993.

[7] R. P. Kurshan, *Computer-aided verification of coordinating processes; the automata-theoretic approach.* Princeton University Press, 1994.

[8] C. Daws and S. Yovine, "Two examples of verification of multirate timed automata with KRONOS," in *Proc. 1995 IEEE Real-Time Systems Symposium, RTSS'95*, (Pisa, Italy), IEEE Computer Society Press, Dec. 1995.

[9] F. Balarin, *Iterative Methods for Formal Verification of Digital Systems.* PhD thesis, University of California, Berkeley, 1994.

[10] F. Balarin, K. Petty, and A. L. Sangiovanni-Vincentelli, "Formal verification of the PATHO real-time operating system," in *IEEE Control and Decision Conference*, pp. 2459–2465, 1994.

[11] A. Puri and P. Varaiya, "Decidability of hybrid systems with rectangular differential inclusions," in *Computer Aided Verification*, pp. 95–104, 1994.

[12] T. Henzinger, P. Kopke, A. Puri, and P. Varaiya, "What's decidable about hybrid automata," in *STOCS*, 1995.

[13] A. Deshpande, *Control of Hybrid Systems.* PhD thesis, Department of Electrical Engineering, University of California, Berkeley, California, 1994.

[14] M. S. Branicky, V. S. Borkar, and S. K. Mitter, "A unified framework for hybrid control: Background, model and theory," Tech. Rep. LIDS-P-2239, Laboratory for Information and Decision Systems, Massachusetts Institute of Technology, 1994.

[15] A. Nerode and W. Kohn, "Multiple agent hybrid control architecture," in *Hybrid System* (R. L. Grossman, A. Nerode, A. P. Ravn, and H. Rischel, eds.), no. 736 in LNCS, pp. 297–316, New York: Springer Verlag, 1993.

[16] J. C. Doyle, K. Glover, P. P. Khargonekar, and B. A. Francis, "State-space solutions to standard H_2 and H_∞ control problems," *IEEE Transactions on Automatic Control*, vol. 34, no. 8, pp. 831–847, 1989.

[17] D. N. Godbole and J. Lygeros, "Longitudinal control of the lead car of a platoon," *IEEE Transactions on Vehicular Technology*, vol. 43, no. 4, pp. 1125–1135, 1994.

[18] J. K. Hedrick, D.McMahon, V. Narendran, and D. Swaroop, "Longitudinal vehicle controller design for IVHS system," in *American Control Conference*, pp. 3107–3112, 1991.

[19] F. M. Callier and C. A. Desoer, *Linear System Theory*. Springer-Verlag, 1991.

[20] T. Basar and G. J. Olsder, *Dynamic Non-cooperative Game Theory*. Academic Press, 2nd ed., 1994.

[21] L. Berkovitz, *Optimal Control Theory*. Springer-Verlag, 1974.

Logic and Mathematical Programming

Sanjoy K. Mitter
(Joint work with V. Borkar, V. Chandru
[both of Indian Institute of Science]
and D. Micciancio [MIT]) *

Department of Electrical Engineering and Computer Science
and
Laboratory for Information and Decision Systems
Massachusetts Institute of Technology
Cambridge, Massachusetts 02139

1 Introduction

A fundamental problem in logic is determining whether a formula is satisfiable, i.e. there exists a valuation for the variables occurring in the formula that makes the whole formula true. Logical deduction can be easily reduced to satisfiability: formula ϕ is a logical consequence of a set of formulas A if and only if the set of formulas $A \cup \{\neg\phi\}$ is unsatisfiable. Therefore algorithms to decide the satisfiability of formulas can immediately be turned into procedures for logical deduction and automated reasoning.

In fact, the first serious studies of spatial embeddings of logic [5, 4] have been primarily aimed to transfer methodologies and algorithms from the field of mathematical programming to the symbolic world of computational logic.

A new perspective on these studies is presented in [3]. In [3] the embedding of logic into mathematical programming is used to prove some well known theorems of first order logic. The novelty of this work is not in the results achieved, but in the approach used: the topological structure of the space logical satisfiability is embedded into is exploited to gain structural insights.

We are interested in logic mainly as a language to describe and reason about computer programs. ¿From this point of view, it would be interesting to see to what extent the spatial embeddings studied for propositional logic can be extended to other logic languages, such as dynamic logic [7] and process logic [6].

Finding embeddings of dynamic logic in the style of [3] is presumably a hard problem because of some non-compactness results that affect that logic.

*This research has been supported by U.S. Army Research Office grant DAAL03-92-G-0115 to the Center for Intelligent Control Systems and by a grant from Siemens AG.

Therefore it seems desirable to explore the feasibility of such embeddings by choosing a fairly simple subset of dynamic logic, namely modal logic.

The rest of this report is organized as follows. In Section 2 we review the definitions and results presented in [3] for propositional and predicate logic. Section 3, 4 and 5 summarize results obtained in [3] In section 6 we show how modal logic can be spatially embedded into a linear systems. Finally, connections of modal logic to the bisimulation relation are described in section 7 together with a simple example.

2 Propositional and Predicate Logic

In propositional logic, formulas are built up from propositional variables through the use of the usual boolean connectives $\lor$, $\land$ and $\neg$. Propositional variables can evaluate to either *True* or *False*. In order to embed a propositional formula into a linear program, we can associate the numbers 0 and 1 to the the symbolic values *False* and *True*. It is then natural to express disjunctions as $0 - 1$ linear inequalities and formulas in conjunctive normal form as $0 - 1$ linear systems.

For example the satisfiability of the logical formula

$$x \lor \neg y \lor z$$

can be embedded as solubility of the inequality

$$x + (1 - y) + z \geq 1$$

where x, y, z are $0 - 1$ variables.

The conjunctive normal formula

$$(x) \land (y \lor \neg z) \land (\neg w \lor x \lor \neg y) \land (w \lor z)$$

is satisfiable if and only if the following system has solution:

$$
\begin{aligned}
x &\geq 1 \\
y + (1 - z) &\geq 1 \\
(1 - w) + x + (1 - y) &\geq 1 \\
w + z &\geq 1 \\
x, y, z, w &= 0 \text{ or } 1
\end{aligned}
$$

It is conventional in mathematical programming to express linear systems in matrix notation:

$$Ax \geq b$$

where A is a matrix of coefficients, x is a vector of unknown and b is a vector of constants. For example the above system can be rewritten as

$$
\begin{bmatrix}
1 & 0 & 0 & 0 \\
0 & 1 & -1 & 0 \\
1 & -1 & 0 & -1 \\
0 & 0 & 1 & 1
\end{bmatrix}
\begin{bmatrix}
x \\
y \\
z \\
w
\end{bmatrix}
\geq
\begin{bmatrix}
1 \\
0 \\
-1 \\
1
\end{bmatrix} .
$$

In general we can associate to each formula in conjunctive normal form ϕ a system $Ax \geq b$ of c linear inequalities over $0 - 1$ variables, where c is the number of clauses in ϕ. The satisfiability of ϕ is equivalent to the solubility of the associated system

$$Ax \geq b, x \in \{0, 1\}^n . \tag{1}$$

This spatial embedding of propositional logic is extended in [3] to predicate calculus, using infinite dimensional linear programming. Briefly, a first order formula is transformed into an infinite conjunction of propositional clauses and the resulting infinite propositional formula is embedded into a system $Ax \geq \beta$ over an infinite set x of $0 - 1$ variables. The passage from predicate calculus to propositional logic uses standard techniques from logic (*Skolemization*). The embedding of infinitary propositional calculus into infinite mathematical programming extends the finitary case in the obvious way. As usual the formula is satisfiable if and only if the associated system has solution. The structure of the resulting system (the *clausal form* and *finite support* property, [3]) and general properties of the product of topological spaces are used in [3] to prove the Herbrand theorem and the existence of a minimal solution for systems of Horn clauses.

We now discuss these results in summary form.

3 Infinite Dimensional $0 - 1$ Linear Programs

Consider a mathematical program of the form

$$\mathcal{D} = \{x \in \{0, 1\}^\infty : A_x \geq \beta\} \tag{2}$$

Each row of the matrix A has entries that are $0, \pm 1$, and each entry of the (uncountably) infinite column β is $1-$ the number of -1's in the corresponding row of A. So this is just an infinite version of (1). The finite support of the rows of A is the important structural property that permits the compactness theorems based on product topologies to go through in the ensuing development. It is a natural restriction in the context of first order logic as it corresponds to the finite "matrix" property of first order formulae. Note that compactness theorems can be pushed through for more general infinite mathematical programs using the so called "weak $*$ topologies" but this shall not concern us.

In discussing Horn Logic, we will encounter the continuous (linear programming) relaxation of our infinite mathematical program (2).

$$\bar{\mathcal{D}} = \{x \in [0, 1]^\infty : Ax \geq \beta\} \tag{3}$$

Let $\{A_\alpha x \geq \beta_\alpha\}_{\alpha \in \mathcal{I}}$ denote a suitable indexing of all finite subfamilies of $\{Ax \geq \beta\}$. And for each α in the uncountable set $\mathcal{I}$ let

$$\begin{aligned}
\mathcal{D}_\alpha &= \{x \in \{0, 1\}^\omega : A_\alpha x \geq \beta_\alpha\} \\
\bar{\mathcal{D}}_\alpha &= \{x \in [0, 1]^\omega : A_\alpha x \geq \beta_\alpha\}
\end{aligned}$$

Thus,

$$\mathcal{D} = \cap_{\alpha \in \mathcal{I}} \mathcal{D}_\alpha$$
$$\bar{D} = \cap_{\alpha \in \mathcal{I}} \bar{\mathcal{D}}_\alpha$$

The analysis of finite dimensional mathematical programs such as (1) is based on elementary techniques from combinatorial and polyhedral theory. The situation in the infinite dimensional case gets more complicated. Constraint qualification is a sticky issue even for semi-infinite mathematical programs. The standard approach in infinite dimensional mathematical programming is to impose an appropriate (weak) topological framework on the feasible region and then use the power of functional analysis to develop the structural theory.

4   A Compactness Theorem

A classical result in finite dimensional programming states that if a finite system of linear inequalities in $\Re^d$ is infeasible, there is a "small" $(d+1)$ subsystem that is also infeasible. This compactness theorem is a special case of the ubiquitous Helly's Theorem. Analogous theorems are also known for linear constraints on integer valued variables. In the infinite dimensional case, we could hope for the "small" witness of infeasibility to simply be a *finite* witness. This is exactly what we prove for mathematical programs of the form (3) and (2).

Let S_γ, $\gamma \in \mathcal{G}$, be copies of a Hausdorff space S. Let $S^{\mathcal{G}} = \Pi_{\gamma \in \mathcal{G}} S_\gamma$. The product *topology* on $S^{\mathcal{G}}$ is the topology defined by a basis $\Pi_\gamma \mathcal{O}_\gamma$ where the $\mathcal{O}_\gamma$ are open in S_γ and $\mathcal{O}_\gamma = S_\gamma$ for all but at most finitely many $\gamma \in \mathcal{G}$. A classical theorem on compact sets with product topology is that of Tychonoff which states that

Theorem 4.1 *Arbitrary (uncountable) products of compact sets with product topology are compact.*

Next we show that $\mathcal{D}_\alpha$ and $\bar{\mathcal{D}}_\alpha$, with product topologies, are also compact for any α in $\mathcal{I}$. This follows form the corollary and the lemma below.

Corollary 4.2 $\{x \in \{0,1\}^\infty\}(\{x \in [0,1]^\infty\})$ *(with product topology) is compact.*

Lemma 4.3 *The set* $\{x : A_\alpha \geq \beta_\alpha\}(\alpha \in \mathcal{I})$ *is closed and hence compact.*

Theorem 4.4 $\mathcal{D}(\bar{\mathcal{D}})$ *is empty if and only if* $\mathcal{D}_\alpha(\bar{\mathcal{D}}_\alpha)$ *is empty for some* $\alpha \in \mathcal{I}$.

5   Herbrand   Theory   and   Infinite $0-1$ Programs

We will assume that the reader has a basic familiarity with Predicate Logic. In particular, we assume that the reader is familiar with the Skoleam Normal Form, the Herbrand Universe and the Horn Formula (see [8]).

Assuming now that H is a Horn formula as defined above, we formulate the following infinite dimensional optimization problem.

$$\inf \left\{ \sum x_j : A_x \geq \beta, x \in [0,1]^\infty \right\} \tag{4}$$

where linear inequalities $A_x \geq \beta$ are simply the clausal inequalities corresponding to the ground clauses of H. The syntactic restriction on Horn clauses translates to the restriction that each row of A has at most one $+1$ entry (all other entries are either 0 or -1's — only finitely many of the latter though). We shall prove now that if the infinite linear program (4) has a feasible solution then it has an integer optimal $(0 - 1)$ solution. Moreover, this solution will be a least element of the feasible space i.e., it will simultaneously minimize all components over all feasible solutions.

Lemma 5.1 *If the linear program (4) is feasible then it has a minimum solution.*

Lemma 5.2 *If x^1 and x^2 are both feasible solutions for (4) then so is $\{\bar{x}_j = \min(x_j^1, x_j^2)\}$.*

Theorem 5.3 *If the linear program (4) is feasible, then it has a unique $0 - 1$ optimal solution which is the least element of the feasible set.*

The interpretation of this theorem in the logic setting is that if a Horn formula H has a model then it has a least model (a unique minimal model). This is an important result in model theory (semantics) of so-called definite logic programs.

6 Modal Logic

Modal logic extends classical logic introducing new quantifiers over formulas, called modalities. The set of modalities may be different from one logic to the other. For example, dynamical logic can be viewed as a modal logic where the modalities are programs. Here we consider a special case of dynamic logic where programs are single atomic actions. More precisely the set of modalities is $\{\Box_a\}_{a \in \Sigma}$ (and their duals $\{\Diamond_a\}_{a \in \Sigma}$) where Σ is a set of symbols.

A model $M = (W, T, \sigma)$ for a modal formula ϕ is given by a set of *worlds* W, a family of transition functions $T = \{t_a : W \to 2^W\}_{a \in \Sigma}$ labeled by the symbols in Σ, and a valuation for the variables $\sigma : V \to 2^W$ that associate to each propositional variable the set of worlds in which the variable is true. Given a model $M = (W, T, \sigma)$ and a world s in W the truth value of a modal formula is defined by induction on the structure of the formula as follows:

- $M, s \models x$ iff $s \in \sigma(x)$,

- $M, s \models \phi \wedge \psi$ iff both $M, s \models \phi$ and $M, s \models \psi$,

- $M, s \models \phi \vee \psi$ iff either $M, s \models \phi$ or $M, s \models \psi$,

- $M, s \models \neg\phi$ iff not $M, s \models \phi$,

- $M, s \models \Box_a \phi$ iff $M, t \models \phi$ for all $t \in t_a(s)$,

- $M, s \models \Diamond_a \phi$ iff $M, t \models \phi$ for some $t \in t_a(s)$.

A formula ϕ is true in a model M, written $M \models \phi$, if ϕ is true in all worlds of M. A formula ϕ is satisfiable iff it is true in some model.

So far, we have defined a logic language that extends propositional logic and we have defined a notion of satisfiability for formulas in that languages. We want to embed the satisfiability of modal formulas into linear problems, as it has been done for propositional logic.

We now propose embedding of modal logic, that preferably preserves the finiteness property of propositional logic. The intuition behind the embedding that we define is to use "timed" linear systems of the form

$$A_0 x(t) + A_1 x(t+1) \geq b$$

where the "time" t is used to express the "dynamics" associated to the modal operators. The above system is of a kind usually encountered in the study of dynamical systems and can be rewritten in a more compact way using a shift operator * as follows:

$$A_0 x + A_1 x^* \geq b.$$

Here the variable x is a function of time t and the action of the shift operator on x is given by $x^*(t) = x(t+1)$.

In order to simplify the presentation in the rest of this section we will take $\Sigma = \{a\}$ so that we have only two modal operators $\Box$ and $\Diamond$ (the subscript a is omitted for brevity). However, everything can be extended to the general case, with $|\Sigma|$ possibly greater than one, with obvious modifications.

Since the transition function t may associate to each world more than one successor (or even none), the dynamics expressed by the modal operators $\Box$ and $\Diamond$ has a branching structure. Therefore *time* is not the right concept to express the modalities. We will consider a notion of *generalized time T*. Variable x is still a function of T, but there are two shift operators $\Box$ and $\Diamond$ that can act over x. Putting it together, we want to embed the satisfiability problem for modal logic into a system of the kind

$$A_0 x + A_1 x^\Box + A_2 x^\Diamond \geq b.$$

where the vector x is a function of T and the action associated to the shift operators $\Box$ and $\Diamond$ is the following. We have said that T can be thought as a "time" in a broad sense (carrying on this analogy, we call *istants* the elements of T). Each istant in T may have more than one immediate successor in T. Let τ be a function that associates to each istant the set of its immediate successors in T. The result $x^\Box$ of applying the $\Box$ shift to x is the set of all possible values that vector x can take after one unit of "time". Analogously the result $x^\Diamond$ of applying the $\Diamond$ shift to x is some of the possible values that vector x can take after one unit of "time".

Now we look at how modal formulas can be represented in this framework. Clearly any propositional formula that doesn't make use of the modalities can be embedded into a system with A_1 and A_2 both equal to the null matrix. Also, formulas that make very simple use of the modalities can be directly embedded. For example the formula

$$(x \vee \neg(\Diamond y \wedge z)) \wedge (\Box z \rightarrow z)$$

can be rewritten as

$$(x \vee \neg z \vee \Box \neg y) \wedge (\Diamond \neg z \vee z)$$

and then represented by the system

$$\begin{bmatrix} 1 & 0 & -1 \\ 0 & 0 & 1 \end{bmatrix} x + \begin{bmatrix} 0 & -1 & 0 \\ 0 & 0 & 0 \end{bmatrix} x^\Box + \begin{bmatrix} 0 & 0 & 0 \\ 0 & 0 & -1 \end{bmatrix} x^\Diamond \geq \begin{bmatrix} -1 \\ 0 \end{bmatrix}$$

Things get harder if the formula makes a more complex use of modal operators. For example there is no direct way to express the formula $\Diamond \Box x$ directly into our system. It seems that the flat structure of the linear system does not allow us to represent nested modal operators. A more subtle problem arises when translating the formula $\Box x \vee \Box y$. One could be tempted to embed this formula into the system

$$\begin{bmatrix} 1 & 1 \end{bmatrix} \begin{bmatrix} x \\ y \end{bmatrix}^\Box \geq 1.$$

At first sight this seems correct but a more careful exam shows that the meaning of the above system is the formula $\Box(x \vee y)$ which is not equivalent to $\Box x \vee \Box y$. In fact, the shift operator $^\Box$ acts on the vector $\begin{bmatrix} x \\ y \end{bmatrix}$ as a whole and therefore we cannot choose $x^\Box$ and $y^\Box$ independently of each other.

We will now illustrate an embedding technique that solves the above problems and allows the encoding of arbitrarily complex formulas into linear systems. The resulting system is finite, and its size is not significantly greater of the starting modal formula. These ideas are due to D. Micciancio [9].

The method is based on the introduction of new variables associated to subexpressions of the logic formula and is defined as a recursive procedure **Embed**(ϕ). On input a formula ϕ of propositional modal logic, **Embed**(ϕ) returns a system of linear equations over the variables of ϕ, plus some fresh variables introduced during the execution of the procedure, whose solubility over $0-1$ variables is equivalent to the satisfiability of the original formula ϕ. First we define a procedure to embed formulas of the form $x \leftrightarrow \phi$:

Embed($x \leftrightarrow \phi$)

- if $\phi = x$, then return $\{x \geq 1\}$,

- if $\phi = \neg\psi$, then introduce a fresh variable z and return

$$\{-x - z \geq -1, x + z \geq 1\} \cup \mathbf{Embed}(z \leftrightarrow \psi)$$

- if $\phi = \psi_1 \wedge \psi_2$, then introduce two fresh variables z_1 and z_2 and return

$$\{-x + z_1 \geq 0, -x + z_2 \geq 0, x - z_1 - z_2 \geq -1\}$$

$$\cup \, \textbf{Embed}(z_1 \leftrightarrow \psi_1) \cup \textbf{Embed}(z_2 \leftrightarrow \psi_2)$$

- if $\phi = \psi_1 \vee \psi_2$, then introduce two fresh variables z_1 and z_2 and return

$$\{x - z_1 \geq 0, x - z_2 \geq 0, -x + z_1 + z_2 \geq 0\}$$

$$\cup \, \textbf{Embed}(z_1 \leftrightarrow \psi_1) \cup \textbf{Embed}(z_2 \leftrightarrow \psi_2)$$

- if $\phi = \Box\psi$, then introduce a fresh variable z and return

$$\{-x + z^{\Box} \geq 0, x - z^{\Box} \geq 0\} \cup \textbf{Embed}(z \leftrightarrow \psi)$$

- if $\phi = \Diamond\psi$, then introduce a fresh variable z and return

$$\{-x + z^{\Diamond} \geq 0, x - z^{\Box} \geq 0\} \cup \textbf{Embed}(z \leftrightarrow \psi)$$

The general case easily follows. Any formula ϕ can be embedded into the linear system $\{z \geq 1\} \cup \textbf{Embed}(z \leftrightarrow \phi)$ where z is a variable not occurring in ϕ. Applying the function **Embed** to the formula $\Diamond\Box x$ we get the system

$$
\begin{aligned}
z_1 &\geq 1 \\
-z_1 + z_2^{\Diamond} &\geq 0 \\
z_1 - z_2^{\Diamond} &\geq 0 \\
-z_2 + z_3^{\Box} &\geq 0 \\
z_2 - z_3^{\Box} &\geq 0 \\
-z_3 + x &\geq 0 \\
z_3 - x &\geq 0
\end{aligned}
$$

Obviously we could have embedded the same formula in the smaller system

$$
\begin{aligned}
z^{\Diamond} &\geq 1 \\
-z + x^{\Box} &\geq 0 \\
z - x^{\Diamond} &\geq 0.
\end{aligned}
$$

However, even if the system obtained by applying **Embed** is not the smallest possible, it can be formally proved the the result of the given procedure is never much bigger than necessary. Namely the system $\textbf{Embed}(\phi)$ has at most $3n + 1$ rows where n is the size of the formula ϕ.

The last system can be written in matrix notation as

$$
\begin{bmatrix} 0 & 0 \\ 0 & -1 \\ 0 & 1 \end{bmatrix} \begin{bmatrix} x \\ z \end{bmatrix} + \begin{bmatrix} 0 & 0 \\ 1 & 0 \\ 0 & 0 \end{bmatrix} \begin{bmatrix} x \\ z \end{bmatrix}^{\Box} + \begin{bmatrix} 0 & 1 \\ 0 & 0 \\ -1 & 0 \end{bmatrix} \begin{bmatrix} x \\ z \end{bmatrix}^{\Diamond} \geq \begin{bmatrix} 1 \\ 0 \\ 0 \end{bmatrix}
$$

Here we see how the introduction of a new variable z allows us to represent a formula with nested modal operators.

Now consider the formula $\Box x \vee \Box y$, we have already remarked that this formula cannot be straightforwardly translated into the system

$$[\,1\quad 1\,]\left[\begin{array}{c}x\\y\end{array}\right]^{\Box} \geq 1$$

which in fact represent a different formula, namely $\Box(x \vee y)$. Let's see how expressions with multiple modal operators in the same clause are handled. The result of applying the embedding function to formula $\Box x \vee \Box y$ is the system

$$
\begin{aligned}
z_1 &\geq 1\\
-z_1 + z_2 + z_3 &\geq 0\\
z_1 - z_2 &\geq 0\\
z_1 - z_3 &\geq 0\\
-z_2 + z_4^{\Box} &\geq 0\\
z_2 - z_4^{\Box} &\geq 0\\
-z_3 + z_5^{\Box} &\geq 0\\
z_3 - z_5^{\Box} &\geq 0\\
-z_4 + x &\geq 0\\
z_4 - x &\geq 0\\
-z_5 + y &\geq 0\\
z_5 - y &\geq 0
\end{aligned}
$$

or with a few simplifications

$$
\begin{aligned}
z + y^{\Box} &\geq 1\\
-z + x^{\Box} &\geq 0\\
z - x^{\Box} &\geq 0.
\end{aligned}
$$

This last example shows how introducing a new variable z we can split a clause with multiple occurrences of the same modal operator into the conjunction of several clauses each of which contains at most one modal operator.

We have showed how any formula of modal logic can be translated into a "small" linear system of the form

$$A_0 x + A_1 x^{\Box} + A_2 x^{\Diamond} \geq b.$$

The equivalence of the system with the modal formula can be easily proved by induction on the size of the formula. The linear system has the same "clausal form" property shown in [3] for the propositional logic embedding. Another property enjoyed by this linear system is that each row of the matrices A_1 and A_2 has at most one non-null entry. It is because of this last property that the shift operators $\Box$ and $\Diamond$ can be applied to the unknown vector x as a whole, as opposed to being applied componentwise.

7   Modal Logic and Bisimulation

The relevance of modal logic in the context of modeling distributed computing is exemplified by its relationship with bisimulation, a widely accepted equivalence relation between labeled transition systems.

A labeled transition system is a graph whose edges are labeled with symbols from some alphabet Σ. Formally a labeled transition system is a tuple (N, E, L) where N is a set of nodes, E is a binary relation on N and L is a function from N to Σ. The nodes N represent the possible internal states of a process or set of processes, the labels Σ are actions the system may perform, and the edges of the graph E express how the internal state of the system changes following the execution of an action. Usually some node $s \in N$ is designated as the starting node, the initial state of the process represented by the transition system.

Two labeled transition systems (N_1, E_1, L_1, s_1) and (N_2, E_2, L_2, s_2) are bisimilar if there exists a binary relation $R \subseteq N_1 \times N_2$ such that

- $(s_1, s_2) \in R$

- for all $(t_1, t_2) \in R$:

    - if $(t_1, t'_1) \in E_1$, then there exists some t'_2 such that $(t_2, t'_2) \in E_2$ and $L_2(t_2, t'_2) = L_1(t_1, t'_1)$.
    - if $(t_2, t'_2) \in E_2$, then there exists some t'_1 such that $(t_1, t'_1) \in E_1$ and $L_2(t_1, t'_1) = L_1(t_2, t'_2)$.

It is natural to view labeled transition systems as models for modal logic. The nodes in the graph are the worlds of the model and the transition relation t_a maps node s to the set $\{t : (s, t) \in E\}$. We can ask when two labeled transition systems can be distinguished by modal formulas. In other words, given two labeled transition systems we look for some formula that is true in one system but false in the other. Two labeled transition systems are considered equivalent if no such formula exists.

It turns out that this notion of equivalence is exactly bisimilarity. Two labeled transition systems are bisimilar if and only if they satisfy the same set of modal formulas. For a formal proof of this statement together with a more accurate description of the relationship between modal logic and bisimulation the reader is referred to [2].

Here we will only illustrate the mentioned result on a simple scheduler example taken from [1]. The scheduler described in [1] communicates with a set $\{P_i\}_i$ of n processes through the actions a_i and b_i $(i = 1, \ldots, n)$. These actions have the following meaning:

- action a_i signals P_i has started executing,

- action b_i signals P_i has finished its performance.

Each process wishes to perform its task repeatedly. The scheduler is required to satisfy the following specification:

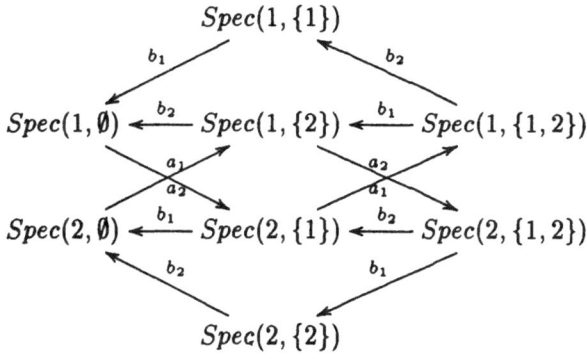

Figure 1: Simple scheduler specification

- actions $a_1, \ldots, a_n$ are performed cyclically starting with a_1,

- action a_i and b_i are performed alternatively.

Informally processes start their task in cyclic order starting with P_1 and each process finish one performance before it begins another.

Then a modular implementation of the scheduler is suggested. The implementation is based on a set of n components $C_1, \ldots, C_n$ connected in cycle that pass a token each other in cyclic order. There is exactly one token, initially owned by C_1, going around. Furthermore, each component C_i performs action a_i after receiving the token and before passing it to $a_{(i \bmod n)+1}$. Then after the token has been passed to $a_{(i \bmod n)+1}$, C_i performs b_i before receiving the token again. For a more accurate description of this example the reader is referred to the original text [1, pages 113–123] where both the specification and the implementation of the scheduler are formally given using the CCS language.

If the number n of processes being scheduled equals two, the specification is given by the labeled transition system shown in Figure 1, while the implementation gives the system described by the labeled transition system in Figure 2.

If the system $[C_1 | \ldots | C_n]$ were a correct implementation of the specification, the two systems in Figure 1 and 2 would not be distinguishable by any modal formula. However this is not the case since formula $s \rightarrow \Box_{a_1} \Box_{a_2} \Diamond_{b_2} t$[1] is true in the system depicted in Figure 1 but not in the one shown in Figure 2. The formula $s \rightarrow \Box_{a_1} \Box_{a_2} \Diamond_{b_2} t$ can be translated into the linear system

$$
\begin{aligned}
-s + x^{\Box_{a_1}} &\geq 0 \\
-x + y^{\Box_{a_2}} &\geq 0 \\
x - y^{\Diamond_{a_2}} &\geq 0 \\
-y + t^{\Diamond_{b_2}} &\geq 0
\end{aligned}
$$

[1]Here s is a predicate true only in the starting state and t is a predicate always true

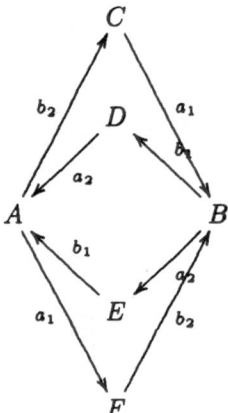

Figure 2: Simple scheduler implementation

$$y - t^{\Box b_2} \geq 0$$
$$t \geq 0$$

which has solution

$$s = \{Spec(1, \emptyset)\}$$
$$x = \{Spec(2, \{1\})\}$$
$$y = \{Spec(1, \{1, 2\})\}$$
$$t = \{Spec(i, S) : i \in \{1, 2\}, S \subseteq \{1, 2\}\}$$

in the model associated to the system in Figure 1 but has no solution in the model associated to the implementation. In conclusion the linear system shows that the proposed implementation of the scheduler does not satisfy the given specification.

References

[1] R. Milner, *Communication and Concurrency.* Prentice Hall, London (1989).

[2] J. van Benthem, J. van Eijck, V. Stebletsova, *Modal logic, transition systems and processes* Math Centrum, CS-R9321 (1993).

[3] V. S. Borkar, V. Chandru, S. K. Mitter, *A Linear Programming Model of First Order Logic* Indian Institute of Science, TR IISc-CSA-95-5 (1995).

[4] R. G. Jeroslow, *Logic-Based Decision Support: Mixed Integer Model Formulation* Annals of Discrete Mathematics 40. North-Holland (Amsterdam 1989).

[5] R. G. Jeroslow, *Computation-oriented reductions of predicate to propositional logic* Decision Support Systems 4 (1988) 183–197.

[6] V. R. Pratt, *Process Logic*, Proc. 6th Ann. ACM Symp. on Principle of Programming Languages (Jan. 1979).

[7] V. R. Pratt, *Dynamic Logic*, Proc. 6th International Congress for Logic, Philosophy, and Methodology of Science, (Hanover, Aug. 1979).

[8] U. Schöning, *Logic for Computer Scientists*, Birkhäuser (1989).

[9] D. Micciancio and S. K. Mitter: Forthcoming LIDS Technical Report, M.I.T.

The Possibility of Indecision in Intelligent Control *

Andrew R. Teel
Department of Electrical Engineering
University of Minnesota
4-174 EE/CS Building
200 Union St. SE
Minneapolis, MN 55455

Abstract

A hybrid system is said to have the hidden possibility of indecision
if the solution of its continuous-valued variables is discontinuous in the
initial data of these variables. This definition is justified in the context
of well-known results concerning ordinary differential equations with discontinuous right-hand side. The hidden possibility of indecision suggests
behavior in the presence of arbitrarily small noise that is not reflected in
the behavior of the classical solutions.

1   Introduction

H. Hermes expressed the view in [12], and it was reiterated by Coron and
Rosier in [8], that it is important to know whether a discontinuous closed-loop
vector field is stable with respect to measurement. Loosely speaking, a vector
field is stable with respect to measurement if arbitrarily small time-varying
disturbances on the right hand side perturb the vector field's classical solutions
by an arbitrarily small amount. Clearly, if the closed-loop vector field is not
stable with respect to measurement then, for certain regions in the state space,
the classical solutions may not always accurately reflect the behavior of the
true system where measurement noise must always be considered.

A discontinuous vector field is guaranteed *not* to be stable with respect to
measurement if its classical solutions are discontinuous in the initial data (see
[12] and [10, sections 2.7,2.8].) So, there are reasons in control problems to
avoid producing closed-loop vectors fields having classical solutions disconti-
uous in the initial data. Unfortunately, some control problem specifications
necessitate discontinuous vector fields having classical solutions discontinuous
in the initial data. Consider the problem for the often studied inverted pendu-
lum/cart system of trying to make the set of zero cart and inverted pendulum

---

*Research supported in part by the NSF under grants ECS-9309523 and ECS-9502034

positions globally asymptotically stable. Fix all initial states in the closed-loop system except the angle and angular velocity of the pendulum. If the classical solutions are continuous in these initial values and the set of inverted positions is stable then each inverted position is attractive to an open set of initial data. From uniqueness, these open sets are disjoint. However, the union of disjoint sets cannot cover $\mathbb{R}^2$. Thus the attractivity cannot be global if the solutions are continuous in the initial data.

2 Decision and indecision

2.1 Hybrid systems

It is likely that many problem specifications in the field of intelligent control will necessitate classical solutions discontinuous in the initial data of the process to be controlled. Indeed, this discontinuous behavior can be associated with the need for decisiveness on the part of the controller, emulating that virtue of the human mind. In the inverted pendulum problem, for example, the controller must decide, based on the initial data, to which stable equilibrium to drive the system. With each requirement for a decision comes the possibility of indecision. For systems with classical solutions discontinuous in the initial data, the possibility of indecision is hidden but can be exposed by arbitrarily small time disturbances. In this paper we briefly discuss the possibility of indecision in finite-dimensional systems having both continuous-valued and discrete-valued variables. Such systems are usually referred to as hybrid systems. For more information on modeling hybrid systems see [3], [2] and the articles in [11].

We will say that a hybrid system having unique classical solutions has the **hidden possibility of indecision** if the classical solution of the continuous-valued variables is discontinuous in the initial data of these variables. We motivate this definition as follows. Quoting from [1, page 11]: "Many evolutionary processes are characterized by the fact that at certain moments of their development they undergo rapid changes. In the mathematical simulation of such processes, it is convenient to neglect the duration of this rapid change and to assume that the process changes its state by jumps." [1] If the hybrid system is a valid approximation, the hidden possibility of indecision will mean the solution of the actual system is discontinuous in the initial data. Consequently, small disturbances in the system that the hybrid system approximates can produce behavior not reflected in the classical solutions of the hybrid system.

2.2 Hidden vs. explicit indecision

Brockett's has shown, when a certain transversality condition is satisfied, that the solution of the continuous-valued variables of his hybrid systems in [4] de-

---

[1] This interpretation for logic variables of a digital computer interacting with a continuous-variable process seems particularly appropriate since the logic states of the computer are aggregated versions of continuous-variable currents and/or voltages evolving in the solid-state circuitry of the machine.

pends continuously on the initial data of these variables. Hence, these systems are examples of hybrid systems that do not have the hidden possibility of indecision.

In general, the hidden solutions exposed by small time disturbances are associated with unstable sliding motions (solutions in the sense of Filippov [10]) [12, Lemma 3], [8, proposition 1.4]. In systems that do not have the hidden possibility of indecision, unstable motions, if they exist, will be explicit, e.g., [4, example 4]. In systems containing explicit, unstable motions it is still possible for small disturbances to induce indecision for motions starting near unstable motions. However, such behavior will be reflected in the classical solutions by the explicit unstable motion. This motivates the modifier "hidden" in our concept of indecision.

2.3   Intrinsic vs. nonintrinsic indecision

As mentioned earlier, hidden indecision is intrinsic to the problem of globally asymptotically stabilizing the set of inverted positions for the cart/pendulum system. The hidden possibility of indecision is not intrinsic to the problem of stabilizing the so-called nonholonomic integrator ($\dot{x}_1 = u_1$, $\dot{x}_2 = u_2$, $\dot{x}_3 = x_1 u_2 - x_2 u_1$). This is because the origin of the nonholonomic integrator can be globally asymptotically stabilized with smooth, time-varying feedback [7]. Since the resulting closed-loop system is smooth, the solutions are unique and continuous with respect to initial data. A hybrid controller for the nonholonomic integrator was discussed in [13, 14]. The solutions of the system under the action of the proposed controller are not continuous in the initial data for x. Briefly, and following the notation used in [13], the solution is discontinuous in the initial data at least at the points on the boundary $x_1^2 + x_2^2 = \pi_2(x_3^2) = 2(1 - e^{-x_3^2})$ satisfying $x_1 = x_2 < 0$ and $\sigma_0 = 1$. Trajectories starting just below the boundary (i.e., $x_1^2 + x_2^2 < \pi_2(x_3^2)$) arrive at the point $x_1^2 + x_2^2 = 0$, $x_3 = x_3^2(0)$ in finite time while trajectories starting just above the boundary, again with $\sigma_0 = 1$, remain forever above the boundary.

The hidden possibility of indecision in this example is to be associated with the possibility, in the presence of noise, that the logic variable σ may be unable to resolve itself into a quantized state when the state of the nonholonomic integrator is near the quantization boundary.

In the area of analog-to-digital converters, the inability of a quantizer to resolve itself into a quantized state in a small amount of time is referred to as metastability. This phenomenon was identified in [5], [6] and [9] as the source of many failures for systems involving asynchronous computers. The probability of metastability for A/D converters has been analyzed (e.g., see [15]) and designs have been proposed to reduce the probability of the occurrence of metastability to an acceptable level (e.g., see [16].)

The fact that A/D converters can be designed to bring down the probability of metastability to acceptable levels in certain applications suggests that, with an appropriate implementation of the logic dynamics, the hybrid control scheme may perform much more reliably than a time-invariant, static discontinuous controller. In fact, from an engineering point of view, the probability may be low enough to warrant retaining the control algorithm for its ease of implementation and fast convergence rate [13]. It would be interesting to investigate these issues further in an attempt to justify hybrid control for this example.

Finally, it is worth noting that if the hybrid system of this example is supposed to truly have impulsive effects, noise that is nonzero only on a set of measure zero can be used to produce (the projection in the x coordinates of) the Filippov solutions with arbitrary precision. For each hidden solution, there is a unique piecewise constant trajectory for $\sigma(t)$ taking values in the set $\{1, 2, 3, 4\}$, only jumping between adjacent values and satisfying $x(t) \in \mathcal{R}_{\sigma(t)}$ (see the notation in [13],[14].) On the other hand, for the impulsive system we can add a disturbance that is nonzero only on a set of measure zero to the state x to produce a $\sigma(t)$ satisfying the above properties almost everywhere. This is because when $x \in \mathcal{R}_i \cap \mathcal{R}_{i+1}$, σ can be reset to either i or $i+1$ at time t by adding to x at time t (only) a disturbance which translates x so that it is in only one of the two, $\mathcal{R}_i$ or $\mathcal{R}_{i+1}$. (Here we are exploiting the fact that σ changes instantaneously.) But this time signal $\sigma(t)$ produces the associated hidden trajectory in the x coordinates.

3 Conclusion

We have defined what is meant by the hidden possibility of indecision in hybrid control systems. The phenomenon that we are emphasizing is the same one that concerned the authors of [12] and [8] in the setting of ordinary differential equations. Depending on the control problem specification, the hidden possibility of indecision may be intrinsic or nonintrinsic. In either case, it would be useful to know the nature of the trajectories resulting from indecision and, in the proper stochastic framework, an acceptable probability of indecision. One nice feature of hybrid control designs is that they seem to focus the problem of minimizing the probability of indecision into the design of the analog-to-digital converter used to interface the continuous-valued and discrete-valued variables.

References

[1] D.D. Bainov and P.S. Simeonov. *Systems with Impulse Effect: stability, theory and applications.* Ellis Horwood Limited, 1989.

[2] M.S. Branicky, V.S. Borkar and S.K. Mitter. A unified framework for hybrid control. In *Proceedings of the 33rd Conference on Decision and Control*, pages 4228–4234, Orlando, FL, December 1994.

[3] R.W. Brockett. Hybrid models for motion control systems. In *Essays on Control: perspectives in the theory and its applications*, H.L. Trentelman and J.C. Willems, eds., Birkhauser, 1993, pp. 29–53.

[4] R.W. Brockett. Language driven hybrid systems. In *Proceedings of the 33rd Conference on Decision and Control*, pages 4210–4214, Orlando, FL, December 1994.

[5] I. Catt. Time loss through gating of asynchronous logic signal pulses. *IEEE Trans. Electron. Comput.*, (Short Notes), vol. EC-15, pp. 108-111, Feb. 1966.

[6] T.J. Chaney and C.E. Molnar. Anomalous behavior of synchronizer and arbiter circuits. *IEEE Transactions on Computers*, vol. 22, pp. 421-422, 1973.

[7] J.-M. Coron. Global asymptotic stabilization for controllable systems without drift. *Mathematics of Control, Signals and Systems*, v. 5 (1992), 295-312.

[8] J.-M. Coron and L. Rosier. A relation between continuous time-varying and discontinuous feedback stabilization. *Journal of Mathematical Systems, Estimation, and Control*, volume 4, number 1, pp. 67–84, 1994.

[9] G.R. Couranz and D.F. Wann. Theoretical and experimental behavior of synchronizers operating in the metastable region. *IEEE Transactions on Computers*, vol. 24, 604-616, 1975.

[10] A.F. Filippov. *Differential Equations with Discontinuous Righthand Sides*. Kluwer Academic Publishers, 1988.

[11] *Hybrid Systems*. R. Grossman, A. Nerode, A. Ravn, and H. Rischel, eds. Lecture Notes in Computer Science, volume 736, Springer-Verlag, New York, 1993.

[12] H. Hermes. Discontinuous vector fields and feedback control. In *Differential Equations and Dynamical Systems*, J.K. Hale and J.P. LaSalle eds., Academic Press, New York, 1967.

[13] J.P. Hespanha. Stabilization of nonholonomic integrators via logic-based switching. Technical report, Yale University, 1995.

[14] A. S. Morse. Control using logic-based switching. In *Trends in Control: a European Perspective*, A. Isidori, ed. pages 69-113, Springer-Verlag, 1995.

[15] H.J.M. Veendrick. The behavior of flip-flops used as synchronizers and prediction of their failure rate. *IEEE J. Solid-State Circuits*, vol. SC-15, pp. 169-176, April 1980.

[16] B. Zojer, R. Petschacher and W.A. Luschnig. A 6-bit 200-MHz full Nyquist A/D converter. *IEEE J. Solid-State Circuits* vol. SC-20, pp. 780-786, June 1985.

A Notion of Discontinuous Feedback

Yu. S. Ledyaev *
Steklov Institute of Mathematics
Moscow 117966, Russia
ledyaev@mian.su

E. D. Sontag[†]
Dept. of Mathematics, Rutgers University
New Brunswick, NJ 08903, USA
sontag@control.rutgers.edu

Abstract

This short note discusses a notion of discontinuous feedback which is appealing from a theoretical standpoint because, on the one hand, solutions of closed-loop equations always exist, and on the other hand, if a system can be stabilized in any manner whatsoever, then it can be also stabilized in our sense. Moreover, the implementation of this type of feedback can be physically interpreted in terms of simple switching control strategies. A recent result obtained by the authors together with Clarke and Subbotin is described, which employs techniques from control-Lyapunov functions, nonsmooth analysis, and differential game theory in order to provide a general theorem on feedback stabilization.

1 Discontinuous Feedback

It is known that, in general, in order to control nonlinear systems one *must* use switching (discontinuous) mechanisms of various types. Of course, time-optimal solutions for even linear systems often involve such discontinuities, but for linear systems most control problems usually admit also (perhaps suboptimal) continuous solutions. However, when dealing with arbitrary systems, discontinuities are unavoidable *even if no optimality objectives are imposed*. This was systematically discussed for one-dimensional systems in the early paper [20] (see also [23]). Perhaps the best-known result regarding the nonexistence of

---

*Visiting Rutgers University. Supported in part by Russian Fund for Fundamental Research Grant 96-01-00219, by NSERC and La Fonde FCAR du Québec, and by the Rutgers Center for Systems and Control (SYCON)

[†]Supported in part by US Air Force Grant AFOSR-94-0293

continuous state feedback is what has become known as the "Brockett necessary condition" (cf. [1]), which imposes severe restrictions on the right-hand
side $f(x, u)$ for it to be true that the origin of the system

$$\dot{x} \; = \; f(x, u) \tag{1}$$

(in any dimension) can be stabilized under continuous feedback, that is, so that
there exist any continuous k so that the origin of

$$\dot{x} \; = \; f(x, k(x)) \tag{2}$$

is asymptotically stable. An elementary exposition of this and related results
can be found in Section 4.8 of the textbook [17] (see also the survey paper [16]).
Faced with the theoretical impossibility of having feedback laws of the form
$u = k(x)$, k a continuous function, various alternatives have been proposed.
(Objectives other than state stabilization, such as tracking and disturbance
rejection, output feedback, or adaptive and robust variants, also lead to the
necessity of discontinuous controllers; the study of stabilization is merely the
first step in the search for general controller structures.)

One way of overcoming the limitations of continuous feedback is to look
for *dynamic* feedback laws, that is to say, to allow for additional (memory)
variables in controllers. This was the approach suggested, and shown to be
powerful enough for all one-dimensional systems, in [20], which demonstrated
that a smooth *time-varying* (and, though not stated explicitly, periodic in t)
controller $u = k(t, x)$ always exists. Far more interesting was the result discov-
ered by Coron in [8], who showed that such feedback stabilizers always exist
(in any dimension) provided that the original system is completely controllable
and has "no drift" ($f(x, 0) = 0$ for all states). (This method is closely related
to the use of classical gradient descent and Newton numerical algorithms for
control, see [18]).) However, for the general case of not exactly-controllable
systems, and possibly with drift, no such dynamic or time-varying solutions
are known, so the interest in *discontinuous* feedback laws $u = k(x)$ arises.

The first positive general theorem along those lines was due to Sussmann,
who established in the early 1980s the existence of a general type of *piecewise
analytic* feedback for the stabilization of controllable analytic systems; see [23].
His work, which was partly motivated by related issues arising in optimal control synthesis via the dynamic programming paradigm, provided an existence
proof for feedback stabilizers for a relatively large class of systems. At around
the same time, one of the coauthors of the present paper proposed a different
approach, based on a general theory of controllers built out of finite automata
and linear systems, in [14], using *constant-rate sampling* to deal with rather arbitrary continuous nonlinear systems. That work took into account constraints
due to partial state measurements, and proposed a synthesis using what would
be today called "hybrid" controllers; an expository survey was recently prepared for the proceedings of a workshop, see [19] and references given there.
However, this approach was not based on true continuous-time feedback, but
rather, as remarked, was basically a discrete-time theory. A general theory was
lacking.

In this note, we outline the new formulation of discontinuous feedback provided in the recent paper [3], stating the relevant definitions and main results, and discuss the implications for control of various types of systems.

2 Theoretical Difficulty: Defining "Solution"

The study of discontinuous feedback leads to a major theoretical difficulty, namely the need to *define the meaning of solution* $x(\cdot)$ of the differential equation (2). As a trivial illustration, consider the one-dimensional equation $\dot{x} = -\text{sign}(x)$, where we write $\text{sign}\,x = 1$ if $x \geq 0$ and $\text{sign}\,x = -1$ if $x < 0$ (this system could have arisen, for example, from using an on/off controller $u(t) \in \{0, 1\}$ for the system $\dot{x} = 1 - 2u$). It is easy to see that there is no possible continuous function $x(t)$ solving this equation when starting at $x(0) = 0$ (in other words, so that $x(t) = -\int_0^t \text{sign}\,x(s)ds$ for small $t > 0$). Thus no "solution" in the classical sense is possible for this example, and in general for discontinuous equations. The most often proposed definition of generalized solution of (2) is due to Filippov, who introduced in [11] an approach based on the solution of a certain associated differential inclusion. It is known, however, that *in general it is impossible to find feedback stabilizers that stabilize in the sense of Filippov*. This negative fact follows from the results in [13, 9], which amount to the statement that the existence of a discontinuous stabilizing feedback in this sense again implies the Brockett necessary condition cited above.

In [3], the following alternative is proposed. It is based on the concept of discontinuous feedback control for positional differential games introduced by Krasovskii and Subbotin in [12]. We want to define the trajectories that arise when using a feedback $k(x)$, that is, a function $k : \mathbb{R}^n \to \mathbb{U}$ from the state-space to the control-value set.

We do so by first considering, for any given partition $\pi = \{t_i\}_{i \geq 0}$ of $[0, +\infty)$, that is, for any infinite sequence consisting of numbers $0 = t_0 < t_1 < t_2 < \ldots$ with $\lim_{i \to \infty} t_i = \infty$, the *$\pi$-trajectory* of (2) *starting from* x_0, obtained as follows: on each interval $[t_i, t_{i+1}]$, the initial state is measured, $u_i = k(x(t_i))$ is computed, and then the constant control $u \equiv u_i$ is applied until time t_{i+1}, when a new measurement is taken. In other words, we use sampled control with switching instants $0 = t_0 < t_1 < t_2 < \ldots$. More precisely, for each i, $i = 0, 1, 2, \ldots$, one solves recursively

$$\dot{x}(t) = f(x(t), k(x(t_i))), \; t \in [t_i, t_{i+1}] \tag{3}$$

using as initial value $x(t_i)$ the endpoint of the solution on the preceding interval, and starting with $x(t_0) = x_0$. The ensuing π-trajectory may fail to be defined on all of $[0, +\infty)$, because of possible finite escape times in one of the intervals, in which case we only have a trajectory defined on some maximal interval, but it is always defined, as in the classical case, for at least small enough intervals.

Technically, we assume in this definition that the control-value set $\mathbb{U}$ is a locally compact metric space (for instance, $\mathbb{U} = \mathbb{R}^m$ for some m) and that the mapping $f : \mathbb{R}^n \times \mathbb{U} \to \mathbb{R}^n : (x, u) \mapsto f(x, u)$ is continuous, and locally

Lipschitz on x uniformly on compact subsets of $\mathbb{R}^n \times \mathbb{U}$. Moreover, when defining feedback, we assume that the function $k : \mathbb{R}^n \to \mathbb{U}$ is locally bounded, meaning that its values on each bounded set are in some a compact subset of $\mathbb{U}$ (there are no "infinite singularities"). For each partition, we consider its *diameter* $d(\pi) := \sup_{i \geq 0}(t_{i+1} - t_i)$.

With these assumptions, and provided the trajectories stay bounded (which our theorems do insure) it is easy to prove that, for any sequence of trajectories corresponding to partitions π_k with diameters converging to zero, there is a convergent subsequence. We think of the limit of any such sequence as a "generalized solution" of the closed-loop equation (2). *Thus solutions always exist in this sense* (though, as with continuous but not necessarily Lipschitz feedback, they may fail to be unique). For ease of reference, we call here such generalized solutions "s-solutions" (for "sampling" or "switching"). It is not difficult to see that if k is continuous, every s-trajectory is a solution in the usual sense, so this indeed generalizes the usual notion.

It is important to emphasize the interpretation of s-solutions: they represent the limits of trajectories arising from high-frequency sampling when using the feedback law $u = k(x)$. This type of interpretation is somewhat analogous, at least in spirit, to the way in which "relaxed" controls are interpreted in optimal trajectory calculations, namely by high-frequency switching of approximating regular controls.

Now that the definition of solution has been given, we can turn to stating the main result of [3]. We present this result in somewhat different but equivalent terms to the formulation in that paper (in the paper, π-trajectories are used instead of the limiting s-solutions). The objective in giving the result, on stabilization, is to show that the notion is useful in that it permits solving an interesting problem. Research in progress deals with the development of extensions to other control problems.

3   A General Result on Stabilization

We define next what we mean by a feedback stabilizer. Intuitively, this is a feedback law which, for fast enough sampling, drives all states asymptotically to the origin and with small overshoot. (The more precise statements in [3] provide very precise estimates on maximal inter-sampling times, overshoot vs accuracy, etc.)

We say that the feedback $k : \mathbb{R}^n \to \mathbb{U}$ *s-stabilizes* the system (1) if for each pair $0 < r < R$, there exist $M = M(R) > 0$ and $T = T(r, R) > 0$ such that, for any initial state x_0 such that $|x_0| \leq R$, every s-solution $x(\cdot)$ of (2) starting from x_0 is defined for all $t \geq 0$ and it holds that:

1. $|x(t)| \leq r$ for all $t \geq T$;

2. $|x(t)| \leq M(R)$ for all $t \geq 0$;

3. $\lim_{R \downarrow 0} M(R) = 0$.

Observe that these three properties represent, respectively, uniform attractiveness, overshoot boundedness, and Lyapunov stability of the closed-loop system.

We also need to recall the definition of global, null-asymptotic controllability.

The system (1) is *asymptotically controllable* if (attractiveness): for each $x_0 \in \mathbb{R}^n$ there exists some control u (a locally essentially bounded measurable function $u : [0, +\infty) \to \mathbb{U}$) such that the trajectory $x(t) = x(t; x_0, u)$ (the solution of (1) at time $t \geq 0$, with initial condition x_0 and control u) is defined for all $t \geq 0$ and $x(t) \to 0$ as $t \to +\infty$, and in addition it holds that (Lyapunov stability): for each $\varepsilon > 0$ there exists $\delta > 0$ such that for each $x_0 \in \mathbb{R}^n$ with $|x_0| < \delta$ there is a control u like this and such that $|x(t)| < \varepsilon$ for all $t \geq 0$. We also assume that there are compact subsets $\mathbb{X}_0$ and $\mathbb{U}_0$ of $\mathbb{R}^n$ and $\mathbb{U}$ respectively such that, if the initial state x_0 satisfies also $x_0 \in \mathbb{X}_0$, then the control can be chosen with $u(t) \in \mathbb{U}_0$ for almost all t. (This last property simply rules out the case in which the only way to control to zero is by using controls that must "go to infinity" as the state approaches the origin.)

The main result in [3] is as follows:

Theorem. *The system (1) is asymptotically controllable if and only if it admits an s-stabilizing feedback.*

The "if" implication is trivial, but the proof of the converse is not, and it is based upon: (a) Techniques from control-Lyapunov functions from [15] and [21]; (b) results of nonsmooth analysis, and in particular the systematic use of proximal subgradients, which are substitutes for the gradient for a nondifferentiable function, and were originally developed in nonsmooth analysis for the study of optimization problems (see [2]); (c) the analysis of viscosity supersolutions of Hamilton-Jacobi equations in the sense of [10], or more precisely, the "proximal supersolutions" studied in [4]) and the interesting connections between these concepts developed in [4] and the book [22]; and finally, (d) methods developed in the theory of positional differential games in [12] (these techniques were used together with nonsmooth analysis tools in the construction of discontinuous feedback for differential games of pursuit in [5] and games of fixed duration in [6]).

4   Remarks

Although we only defined s-solutions for closed-loop systems (2), in which k is arbitrary but f is locally Lipschitz, one can extend the notion to deal with *general interconnections of systems operating at discrete and continuous time scales*. For example, if either the controller or the plant incorporate variables whose update occurs at discrete instants $s_0, s_1, \ldots$, and if the sequence of s_i's does not have any finite limit points, then one may define π-trajectories by first obtaining a common refinement of the partition π and the partition given by the s_i's. We omit the straightforward details.

References

[1] Brockett, R.W., "Asymptotic stability and feedback stabilization," in *Differential Geometric Control theory* (R.W. Brockett, R.S. Millman, and H.J. Sussmann, eds.), Birkhauser, Boston, 1983, pp. 181-191.

[2] Clarke, F.H., *Methods of Dynamic and Nonsmooth Optimization.* Volume 57 of *CBMS-NSF Regional Conference Series in Applied Mathematics*, S.I.A.M., Philadelphia, 1989.

[3] Clarke, F.H., Yu.S. Ledyaev, E.D. Sontag, and A.I. Subbotin, "Asymptotic controllability implies feedback stabilization," submitted[1]. (Summarized version in Proc. Conf. Inform. Sci. and Systems, Princeton, March 1996.)

[4] Clarke, F.H., Yu.S. Ledyaev, R.J. Stern, and P. Wolenski, "Qualitative properties of trajectories of control systems: A survey," *J. Dynamical and Control Systems* 1(1995): 1-48 .

[5] Clarke, F.H., Yu.S. Ledyaev, and A.I. Subbotin, "Universal feedback strategies for differential games of pursuit," *SIAM J. Control*, (in press).

[6] Clarke, F.H., Yu.S. Ledyaev, and A.I. Subbotin, "Universal feedback via proximal aiming in problems of control and differential games," preprint, U. de Montréal.

[7] Clarke, F.H., Yu.S. Ledyaev, and P. Wolenski, "Proximal analysis and minimization principles," *J. Math. Anal. Appl.* **196** (1995): 722-735

[8] Coron, J-M., "Global asymptotic stabilization for controllable systems without drift," *Math of Control, Signals, and Systems* 5(1992): 295-312.

[9] Coron, J.-M., and L. Rosier, "A relation between continuous time-varying and discontinuous feedback stabilization," *J.Math. Systems, Estimation, and Control* 4(1994): 67-84.

[10] Crandall, M., and P.-L. Lions (1983) "Viscosity solutions of Hamilton-Jacobi equations," *Trans. Amer. Math. Soc.* **277**(1983): 1-42.

[11] Filippov, A.F., "Differential equations with discontinuous right-hand side," *Matem. Sbornik* 5(1960): 99-127. English trans. in *Amer. Math. Translations* **42**(1964): 199-231.

[12] Krasovskii, N.N., and A.I. Subbotin, *Positional differential games*, Nauka, Moscow, 1974 [in Russian]. French translation *Jeux differentiels*, Editions Mir, Moscou, 1979. Revised English translation *Game-Theoretical Control Problems*, Springer-Verlag, New York, 1988.

[13] Ryan, E.P., "On Brockett's condition for smooth stabilizability and its necessity in a context of nonsmooth feedback," *SIAM J. Control Optim.* **32**(1994): 1597-1604.

---

[1]Available via the WWW at **http://www.math.rutgers.edu/˜sontag**

[14] Sontag, E.D., "Nonlinear regulation: The piecewise linear approach," *IEEE Trans. Autom. Control* **AC-26**(1981): 346-358.

[15] Sontag E.D., "A Lyapunov-like characterization of asymptotic controllability," *SIAM J. Control and Opt.* **21**(1983): 462-471.

[16] Sontag E.D., "Feedback stabilization of nonlinear systems," in *Robust Control of Linear Systems and Nonlinear Control* (M.A. Kaashoek, J.H. van Schuppen, and A.C.M. Ran, eds,.) Birkhäuser, Cambridge, MA, 1990.

[17] Sontag E.D., *Mathematical Control Theory, Deterministic Finite Dimensional Systems*, Springer-Verlag, New York, 1990.

[18] Sontag E.D., "Control of systems without drift via generic loops," *IEEE Trans. Autom. Control* **40**(1995): 1210-1219.[1]

[19] Sontag E.D., "Interconnected automata and linear systems: A theoretical framework in discrete-time," in Hybrid Systems III: Verification and Control (R. Alur, T. Henzinger, and E.D. Sontag, eds.), Springer, NY, 1996, to appear.[1]

[20] Sontag, E.D., and H.J. Sussmann, "Remarks on continuous feedback," in *Proc. IEEE Conf. Decision and Control, Albuquerque, Dec. 1980*, IEEE Publications, Piscataway, pp. 916-921.

[21] Sontag, E.D., and H.J. Sussmann, "Nonsmooth control-Lyapunov functions," *Proc. IEEE Conf. Decision and Control, New Orleans, Dec. 1995*, IEEE Publications, 1995, pp. 2799-2805.[1]

[22] Subbotin, A.I., *Generalized Solutions of First-Order PDEs: The Dynamical Optimization Perspective*, Birkhaüser, Boston, 1995.

[23] Sussmann, H.J., "Subanalytic sets and feedback control," *J. Diff. Eqs.* **31**(1979): 31-52.

Multimode Regulators for Systems with State & Control Constraints and Disturbance Inputs

Ilya Kolmanovsky and Elmer G. Gilbert
Department of Aerospace Engineering
The University of Michigan
Ann Arbor, Michigan 48109-2118

1 Introduction

This paper considers multimode regulators (Figure 1) that utilize logic-based switching of fixed-structure controllers. Their purpose is to resolve a common conflict in the design of state regulators: meeting performance objectives such as fast response and good disturbance rejection while simultaneously avoiding violation of specified, pointwise-in-time constraints on the state and control of the plant. The family of switched controllers is designed to have increasing levels of performance with the highest level controller being the desired controller. Lower level controllers sacrifice performance for improved safety - the capability of the corresponding closed-loop systems to avoid constraint violations resulting from larger sets of initial plant states. Supervisory logic selects the highest level controller that is "safe" for the current state of the plant. It also initializes the controllers when they are newly selected. Reasonable assumptions on the set of initial plant states, the disturbance input and properties of the controllers guarantee that the logic ultimately selects the highest level controller.

Of course, the idea of controller switching has been used before in situations ranging from gain scheduling to adaptive control [1]. To the best of our knowledge, Tan [2], [3] was the first person to have used a multimode system for the purpose we have described. He was primarily concerned with discrete-time, disturbance-free, linear systems with linear state feedback. His basis for predicting "safe" operation was the maximal constraint admissible set: the largest set of closed-loop initial states that assure subsequent satisfaction of the constraints. Sets of this nature have been studied in a variety of contexts; see, for example, [4]-[9]. For a wide class of linear systems these sets can be expressed concretely by a finite set of functional inequalities [7], [9]. It is this feature that led Tan to successful implementations of multimode regulators.

Section 2 puts forth a more general setting for multimode regulators. Attention is limited to discrete-time systems, since deciding whether or not a given state is "safe" is much more difficult in the continuous-time case [7].

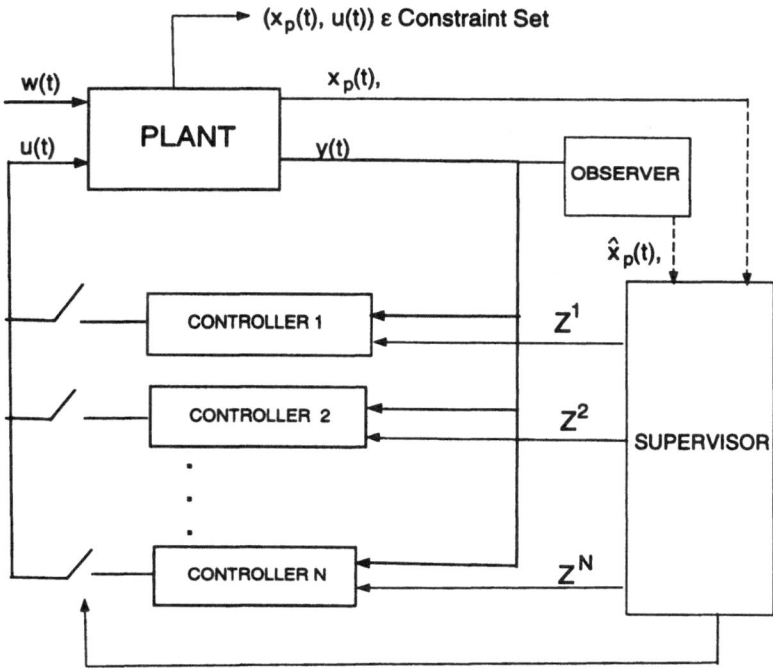

Figure 1: Multimode Regulator.

The generalizations include: consideration of nonlinear plants and controllers, addition of disturbance inputs, use of dynamic controllers and means for their initialization, simplification of switching criteria, new conditions which guarantee ultimate selection of the highest level controller and treatment of errors in the plant model and the determination of plant state. Technical details are not deep and are left for the reader. We do not pretend that all issues of interest have been covered or that practical implementations of the multimode controllers are always feasible. It is hoped that our new ideas may lead to fruitful lines of inquiry. Section 3 considers the special case of linear plants and controllers where both theory and computational methods are in relatively good shape. Section 4 provides a simple illustrative example. Concluding remarks appear in Section 5.

Most of our mathematical notations are standard and should be clear from the context. It is necessary to be precise about a few set notations. Let $A, B \subset \mathbb{R}^n$ and $\alpha \in \mathbb{R}$. Then $A + B = \{x + y : x \in A, y \in B\}$ and $\alpha A = \{\alpha x : x \in A\}$. We call $A \sim B = \{z : z + B \subset A\}$ the P-difference in recognition of Pontryagin who appears to have been the first to have used it in a control application [10]. Other papers where it has been used, either explicitly or implicitly, include [4], [5], [8] [9]. Some of the properties of P-subtraction will be exploited without special comment; see [8], [9] for more details. When A and B are polyhedra expressed by a set of linear inequalities, there exist constructive procedures for determining $A + B$ and $A \sim B$ (also polyhedra) as sets of linear inequalities [9], [11], [12].

2   Formulation of the Multimode Controller

The given discrete-time nonlinear plant is described by

$$x_p(t+1) = f(x_p(t), w(t), u(t)), \quad y(t) = h(x_p(t), w(t)), \tag{1}$$

where $x_p(t) \in \mathbb{R}^p$, $w(t) \in \mathbb{R}^q$, $u(t) \in \mathbb{R}^m$ and $y(t) \in \mathbb{R}^r$. There are no explicit assumptions on the functions $f : \mathbb{R}^{p+q+m} \to \mathbb{R}^p$ and $h : \mathbb{R}^{p+q} \to \mathbb{R}^r$. It is assumed that the input $w(t)$, which accounts for plant disturbances, measurement noise and time-varying parameters, is confined to a specified set W,

$$w(t) \in W \quad \forall t \in Z^+, \tag{2}$$

but is otherwise unknown. Sequences, $w : Z^+ \to W$, are denoted by $w \in \mathcal{W}$.

The operation of the multimode controller is based on switchings between fixed structure controllers represented by

$$x_c^i(t+1) = f^i(x_c^i(t), y(t)), \quad u(t) = h^i(x_c^i(t), y(t)), \tag{3}$$

where $i = 1, \cdots, N$, $x_c^i(t) \in \mathbb{R}^{c_i}$, $f^i : \mathbb{R}^{c_i+r} \to \mathbb{R}^{c_i}$ and $h^i : \mathbb{R}^{c_i+r} \to \mathbb{R}^m$. Controller state dimension, c_i, can change with i. Nondynamic controllers are also allowed; they are represented by $c_i = 0$ and $u(t) = h^i(y(t))$. It is assumed that the N controllers are arranged in a sequence of improving performance with the Nth controller providing the desired best performance. Assumptions on the functions f^i and h^i and on the definition of "performance" are left open. Specific numerical measures of performance may impose, implicitly, special assumptions on f, f^i and h^i, such as closed loop stability. Qualitative, highly subjective notions of performance may require few of any additional assumptions.

State and control constraints for the plant (1) result in *state* constraints for each of the N closed loop systems:

$$(x_p(t), x_c^i(t)) \in X^i \ \forall t \in Z^+, \qquad i = 1, \cdots, N, \tag{4}$$

where $X_i \subset \mathbb{R}^{p+c_i}$ are specified sets. Typically, the sets $X^i \subset \mathbb{R}^{p+c_i}$ can be represented by systems of linear or nonlinear inequalities. For instance, a control constraint $||u(t)||_\infty \leq 1$ implies $X^i = \{(x_p, x_c^i) \in \mathbb{R}^{p+c_i} : ||h^i(x_c^i, h(x_p, w))||_\infty \leq 1 \ \forall w \in W\}$.

Before continuing it is necessary to introduce some notations concerning the dynamics of the closed loop system (1), (3). Let

$$\phi_p^i(t, x_p(0), x_c^i(0), w), \phi_c^i(t, x_p(0), x_c^i(0), w)$$

denote the solution of (1)-(3) for initial conditions $x_p(0)$ and $x_c^i(0)$ and a sequence $w \in \mathcal{W}$. Define

$$
\begin{aligned}
O_\infty^i \;=\; & \{(x_p, x_c^i) \in \mathbb{R}^{p+c_i} : (\phi_p^i(t, x_p, x_c^i, w), \phi_c^i(t, x_p, x_c^i, w)) \in X^i \\
& \forall t \in Z^+ \text{ and } \forall w \in \mathcal{W}\}.
\end{aligned}
\tag{5}
$$

We assume hereafter that problem data are such that $O_\infty^i \neq \emptyset$ for $i = 1, \cdots, N$. Clearly, O_∞^i is the maximal "safe" set for (1)-(3). That is, $(x_p(\tau), x_c^i(\tau)) \Rightarrow (x_p(t), x_c^i(t)) \in X^i$ for all $t \geq \tau$ and $w \in W$ if and only if $(x_p(\tau), x_c^i(\tau)) \in O_\infty^i$. Definitions related to (5) appear in [4]-[9]. Note O_∞^i is positively invariant for (1)-(3); that is, $(x_p(t), x_c^i(t)) \in O_\infty^i$ and $w(t) \in W$ imply $(x_p(t+1), x_c^i(t+1)) \in O_\infty^i$.

Finally, we use $H^i(T) \subset \mathbb{R}^{p+c_i}$ to denote any set which has the following property:

$$(\phi_p^i(t, x_p, x_c^i, w), \quad \phi_c^i(t, x_p, x_c^i, w)) \in H^i(T)$$

$$\forall (x_p, x_c^i) \in O_\infty^i, w \in W \text{ and } t \geq T. \tag{6}$$

Clearly, $H^i(T)$ is a subset of O_∞^i which serves as an ultimate bound on the safe solutions of (1)-(3). Generally, the bound can be made tighter as T increases. If $H^i(T)$ rapidly shrinks in T toward a small set which is well inside O_∞^i, the ith closed loop system has good disturbance rejection and stability properties. Often, it is impossible to find a single controller $i = 1 = N$ which simultaneously makes O_∞^1 large and $H^1(T)$ acceptably small. Flexibility of the multimode controller circumvents the conflict: the $i = 1$ controller achieves large O_∞^1 while the $i = N$ controller yields small $H^N(T)$.

Mechanizations of the multimode controller are based on sets $G^i \subset O_\infty^i$, $i = 1, \cdots, N$. Let $\Pi^i = [I_p : 0] \in \mathbb{R}^{p+c_i}$ denote the projection operator that deletes the last c_i coordinates in $\mathbb{R}^p \times \mathbb{R}^{c_i}$. In particular, $\Pi^i G^i = \{x_p \in \mathbb{R}^p : \exists x_c^i \in \mathbb{R}^{c_i}, (x_p, x_c^i) \in G^i\}$. Assume that $Z^i : \Pi^i G^i \rightarrow \mathbb{R}^{c_i}$ is a function that satisfies the condition

$$(x_p, Z^i(x_p)) \in G^i. \tag{7}$$

Then $x_p(\tau) \in \Pi^i G^i$ and $x_c^i(\tau) = Z^i(x_p(\tau))$ imply $(x_p(t), x_c^i(t)) \subset O_\infty^i \subset X^i$ for all $t \geq \tau$ and $w \in W$. Thus, $\Pi^i G^i$ is a "safe set" of plant initial conditions. Obviously, $\Pi^i O_\infty^i$ is the largest safe set and it is obtained by setting $G^i = O_\infty^i$ and initializing the controller by $x_c^i(\tau) = Z^i(x_p(\tau))$. Generally, many choices for Z^i are available; some have preferred properties such as reduced control effort or quicker closed loop response.

The logic of the multimode controller selects the current controller by choosing i so that $x_p(t) \in \Pi^i O_\infty^i$. Specifically, at time t the controller $i(t)$ is active or appropriately initialized:

$$x_c^{i(t)}(t) = f^{i(t)}(x_c^{i(t)}(t-1), y(t-1)), \quad i(t) = i(t-1), t > 0,$$

$$= Z^{i(t)}(x_p(t)), \quad t = 0, \text{ or } i(t) \neq i(t-1), \tag{8}$$

$$u(t) = h^{i(t)}(x_c^{i(t)}(t), y(t)). \tag{9}$$

For nondynamic controllers $u(t) = h^i(y(t))$, Z^i is not needed and Π^i is the identity matrix. Transitions in $i(t)$ are determined by the set of safe indices

$$I(x_p) = \{i \in \{1, \cdots, N\} : x_p \in \Pi^i G^i\}. \tag{10}$$

To avoid potential up-down cycling of $i(t)$, caused by $x_p(t)$ leaving $\Pi^{i(t)} G^{i(t)}$ while it remains in $\Pi^{i(t)} O_\infty^{i(t)}$, it is necessary to guarantee separately that

i : $Z^+ \rightarrow \{1, \cdots, N\}$ is non-decreasing. Subject to this constraint $i(t)$ is maximum over the set of safe indices.

Algorithm 2.1: Set $i(0) = \max I(x_p(0))$. For $t > 0$ apply the following procedure: if $I(x_p(t)) = \emptyset$ or $\max I(x_p(t)) < i(t-1)$, set $i(t) = i(t-1)$; else, set $i(t) = \max I(x_p(t))$.

Theorem 2.1: Assume $G^i \subset O^i_\infty$, $i = 1, \cdots, N$, and $x_p(0) \in \bigcup_{i=1,\cdots,N} \Pi^i G^i$. Let $i(t)$ and $u(t)$ be defined respectively by Algorithm 2.1 and (8)-(9). Then i : $Z^+ \rightarrow \{1, \cdots, N\}$ is nondecreasing and the constraints (4) are satisfied for $i = i(t)$, $t \in Z^+$. Assume further, that for each $i \in \{1, \cdots, N-1\}$ there exist $\bar{i} > i$ and $T_i \in Z^+$ such that $\Pi^i H^i(T_i) \subset \Pi^{\bar{i}} G^{\bar{i}}$. Then $i(t) = N$ for all t sufficiently large.

Remark 2.1: The allowed set of initial plant states can be enlarged by changing the $t = 0$ step in Algorithm 2.1 and (8): if $I(x_p(0)) = \emptyset$, set $i(0) = 1$ and $x^1_c(0) = 0$; else, set $i(0) = \max I(x_p(0))$ and $x^{i(0)}_c(0) = Z^i(x_p(0))$. Then the results of Theorem 2.1 hold for $x_p(0) \in \{\zeta \in \mathbb{R}^p : (\zeta, 0) \in O^1_\infty\} \bigcup (\bigcup_{i=1,\cdots,N} \Pi^i G^i)$.

Remark 2.2: Suppose the sets G^i are positively invariant. Then $x_p(\tau) \in \Pi^i G^i$ implies $x_p(t) \in \bigcup_{j=1,\cdots,N} \Pi^j G^j$ for all $t \geq \tau$ and i : $Z^+ \rightarrow \{1, \cdots, N\}$ is *automatically* nonincreasing. Thus, for $t > 0$, Algorithm 2.1 can be simplified to $i(t) = \max I(x_p(t))$. The simplification has an important consequence: the logic for determining $i(t)$ depends only on the *current* plant state. Hence, the multimode controller functions properly in the presence of unknown "jumps" in plant state. It is only necessary that after a jump, $x_p(t) \in \bigcup_{i=1,\cdots,N} \Pi^i G^i$.

Remark 2.3: The best choices for the G^i appear to be $G^i = O^i_\infty$. Then the G^i are all positively invariant and the set of safe plant states is maximized. There are, however, certain practical advantages in choosing the G^i to be proper subsets of the O^i_∞. They may be much simpler sets which are easier to determine. For example, if $G^i \subset X^i$ is a level set of a Lyapunov function for the system (1)-(3), then it is positively invariant and meets the requirement $G^i \subset O^i_\infty$. Alternatively, the property of positive invariance may be abandoned in favor of defining G^i by several simple inequalities. This speeds on-line testing of $x_p \in \Pi^i G^i$ and the determination of $I(x_p)$. For linear systems numerical techniques are available for simplifying G^i and verifying the inclusion $G^i \subset O^i_\infty$. For complex, nonlinear systems where such techniques do not exist, simulation or experiments may be useful in identifying appropriate sets G^i.

Remark 2.4: In principle, the results of Theorem 2.1 can be made robust to errors in the modeling of plant dynamics. The sets O^i_∞ and $H^i(T)$ depend on f and h. If the inclusions $G^i \subset O^i_\infty$ and $\Pi^i H^i(T_i) \subset \Pi^{\bar{i}} G^{\bar{i}}$ hold over all the expected variations in f and h, the multimode controller will perform as predicted. Of course, the choices for G^i are more restricted and it is unlikely that they will be positively invariant for each f and h in the class of plant models.

Remark 2.5: An important issue in the design of the controllers is the nesting of the sets, $\Pi^i H^i(T_i) \subset \Pi^{\bar{i}} G^{\bar{i}}$. In the absence of disturbances ($W = \{0\}$) the nesting is achieved quite simply. It follows if, for $i = 1, \cdots, N$, $(x_p, x^i_c) = 0$ is an asymptotically stable equilibrium point of (1), (3) and G^i is a region of attraction. In the presence of disturbances the controller designs are coupled together by $\Pi^i H^i(T_i) \subset \Pi^i G^i$. Generally, the resulting complexity of the design process will be eased if the i and $i+1$ controllers do not differ greatly.

Remark 2.6: Our formulation of the multimode controller is based on $I(x_p)$, which assumes that the full state of the plant is known exactly. Suppose instead that $x_p(t) = \hat{x}_p(t) + e(t)$, where $\hat{x}_p(t)$ is obtained from a state estimator which processes the input and output sequences of the plant. If the initialization error for the state estimator is bounded and W is bounded, it is not unreasonable to expect, for the ith controller, that $e(t)$ is confined to a known compact set, E^i. Using this and the P-difference shows that $\hat{x}_p(t) \in \Pi^i G^i \sim E^i$ implies $x_p(t) \in \Pi^i G^i$. Assume that E^i is relatively small so $\Pi^i G^i \sim E^i \neq \emptyset$, $i = 1, \cdots, N$. Then the results of Theorem 2.1 still hold if the safe indices are redefined by $I_p(\hat{x}_p) = \{i \in \{1, \cdots, N\} : \hat{x}_p \in \Pi^i G^i \sim E^i\}$ and the nesting conditions become $\Pi^i H^i(T_i) \subset \Pi^i G^i \sim E^i$, $i = 1, \cdots, N-1$. Note $0 \in E^i$ implies $\Pi^i G^i \sim E^i \subset \Pi^i G^i$. In this case it is clear that transitions of $i(t)$ are slowed and it is harder to satisfy the nesting conditions.

Remark 2.7: Algorithm 2.7 and (8)-(9) require instantaneous computation of $I(x_p(t))$ and (8)-(9) once the value of $x_p(t)$ becomes available. Thus, it may be necessary to modify the construction of the multimode controller to account for a computational delay. Several options are available. One is to select simpler sets G^i and functions Z^i, f^i and h^i. Another is to add a delay to the plant model (1) (e.g., $u(t) = v(t-1)$, where v is a fictitious control input) and design the controller family (3) for the augmented plant. Then the value of $u(t) = v(t-1)$ is generated from the values of $x_p(t-1)$, $u(t-1)$ and $x_c^{i(t-1)}(t-1)$. A final option is to predict $(x_p(t), x_c^{i(t)}(t))$ from $(x_p(t-1), x_c^{i(t-1)}(t-1))$ using (1) and (3). Notational details are a bit complicated and are omitted.

Remark 2.8: If all the controllers have the same dimension, $c = c_i$, it is possible to base the determination of $i(t)$ on G^i instead of $\Pi^i G^i$. Let $x_c(t) = x_c^i(t) \in \mathbb{R}^c$ and replace (8), (9) by: $x_c(0) = 0$ and $x_c(t) = f^{i(t-1)}(x_c(t-1), y(t-1))$, $y(t) = h^{i(t)}(x_c(t), y(t))$. Define $\bar{I}(x_p, x_c) = \{i \in \{1, \cdots, N\} : (x_p, x_c) \in G^i\}$ and modify Algorithm 2.1 by replacing $I(x_p(t))$ by $\bar{I}(x_p(t), x_c(t))$. Then the results of Theorem 2.1 hold if: $x_p(0) \in \bigcup_{i=1,\cdots,N}\{x_p \in \mathbb{R}^p : (x_p, 0) \in O_\infty^i\}$ and Π^i is the identity matrix. Similar adjustments can be made in Remarks 2.2-2.7. The changes in logic and controller initialization have potential advantages and disadvantages. Since $\bar{I}(x_p(t), x_c(t)) \subset I(x_p(t))$, transitions in $i(t)$ occur more slowly. The set $\bigcup_{i=1,\cdots,N}\{x_p \in \mathbb{R}^p : (x_p, 0) \in O_\infty^i\}$ may or may not be larger than $\bigcup_{i=1,\cdots,N} G^i$. The function Z^i is eliminated and, when h^i does not change with i, jumps in $u(t)$ are avoided. Even though G^i is in a space of greater dimension it may have a simpler characterization than $\Pi^i G^i$. For example, if G^i is defined by several simple inequalities it does not follow that $\Pi^i G^i$ is defined by several simple inequalities.

3   Linear Systems

In this section we assume that the plant and the controllers are linear:

$$x_p(t+1) = Ax_p(t) + B_u u(t) + B_w w(t), \quad y(t) = Cx_p(t) + Dw(t), \qquad (11)$$

$$x_c^i(t+1) = A_c^i x_c^i(t) + B_c^i y(t), \quad u(t) = C_c^i x_c^i(t) + D_c^i y(t). \qquad (12)$$

This special case has great practical interest and it is susceptable to computational techniques. To simplify subsequent notations we represent the ith closed loop system (11),(12) by

$$x^i(t+1) = A^i x^i(t) + B^i w(t), \qquad (13)$$

where $x^i = (x_p, x_c^i) \in \mathbb{R}^{p+c_i}$ and A^i and B^i are appropriately defined matrices. State and control constraints on plant and controller variables take the form

$$y^i(t) = C^i x^i(t) + D^i w(t) \in Y^i \subset \mathbb{R}^{d_i}. \qquad (14)$$

For instance, suppose the only constraint is on the control: $u(t) \in U \subset \mathbb{R}^m$. Then, for $i = 1, \cdots, N$, $Y^i = U$, $C^i = [D_c^i C \; : \; C_c^i]$ and $D^i = D_c^i D$. Note that $X^i = \{x^i \in \mathbb{R}^{p+c_i} \; : \; C^i x^i + D^i w \in Y^i \; \forall w \in W\}$ is not necessarily bounded, even when Y^i is compact.

   We make the following assumptions for $i = 1, \cdots, N$: (A1) the system (13) is asymptotically stable (eigenvalues of A^i inside the unit circle), (A2) W is compact and $0 \in W$; (A3) Y^i is compact and $0 \in int Y^i$, (A4) the pairs (C^i, A^i) are observable. These assumptions, while somewhat stronger than needed, are reasonable and simplify the exposition and the notations.

   With the preceding notations the expression for the maximal constraint admissible set, (5), becomes

$$
\begin{aligned}
O_\infty^i \;&=\; \{x^i \in \mathbb{R}^{p+c_i} \; : \; C^i x^i + D^i W \subset Y^i\} \\
&\bigcap \; \{x^i \in \mathbb{R}^{p+c_i} \; : \; C^i A^i x^i + C^i B^i W + D^i W \subset Y^i\} \bigcap \cdots \\
&\bigcap \; \{x^i :\in \mathbb{R}^{p+c_i} \; : \; C^i (A^i)^t x^i + \sum_{\tau=0}^{t-1} C^i (A^i)^\tau B^i W + D^i W \subset Y^i\} \\
&\bigcap \; \cdots .
\end{aligned}
\qquad (15)
$$

Using this characterization and the P-difference shows

$$O_\infty^i = \bigcap_{t=1}^{t=\infty} O_t^i, \qquad (16)$$

where O_t^i is defined by the following recursions:

$$O_{t+1}^i = O_t^i \bigcap \{\psi \in \mathbb{R}^{p+c_i} \; : \; C^i (A^i)^t \psi \in Y_t^i\},$$
$$O_0^i = \{\psi \in \mathbb{R}^{p+c_i} \; : \; C^i \psi \in Y_0^i\}, \qquad (17)$$

$$Y_{t+1}^i = Y_t^0 \sim C^i(A^i)^t B^i W, \quad Y_0^i = Y^i \sim D^i W. \tag{18}$$

The properties of O_∞^i are summarized in the following theorem.

Theorem 3.1[8], [9]: Suppose $0 \in int Y_t^i$ for all $t \in Z^+$. Then: (i) O_∞^i is nonempty; (ii) O_∞^i is finitely determined, i.e., there exists $t^* \in Z^+$ such that $O_\infty^i = O_t^i$ for all $t \geq t^*$; (iii) O_∞^i is compact and $0 \in int O_\infty^i$; (iv) if Y^i is convex then O_∞^i is convex; (v) if Y^i is polyhedral then O_∞^i is polyhedral.

Remark 3.1: Property (ii) provides the basis for computing O_∞^i. If the set Y^i is polyhedral, the recursions (17), (18) can be easily implemented algorithmically by solving linear programming problems [7], [9]. If for some t, $O_t^i = O_{t+1}^i \neq \emptyset$ then $O_\infty^i = O_t^i$ and the algorithm stops. If for some t, $O_t^i = \emptyset$ then $O_\infty^i = \emptyset$ and the algorithm stops.

Remark 3.2: Finite determination has another important advantage: O_∞^i is expressed by a finite number of inclusions rather than infinitely many inclusions. Thus, it is numerically feasible to test if points or sets, such as G^i, belong to O_∞^i.

Remark 3.3: Property (iv) provides an effective way of obtaining subsets $G^i \subset O_\infty^i$ *without* computing O_∞^i. By performing simulations or experiments one may be able to identify several points in O_∞^i; then the convex hull of these points is a polyhedron $G^i \subset O_\infty^i$.

Remark 3.4: When O_∞^i is polyhedral, and is expressed by a finite number of linear inequalities, subsets G^i expressed by fewer linear inequalities can be generated using a procedure described in the next Section. The projection $\Pi^i G^i$, and the function Z^i needed in the implementation of the multimode controller, can be determined using Fourier elimination [11, 12].

Implicit in the theorem and remarks are numerical procedures for determining O_∞^i and obtaining concrete representations of the sets $G^i \subset O_\infty^i$. These representations lead to on-line implementation of Algorithm 2.1.

Progress can also be made on the determination of sets $H^i(T)$ needed for the set nesting requirement of Theorem 2.1. There exists a minimal, compact limit set, $F^i \subset \mathbb{R}^{p+c_i}$, for the trajectories of (13) with $w \in \mathcal{W}$ [8], [9]. It is given by the infinite set series,

$$F^i = \sum_{t=0}^{\infty} (A^i) B^i W, \tag{19}$$

which is convergent in Hausdorff-norm sense. Let $\hat{F} \subset \mathbb{R}^{p+c_i}$ be any set such that $F^i \subset int \hat{F}$. Then there exists $T \in Z^+$ such that $H^i(T) = \hat{F}$. Since (19) is an infinite series it is generally not possible to obtain a concrete representation of F^i. However, there is an algorithmic approach [9] for testing whether or not a given set contains F^i. The algorithm takes the following form: Set $X^i = \hat{F}$ and compute recursively the corresponding maximal constraint admissible set, $\hat{O}_\infty^i$, for (13) with $w \in \mathcal{W}$. Then $F^i \subset \hat{F}$ if and only if $\hat{O}_\infty^i \neq \emptyset$.

The set nesting condition may be avoided entirely if an additional assumption is made on the sequences $w \in \mathcal{W}$. We say $w : Z^+ \to \mathbb{R}^q$ is *origin-persistent* if for any $\epsilon > 0$ and for any $T, L \in Z^+$ there exists $\tilde{t} \geq T$ such that $w(\tilde{t} - k) \in \{\zeta \in \mathbb{R}^q : \|\zeta\| \leq \epsilon\}$ for $k = 0, 1, \cdots, L$. Origin persistent sequences must ultimately have long runs where $w(t)$ is arbitrarily near the origin. For example, a disturbance sequence is origin-persistent with probability one when the elements of $\{w(t),\ t \in Z^+\}$ are statistically independent random variables and for any $\epsilon > 0$ there exists $\alpha > 0$ such that for all $t \in Z^+$ the probability that $w(t) \in \{\zeta \in \mathbb{R}^q : ||\zeta|| \leq \epsilon\}$ is greater than α.

Theorem 3.2:   Assume $w \in \mathcal{W}$ is origin-persistent and G^i, $i = 1, \cdots, N$, satisfies the conditions: $G^i \subset O^i_\infty$, $0 \in intG^i$. Let $i(t)$ and $u(t)$ be defined respectively by Algorithm 2.1 and (8)-(9). Then $i(t) = N$ for all t sufficiently large.

Finally, it is easier to be more specific about the treatment of observer errors when the plant is linear. For simplicity consider the disturbance-free case, $W = \{0\}$, where a classical linear observer operates on $u(t)$ and $y(t)$ to produce an estimate, $\hat{x}_p(t)$, of the plant state. Under the usual assumptions, the estimator error in Remark 2.6 is given by $e(t+1) = He(t)$, $e(0) \in E_0$, where $H \in \mathbb{R}^{p \times p}$ is asymptotically stable and E_0 is the set of all possible errors in initializing the observer. Clearly,

$$\hat{x}_p(t) \in \Pi^i G^i \sim H^t E_0 \tag{20}$$

implies $x_p(t) \in \Pi^i G^i$. Thus, following the idea outlined in Remark 2.6, we redefine the set of safe indices to be $I(\hat{x}_p, t) = \{i \in \{1, \cdots, N\} : \hat{x}_p \in \Pi^i G^i \sim H^t E_0\}$. If $\Pi^i G^i$ is polyhedral it is not difficult to determine $\Pi^i G^i \sim H^t E_0$ as a set of t-dependent linear inequalities.

4   An Example

We now consider a simple second order example. Its purpose is to illustrate in a clearly visible way the main ideas. High-order systems can be handled as well. For example, Tan [2] considers a $p = 9, c_i = 0, m = 4$ helicopter control system where $G_i = O^i_\infty$ and there is no disturbance input ($W = \{0\}$).

Our plant is an unstable system with the control input generated by a zero order hold. The input to the zero order hold, $u(t) \in U = [-1, +1]$ is corrupted by an additive disturbance $w(t) \in W = [-0.03, 0.03]$. The continuous time model of the plant is $\ddot{\theta} - \theta = \alpha$, where α is the output of the zero-order hold. This equation can represent, for example, an inverted pendulum driven by a dc motor, linearized in a neighborhood of the unstable equilibrium. The discrete-time model, (11), is obtained by assuming a sampling period of 0.1 seconds and by defining the components of $x_p(t)$ by $x_{p1}(t) = \theta(0.1t)$, $x_{p2}(t) = \dot{\theta}(0.1t)$. The controllers are implemented by static state feedback, so: $C_i = 0$, $\Pi^i = I_2$, $A^i = A + B_u D^i_c$, $B^i = B_u = B_w$ and $x(t) = x^i(t) = y_p(t) = x_p(t)$ is independent of i. There are four controller designs: $D^i_c = [(-1-\Omega_i)^2 - \Omega_i)]$, $\Omega_i = \frac{5}{16} 2^i$, $i = 1, \cdots, 4$. These gains have a simple interpretation: applied to the continuous-time plant they yield a second order closed loop system with a fixed damping ratio of 0.5 and natural frequencies, $\Omega_i = \frac{5}{8}, \frac{5}{4}, \frac{5}{2}, 5$ rad/sec.

The resulting sets O^i_∞ are shown in Figure 2, with the sets decreasing in size as i increases. They are polygons expressed by M^i linear inequalities, where for increasing i, $M^i = 60, 30, 16, 8$.

Because the M^i are large, it is of interest to form polygons, $G^i \subset O_\infty^i$, which approximate O_∞^i and are expressed by fewer inequalities. We have developed an automatic numerical procedure for doing this. Inequalities which have relatively little effect in the characterization are successively eliminated until only dominant inequalities remain. The dominant inequalities then characterize a set $\hat{G}^i$ which is a superset of O_∞^i. The set G^i is formed by shrinking $\hat{G}^i$ until it is a subset of O_∞^i: $G^i = \alpha_i \hat{G}^i \subset O_\infty^i$, $\alpha \hat{G}^i \not\subset O_\infty^i$ for $\alpha > \alpha^i$. The values α^i are found by bisection on α, with the truth of $\alpha \hat{G}^i \subset O_\infty^i$ tested by solving a sequence of linear programming problems. The "error" of the approximation is measured by the size of $1 - \alpha^i$.

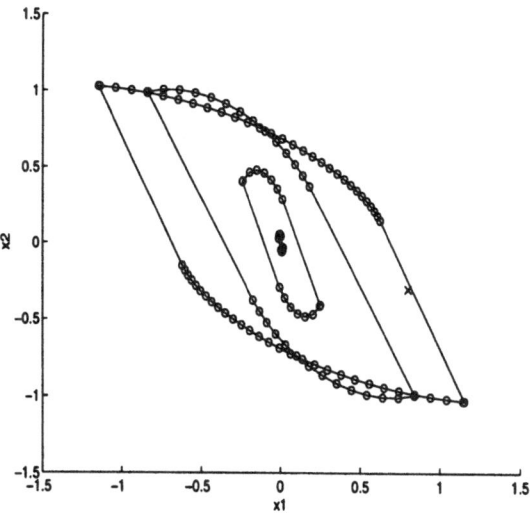

Figure 2: The sets O_∞^i.

Figure 3 shows a family of sets, G^i, generated in this manner: α^i and the number of inequalities, $\hat{M}_i$, are given by

$$(\hat{M}_i, \alpha^i) = (8, 0.95), (6, 0.87), (4, 0.80), (4, 0.75).$$

The computational effort required for evaluation of $I(x_p)$ is greatly reduced: for $G^i = O_\infty^i$ a total of 114 inequalities must be tested, for the approximate G^i the total is 22.

Results of simulation studies are shown in Figure 4 and Figure 5. The initial condition is the cross shown in Figures 2 and 3 and the disturbance is a sine function of frequency 0.31 rad/sec and amplitude 0.03. Note $0.31 < \Omega_i$, $i = 1, \cdots, 4$. Thus the disturbance frequency is well within the pass band of each closed-loop design. The solid curves are for the multimode controller and $G^i = O_\infty^i$; the dashed curves are for the G^i described in the preceding

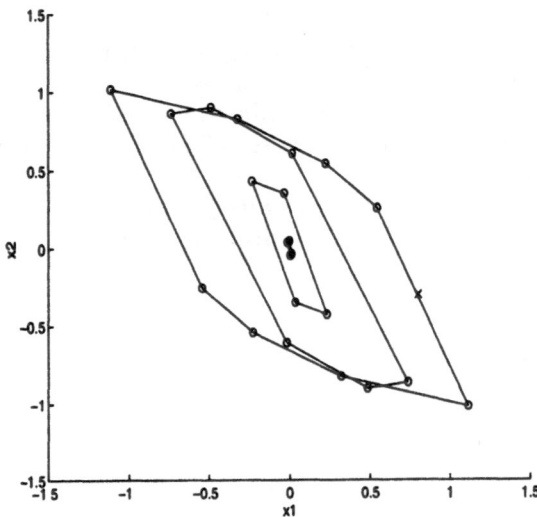

Figure 3: The sets G^i.

paragraph. Transition times for the controller index can be detected from the points of slope discontinuity in $x_2(t)$.

Clearly, the multimode controller gives much faster responses than the fixed, $i = 1$, controller (see dash-dot line in Figure 4). Note that the $i = 1$ controller is the only safe linear controller in the family for the specified initial condition. Performance loss in going from the $G^i = O_\infty^i$ to the simplified $G^i \subset O_\infty^i$ is quite small, considering the over 5 to 1 reduction in on-line computational effort. It is obvious from Figure 5 that the multimode controller has a much smaller disturbance response than the $i = 1$ controller. Actually, the reduction in x_{p1} is about 15 to 1. It is also evident that the relative size of safe initial condition sets for the multimode controller $(O_\infty^1 \bigcup G^2 \bigcup G^3 \bigcup G^4 = O_\infty^1)$ and the high performance controller (O_∞^4) is very different. The ratio of their areas is over 16,000 to 1. This vividly demonstrates the power of nonlinear multimode controllers to achieve a better balance between disturbance rejection and safe recovery from large initial conditions.

5   Conclusion

Multimode control is a novel paradigm for the design of regulator systems with pointwise-in-time constraints on state and control. It overcomes the common design dilemma associated with fixed controllers: achievement of high performance while maintaining a large region of safe attraction.

In the important case of linear systems, much can be said. Basic computational tools needed for the characterization of the G^i exist and can be applied to systems of fairly high order (> 10). Simplifications of the G^i, described in Section 4, can significantly reduce the on-line computational effort associated with the determination of $I(x_p)$. The potential for reduced computational effort is particularly striking when the plant is a sampled-data system with a high sampling rate, because then the characterization of O_∞^i often requires many functional inequalities. Since the plant and its controllers are linear, powerful methods exist for designing a suitable family of controllers. For instance, when the constraints are on the control, $u(t)$, it is effective to use varying control weights in a sequence of optimal linear-quadratic designs [2], [3]. Other design strategies and issues, such as those suggested in the Remarks of Section 2, need further investigation and elaboration.

It should also be possible to extend the multimode concept to linear, input-tracking systems which have pointwise-in-time constraints. The extension would be based on the error and reference governors described in [7], [13], [14]. These nonlinear devices exploit a single "safe" set. Switching over a family of safe sets, corresponding to a family of closed-loop designs, would make it possible for the governors to operate properly over a larger set of initial plant states.

Little of specific nature can be said about nonlinear systems. Theory and algorithms for determining the sets O_∞^i or for constructing of the sets G^i are lacking and methods for controller design are limited. However, the design paradigm should remain valid, at least in some situations of interest. Speculating freely, one can imagine a family of heuristically designed controllers with the safe sets G^i characterized by neural nets which are trained by the results of extensive simulations of the closed-loop systems.

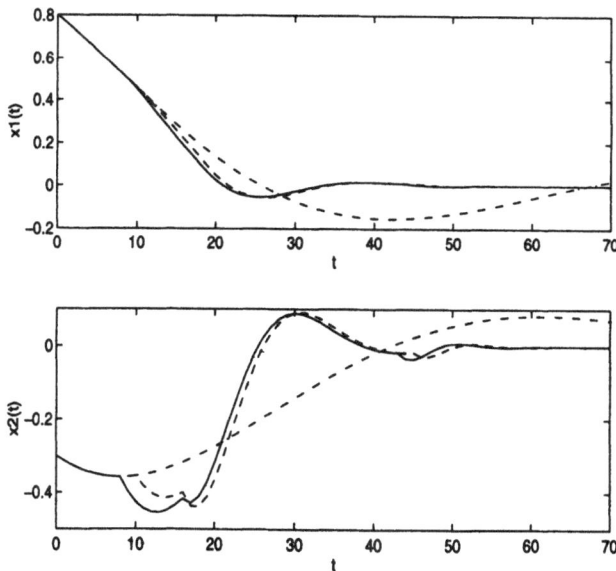

Figure 4: The state responses of the three controllers.

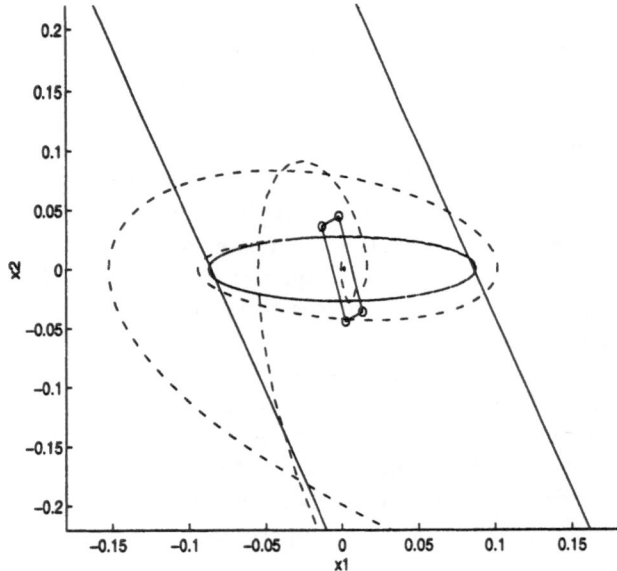

Figure 5: Phase plane responses of the $i = 1$ linear controller and $G^i \subset O^i_\infty$ multimode controller. Solid curves are boundaries of $G^3 \subset O^4_\infty$ and $G^4 \subset O^4_\infty$.

References

[1] Morse A S. Control using logic-based switching. In: Isodori A. (ed) Trends in control. Springer-Verlag, New York, 1995, pp. 69-114

[2] Tan K T. Maximal output admissible sets and the nonlinear control of linear discrete-time systems with state and control constraints. PhD thesis, University of Michigan, 1991

[3] Tan K T. Multimode controllers for linear discrete-time systems with general state and control constraints. In: Optimization: techniques and applications, World Scientific Publishing Company, 1992, pp. 433-442

[4] Glover J D, Schweppe F C. Control of linear dynamic systems with set constrained disturbances. IEEE Trans Automat Contr 1971; AC-16:411-423

[5] Bertsekas D P, Rhodes I B. On the minimax reachability of target sets and target tubes. Automatica 1971; 7:233-247

[6] Blanchini F. Control synthesis for discrete-time linear systems with control and state bounds in the presence of disturbances. J Opt Theory and Appl 1990; 65:29-40

[7] Gilbert E G, Tan K T. Linear systems with state and control constraints: the theory and application of maximal output admissible sets. IEEE Trans Automat Contr 1991; 36:1008-1020

[8] Blanchini F. Constrained control for systems with unknown disturbances. In: Control and dynamic systems, vol. 51. Academic Press, 1992

[9] Kolmanovsky I, Gilbert E G. Maximal output admissible sets for discrete-time systems with disturbance inputs. In: Proceedings of the American Control Conference 1995; 1995-2000

[10] Pontryagin L S. Linear differential games. Soviet Math Dokl 1967; 8:769-771

[11] Keerthi S S, Gilbert E G. Computation of minimum time feedback control laws for discrete-time linear systems with state-control constraints. IEEE Trans Automat Contr 1987; AC-32:432-435

[12] Keerthi S S, Sridharan K. Solution of parametrized linear inequalities by Fourier elimination and its applications. J Opt Theory and Appl 1990; 65: 161-169

[13] Gilbert E G, Kolmanovsky I, Tan K T. Discrete-time reference governors and the nonlinear control of systems with state and control constraints. International J of Robust and Nonlinear Contr 1995; 5:487-504.

[14] Gilbert E G, Kolmanovsky I. Reference governors for systems with state and control constraints and disturbance inputs. In: Proceedings of the 34th IEEE Conference on Decision and Control 1995; 1189-1194

Switching Control of Constrained Linear Systems

D. Q. Mayne [*†]

Abstract

Hybrid controllers commonly consist of a supervisor, whose state is discrete, and a set of conventional controllers with continuous states. Usually, both the supervisor and the conventional controllers are dynamic, but each or both may be static (memoryless). Attention is focussed in this paper on the control of linear systems with hard constraints on both states and controls and subject to a bounded disturbance. To simplify derivation of a controller, it is assumed that the state of the linear system is accessible. Efficient control of constrained systems necessitates nonlinear control, even if the system being controlled is linear. The state space is partitioned into a set of overlapping polytopes $\{X_i\}$ or into a set of simplices $\{X_{ij}\}$ with disjoint interiors; the polytopes are nested ($X_i \subset X_{i+1}$ for all i) and the simplices partition the polytopes ($X_i = \cup_j X_{ij}$). The set X_0 is robust control invariant, i.e. there exists a feedback controller which maintains the state of the system in X_0 (despite the disturbance) if the initial state is in X_0. Two nonlinear controllers are described. The first, which uses the polytope partition, determines the least integer i such that the current state x lies in X_i, and determines (by solving an optimization problem) an admissible control which steers x to X_{i-1}. The second uses a precomputed affine controller for each simplex; the supervisor merely selects the current controller to be that corresponding to the simplex in which the current state resides. Both controllers steer any initial state to the target set X_0 in minimum time (despite the disturbance). Both components (discrete and continuous) of the hybrid controller are static (memoryless).

1 Introduction

Hybrid control, which is defined here to be control of a process by means of a controller with continuous and discrete subsystems, has a long history. Indeed, virtually every controller incorporates a discrete subsystem, usually devoted to ensuring safe operation under fault conditions. An early example in identification is the use of hysteresis for consistent selection of parameterizations [1]. The model set S is a real-analytic manifold of dimension $2np$, where n is the McMillan degree of the system and p the dimension of the output; $U_\alpha \subset S$,

---

[*]Department of Electrical and Computer Engineering, University of California, Davis
[†]Research supported by National Science Foundation under grant ECS-93-12922

where α is a (nice) multi-index, is the set of s such that rows of the Hankel matrix (of s) specified by α are linearly independent. Each $s \in U_\alpha$ may be expressed in a canonical form (which depends on α).The set S is covered by the family $\{U_\alpha\}$ of (overlapping) canonical forms. Identification yields a sequence of estimates $\{s_k\}$ assumed to converge almost surely to s^0. A coordinate selection structure of the form $\alpha_k = f(s_k)$ cannot be guaranteed to converge (in the discrete topology, i.e. in a finite number of steps). Instead, Clark [1] establishes consistency for a hysteresis selection algorithm which employs a family of continuous functions $d_\alpha(s) := |\det(Q^\alpha)|/\mu$ with $\mu := \max_\alpha |\det(Q^\alpha)|$ where Q^α denotes the matrix whose rows are those rows of the observability matrix (of any realization of s) indexed by α; $d_\alpha(s) = 1$ indicates that α is a 'good' parameterization for the system s whereas $d_\alpha(s) = 0$ indicates that $s \notin U_\alpha$ so that this parameterization is inappropriate. The coordinate selection algorithm is:

$$\begin{aligned}
\alpha_k &= \alpha_{k-1} \text{ if } d_{\alpha_{k-1}}(s_k) > \nu \\
&= \text{any } \alpha \text{ such that } d_\alpha(s_k) = 1 \text{ otherwise}
\end{aligned}$$

where $\nu \in (0,1)$. This algorithm employs a discrete subsystem (with memory) whose state at time k is α_k. The 'obvious' strategy of choosing $\alpha_k = $ any i such that $d_i(s_k) = 1$ does not guarantee convergence of $\{\alpha_k\}$ in finite time.

An early application of hysteresis to adaptive control is reported in [2]. Here a number of constrained parameter estimators, indexed by $i \in I$, are employed, each covering a region P_i of parameter space in which the model is stabilizable. The unknown parameter lies in P_{i^0} but the index i^0 is not known. Use is made of the modular structure [3] of adaptive control; if the parameter estimator and the certainty equivalence controller have, independently, certain properties, then the adaptive controller is stabilizing. Assume, for the purposes of exposition, that only one property of the estimator is required, namely that its normalized estimation error is in L_2. The adaptive controller computes $z^i(t) := \int_0^t e^i(s)^2 ds$ for each $i \in I$ where $e^i(t)$ is the (normalized) estimation error at time t of the ith estimator. The estimator $i(t)$ employed, at time t, in the certainty equivalence controller is selected according to the hysteresis algorithm:

$$\begin{aligned}
i(t+) &= i(t) \text{ if } z^{i(t)}(t) < z^0(t) + \delta \\
&= \text{any } i \text{ such that } z^i(t) = z^0(t), \text{ otherwise}
\end{aligned}$$

where $z^0(t) := \min_i z^i(t)$. It can be shown that only a finite number of 'switches' are made, i.e. $i(t) \to i^*$ in finite time. The estimator finally accepted (i^*) is not necessarily the true estimator i^0 but has the requisite property (its normalized error lies in L_2) to ensure stabilization. See [4, 5] for analysis and further applications of hysteresis switching to adaptive control.

Another example occurs in gain scheduled control. Here the state space partitioned into a set $\{X_i\}$ of regions with disjoint interiors and a different (affine) controller assigned to each region. To avoid chattering of the control (between two controllers) on the boundary between regions, hysteresis may be employed.

The first two examples involve uncertainty in a discrete parameter of the system being identified or controlled; all three examples use hysteresis to avoid infinite switching or chattering. Alternative strategies, such as cyclic switching [6] may be employed.

We consider the problem of controlling linear systems with hard constraints on controls and states. Such constraints occur in many applications and this has led to the widespread adoption by industry of ad-hoc methods which have little theoretical justification. In the petro-chemical industry, where processes are relatively slow and satisfaction of constraints economically important, model predictive control, which has the ability to handle state and control constraints, has been widely adopted. Computing requirements inhibit the use of this method for controlling fast processes.

An important, and useful, methodology for coping with state and control constraints is described by Gilbert and Tan [7] who introduce the concept of a maximal output admissible set $\mathcal{O}$, provide methods for constructing this set, and show that, in the discrete-time case, it can be determined in a finite number of iterations. They employ this concept in [7] to construct a nonlinear controller of the form $u = K_{i(x)}x$ where $\{u = K_i x \mid i = 1, \ldots, N\}$ is a sequence of stabilizing controllers, ranging from 'aggressive' to 'conservative' and $i(x)$ is the smallest integer in $\mathbf{N} := \{1, \ldots, N\}$ satisfying $x \in \mathcal{O}_i(x)$ where $\mathcal{O}_i(x)$ is the output admissible set associated with controller $u = K_i x$. This approach has recently been extended to deal with bounded disturbances [8]. In [9] the output admissible set is used to obtain a nonlinear controller for tracking.

Under reasonable assumptions, the value function associated with a robust, time-optimal, control problem, can be finitely parameterized. Consequently, a version of dynamic programming which computes, in reverse time, the level sets $\{X_k\}$ of the value function, can be obtained. These level sets, which are polytopes, can be employed to construct nonlinear controllers. The sets are used to partition the state space into a set of regions, in each of which a different, memoryless controller is employed. Memory, either in the controller associated with each region, or in the switching mechanism, does not bring any obvious advantage. This approach to the problem of controlling constrained linear systems, developed more fully in [10], is summarized below. For related earlier work see [11, 12, 13]

Notation: Given any set $X \subset I\!\!R^p$, int(X) denotes its interior, ri(X) its relative interior, closure(X) its closure, and $\partial(X)$ its boundary. Given two sets A and B, both subsets of $I\!\!R^n$, $A + B := \{a + b \mid a \in A, b \in B\}$, $A - B := A + (-B) = \{a - b \mid a \in A, b \in B\}$, and $A \ominus B := \{a \in I\!\!R^n \mid a + B \subset A\}$. We have [14]:

$$A + B = \bigcup_{b \in B}(A + b), \quad A \ominus B = \bigcap_{b \in B}(A - b)$$

Subscripts denote elements of a sequence; superscripts denote components of a vector. For any integer q, Σ_q is the convex set defined by:

$$\Sigma_q = \{\mu \in \mathbb{R}^{q+1} \mid \mu^i \geq 0, \sum_{i=0}^{q} \mu^i = 1\} \tag{1}$$

If $v \in \mathbb{R}^p$ and $w \in \mathbb{R}^q$, are column vectors, then (v, w) denotes the column vector $(v^T, w^T)^T \in \mathbb{R}^{p+q}$. If j is an integer, $\mathbf{j}$ denotes the set $\{1, \ldots, j\}$.

2   Problem Statement

We wish to determine a state feedback controller for the linear, discrete-time system described by

$$x[k+1] = Fx[k] + Gu[k] + w[k] \tag{2}$$

where $x[k] \in \mathbb{R}^n$ is the state, $u[k] \in \mathbb{R}^m$ the control, and $w[k] \in \mathbb{R}^n$ the disturbance at time k. The control and state are subject to the following hard constraints

$$u[k] \in \Omega, \quad x[k] \in E \tag{3}$$

where $\Omega \subset \mathbb{R}^m$ is a polytope (therefore, convex and compact) and $E \subset \mathbb{R}^n$ a polyhedron (therefore, convex and closed); each set contains the origin in its interior. The disturbance $w[k]$ takes values in W which is assumed to be a polytope in $\mathbb{R}^n$, also containing the origin in its interior; $\mathcal{W}$ denotes the class of admissible disturbance sequences $\mathbf{w} = \{w[k]\}$, i.e. those sequences $\mathbf{w}$ satisfying $w[k] \in W$ for all k. We assume

A1: The pair (F, G) is controllable.

A2: The matrix G has maximal rank m.

A3: F is invertible.

The last assumption is made for simplicity; it holds, of course, when $x[k+1] = Fx[k] + Gu[k]$ is the discrete time version of the continuous time system $\dot{x} = Ax + Bu$. We seek an *admissible* feedback controller $u = h(x)$ which steers any initial state in a specified set to a robust, control-invariant set X_0 in minimum time despite the disturbance (an admissible controller satisfies the state and control constraints and an admissible disturbance satisfies the disturbance constraint)).

Definition 1 *The set $X \subset E$ is said to be* robust control invariant *for the discrete time system* (F, G, Ω, W, E) *if, for every $x \in X$, there exists a $u \in \Omega$ such that $Fx + Gu + W \subset X$.*

For every initial state $x \in X$, where X is robust control invariant, there exists a control $u = h(x) \in \Omega$ such that the successor state $Fx + Gh(x) + w \in X$ for all $w \in W$. Hence, if X is robust control invariant, there exists a state feedback controller $u = h(x)$ such that $x^{h,w}[k; x, 0] \in X$ for all $x \in X$, all

$k \geq 0$, and all admissible disturbance sequences $\mathbf{w} = \{w[0], w[1], \ldots\}$; here $x^{h,w}[k; x, 0]$ denotes the solution of

$$x[k+1] = Fx[k] + Gh(x[k]) + w[k] \tag{4}$$

at time k, due to a disturbance sequence $\mathbf{w} = \{w[k]\}$, given the initial state is $x[0] = x$. Since $x \in E$ and $h(x) \in \Omega$ for all $x \in X$, the controller is admissible in X. We seek, therefore, the solution of the robust minimum time problem

$$V^0(x) = \inf_h \sup_w \{k \mid x^{h,w}[k; x, 0] \in X_0\} \tag{5}$$

$$= \inf_h \{k \mid x^{h,w}[k; x, 0] \in X \; \forall \mathbf{w} \in \mathcal{W}\} \tag{6}$$

The disturbance prevents control to the origin; hence part of the problem is determination of a (preferably small) robust control invariant set to which the state may be steered; once this set is reached, a local controller maintains the state in this set.

The value function V^0 is integer valued. For every positive integer k, $X_k :$ $= \{x \mid V^0(x) \leq k\}$ (a level set of V^0) is the set of states which can be *robustly* steered to X_0 in no more than k steps (x can be robustly steered to X_0 in k steps if there exists a feedback controller $h(\cdot)$ satisfying $x^{h,w}[k; x, 0] \in X_0 \subset E$, $x^{h,w}[j; x, 0] \in E$, and $h(x^{h,w}[j; x, 0]) \in \Omega$ for all $j \in \{0, \ldots, k-1\}$ and all $\mathbf{w} \in \mathcal{W}$. Hence the problem we address is solved once the level sets $X_k, k = 1, 2, \ldots$ are determined; these sets characterize the value function and, therefore, the optimal feedback controller.

We address, in turn, the following problems:

(i) The determination of a *'small'* robust control invariant set X_0 for the system (2), and a controller which maintains the state in this set.

(ii) The determination of a control law $h : X_N \to \Omega$ which steers any point in X_N to X_0 in minimum time (N is a prespecified integer).

3   The robust control invariant set X_0

It is necessary to specify how the robust control invariant set X_0 may be obtained. To do this we require the following elementary results [14]. Let $A, B \in \mathbb{R}^{n \times n}$ be arbitrary, and let X, Y be arbitrary, non-empty, convex subsets of $\mathbb{R}^n$. Then:

$$(X \ominus Y) + Y \quad \subset \quad X \tag{7}$$

$$(A + B)X \quad \subset \quad AX + BX \tag{8}$$

$$A(X + Y) \quad = \quad AX + BY \tag{9}$$

It is desirable that X_0 be 'small' (to contain the effect of the disturbance), but a smallest set (with respect to inclusion) does not necessarily exist. However, the set X_0 specified in the next result is robust control invariant and does have the desirable property that $X_0 \to \{0\}$ as $W \to \{0\}$ (in the Hausdorf metric).

Theorem 1 *Suppose the controller* $u = Kx$ *steers any state of the uncon-strained discrete time system to the origin in* $s \leq n$ *steps. Let* $F_K := F + GK$. *Then the set*

$$X_0 := W + F_K W + \ldots + F_K^{s-1} W$$

is a robust control invariant set for (F, G, Ω, W, E) *if*

$$X_0 \subset E \text{ and } KX_0 \subset \Omega$$

and contains the origin in its interior. The controller $u = Kx$ *maintains the state of the perturbed system in* X_0 *if its initial state is in* X_0.

Since $F_K^s = 0$, $F_K X_0 + W = X_0$, so that for all $x \in X_0$ there exists a control $u = Kx \in \Omega$ such that $Fx + Gu + W = F_K x + W \subset X_0$. Since W contains the origin in its interior, so does X_0. The set X_0 is unique if $m = 1$ because, then, $s = n$ and K is unique. If $m > 1$, neither K nor X_0 are unique. By virtue of (7), $F_K X_0 = X_0 \ominus W$.

4   The level sets $\{X_k\}$

Level sets of the value function for this problem are generated by Algorithm 1 below:

Algorithm 1

Data: Polytope X_0.
Step 1:   Set k=0.
Step 2:   Compute the set
          $X_k' = X_k \ominus W$.
Step 3:   Compute the set
          $X_{k+1} = \{F^{-1} X_k' - F^{-1} G\Omega\} \cap E$.
Step 4:   Set $k = k + 1$. If $k = N$, stop.
          Else, go to Step 2.

The properties of Algorithm 2 are summarized in:

Theorem 2 *Suppose* $X_0 \subset E$ *is a robust control invariant polytope containing the origin in its interior. Then, (i) Algorithm 2 is well defined and generates a non-decreasing sequence* $\{X_k\}$ *of robust control invariant polytopes for the system* (F, G, Ω, W, E), *each lying in* E *and containing the origin in its interior (ii) for all* k, *all* $x \in X_k$, *there exists a* $u \in \Omega$ *such that* $Fx + Gu + W \subset X_{k-1}$, *and (iii) there exists a control law which satisfies the state and control constraints and which robustly steers the discrete time system from any initial state in* X_k *to the set* X_0 *in (no more than)* k *steps, subsequently maintaining the state in* X_0.

By assumption, $X_0 \subset E$ is a robust control invariant polytope; this implies the existence of (a non-empty) X_0' such that $X_0' + W \subset X_0$; also, $X_0' \subset X_0 \subset E$ since $0 \in W$. By definition, for all $x \in X_0 \subset E$, there exists a $u \in \Omega$ such that $Fx + Gu \in X_0'$. By step 3, X_1 is the set of x in E for which there exists a $u \in \Omega$ such that $Fx + Gu \in X_0'$. Hence $X_1 \supset X_0$ and contains the origin in its interior. The polytope property is invariant under the operations of set addition, intersection, matrix multiplication and the contraction operation $X \mapsto X \ominus \Omega$. Since X_0 is a polytope, both X_0' and X_1 are polytopes. It follows that $\{X_k\}_{k=0}^N$ is a sequence of nested polytopes ($X_{k+1} \supset X_k$ for all k), each lying in E and containing the origin in its interior. From Step 3, for any k and any point $x \in X_k$, there exists a $u \in \Omega$ such that $Fx + Gu \in X_{k-1}'$; hence $Fx + Gu + w \in X_k$ for all $w \in W$. Thus any point $x \in X_k$ can be robustly steered to X_{k-1} in one step, and, hence to X_0 in k steps.

It is a direct consequence that Algorithm 2 solves the min-max problem defined at the outset of this section, in the sense that it yields the level sets of the value function and, hence, the value function itself.

We next turn our attention to controller design.

5 Nonlinear controllers

The information encoded in the sets $\{X_k\}$, the level sets of the value function for the minimum time and robust minimum time problems, is sufficient to construct a controller. We present below two alternatives for the robust minimum time problem.

5.1 One step model predictive control

One possibility is a controller which determines the minimum integer k such that the current state $x \in X_k \setminus X_{k-1}$, and the control $h(x)$ by solving:

$$u(x) = \arg \min_u \{\|u\| \mid u \in \Omega, \ Fx + Gu \in X_{k-1}'\}$$

If either the one-norm or max-norm is employed, this problem may be expressed as a linear program; if the two-norm is used, it may be expressed as a quadratic program. Excellent software exists for either alternative. The dimension of the program is low. This model predictive controller $u(\cdot)$ robustly steers the discrete time system from any $x \in X_k \setminus X_{k-1}$, any $k \in \{1, \dots N\}$, into X_{k-1} in one step. This controller controller steers any $x \in X_k \setminus X_{k-1}$ to the boundary of X_{k-1}'. It might be preferable to steer into the interior of X_{k-1}'. This may be done by solving the following program:

$$(u(x), v(x)) = \arg \min_{u, v \leq 0} \{v \mid u \in \Omega, \ A_{k-1}[Fx + Gu] \leq b_{k-1} + va\}$$

where, for all k, $X_k' = \{x \mid A_k x \leq b_k\}$, $v \in \mathbb{R}$, and $a := (1, 1, \dots, 1)'$. This may be formulated as a linear program.

5.2 Simplicial controller

This controller makes use of the following result:

Lemma 1 *Suppose, for $i = 0, 1, \ldots s$, that control $\{u[k] \mid k = 0, \ldots, N\}$ steers the system $x[k+1] = Ax[k] + Bu[k]$ from state x_i to state z_i along a trajectory that satisfies both state and control constraints, and that $x \in S_1$, where $x_0 \ldots x_s$ and $z_0 \ldots z_s$ are affinely independent sets and simplices S_1 and S_2 are defined by:*

$$S_1 := \text{co } \{x_i \mid i \in \{0 \ldots s\}\}$$
$$S_2 := \text{co } \{z_i \mid i \in \{0 \ldots s\}\}.$$

Then $x = \sum_{i=0}^{s} \mu^i x_i$ for some $\mu \in \Sigma_s$ and the control $\{u[k]\} = \sum_{i=0}^{s} \mu^i \{u[k]\}$ steers the system from state $x \in S_1$ to the the state $z = \sum_{i=0}^{s} \mu^i z_i \in S_2$ along a trajectory that satisfies both state and control constraints.

This result is a direct consequence of the linearity of the system and the convexity of the constraint sets Ω and E. For every integer j_k let $\mathbf{j}_k$ denote the set $\{1, 2, \ldots j_k\}$. Each polytope X_k is decomposed into a sequence $\{X_{kj} \mid j \in \mathbf{j}_k\}$ of possibly overlapping simplices whose union covers $X_k \setminus X_{k-1}$. The simplicial controller determines the simplex X_{kj} in which the current state x lies (k is, as before, the least integer such that $x \in X_k \setminus X_{k-1}$), determines $\mu \in \Sigma_n$ and $u(x)$ to satisfy:

$$x = \sum_{i=0}^{s} \mu^i x_{kji}, \quad u(x) = \sum_{i=0}^{s} \mu^i u_{kji},$$

and applies control $u(x)$ to the discrete or continuous time system for time Δ. Here x_{kji}, $i = \{0 \ldots s\}$ are the vertices of X_{kj}, and, for each i, u_{kji} steers the discrete time system from x_{kji} to X_{k-1}; these controls exist by construction, and may be obtained as a byproduct of Algorithm 2. By Lemma 1, control $u(x)$ steers either system from state $x \in X_{kj} \subset X_k$ to the set X_{k-1}. It is easily established that the control is affine in each simplex, i.e.

$$u(x) = c_{kj} + K_{kj} x$$

for all $x \in X_{kj}$. The coefficients c_{kj} and K_{kj} can be pre-computed so that the main task of the controller is to determine in which simplex the current state lies.

There are many ways to construct the simplices X_{kj}; it is merely required that, for each k, $X_k \setminus X_{k-1} \subset \cup\{X_{kj} \mid j \in \mathbf{j}_k\}$. A simple option is to choose the simplices such that $X_k = \cup\{X_{kj} \mid j \in \mathbf{j}_k\}$; one vertex of each simplex is at the origin and the remaining vertices are appropriate vertices of the polytope X_k. With this construction, the control is linear in each simplex ($c_{kj} = 0$ for all k, j).

5.3 Properties of the two controllers

Theorem 3 *The one step model predictive and simplicial controllers steer the perturbed discrete time system from any initial state $x \in X_j \subset X_N$ to X_0 in no more than j steps, and then maintains the state in X_0, satisfying, for all k, the state and control constraints.*

This result is a direct consequence of the properties of the level sets $\{X_k\}$, established in Theorem 2, and the fact that each controller steers a state $x \in X_k \setminus X_{k-1}$ into the set X_{k-1} in one step.

6 Conclusions

The controller is nonlinear. In the first implementation, the control is determined by solving a simple optimization problem which depends on the set $X_k \setminus X_{k-1}$ in which the current state resides. In the second, the control is affine (even linear) in each simplex X_{kj}.

The controller can be extended to deal with the tracking problem in which it is desired that an output $Cx[k]$ of the system tracks the output $Dz[k]$ of an exogenous system. The composite system (plant plus disturbance) is described by

$$x[k+1] = Fx[k] + Gu[k] + w[k] \tag{10}$$
$$z[k+1] = Tz[k] \tag{11}$$
$$e[k] = Cx[k] - Dz[k] \tag{12}$$

in which $z[k]$ is the state of the exogenous system generating the signal to be tracked. The resultant control is a nonlinear function of the composite state (x, z); the controller steers this state robustly to a neighbourhood N of the zero error manifold in the composite state space (a manifold on which $e = Cx - Dz = 0$). This neighbourhood is robust control invariant and there exists a linear controller $u = K_1 x + K_2 z$ which keeps the state of this system in N (despite the disturbance) if the initial state is in N. The neighbourhood shrinks to the zero error manifold as W shrinks to $\{0\}$.

The resultant controller is switched, in that the actual control law selected depends only on the region of the state space in which the current state lies; In the first implementation the control law is nonlinear in each 'layer' $X_k \setminus X_{k-1}$ and affine (or linear) in each simplex X_{kj} in the second implementation. Both the discrete and continuous state subsystems of the controller are, therefore, memoryless.

The complexity of the sets X_k (number of inequalities required in their specification) increases rapidly with k, but there exists considerable scope for simplification [10].

It is not clear that incorporating memory is advantageous in this situation. In the tracking case, it is possible that the trajectory may remain close to the boundary between two regions in state space with differing controllers. The potential for chattering can be removed by incorporating hysteresis.

References

[1] J. M. C. Clark. The consistent selection of parameterizations in system identification. In *Proceedings of Joint Automatic Control Conference*, West Lafayette, Indiana, 1976.

[2] R. H. Middleton, G. C. Goodwin, D. J. Hill, and D. Q. Mayne. Design issues in adaptive control. *IEEE Transactions on Automatic Control*, 33:50–58, 1988.

[3] G. C. Goodwin and D. Q. Mayne. A parameter estimation perspective of continuous time model reference adaptive control. *Automatica*, 23:57–70, 1987.

[4] A. S. Morse, D. Q. Mayne, and G. C. Goodwin. Applications of hysteresis switching in parameter adaptive control. *IEEE Transactions on Automatic Control*, 37:1343–1354, 1992.

[5] S. R. Weller and G. C. Goodwin. Hysteresis adaptive control of linear multivariable systems. In *Proceedings 31st IEEE Conference on Decision and Control*, pages 1731–1736, Tucson, Arizona, 1992.

[6] F. M. Pait and A. S. Morse. A cycling switching stategy for parameter adaptive control. *IEEE Transactions on Automatic Control*, 39(6):1172–1183, 1994.

[7] E. G. Gilbert and K. T. Tan. Linear systems with state and control constraints: the theory and application of maximal output admissible sets. *IEEE Transactions on Automatic Control*, AC-36:1008–1020, 1991.

[8] E. C. Gilbert and I. Kolmanovsky. Discrete-time reference governors for systems with state and control constraints and disturbance inputs. In *Proceedings of the 34th IEEE Conference on Decision and Control*, pages 1189–1194, New Orleans, Louisiana, 1995.

[9] K. T. Tan. *Maximal output admissible sets and the nonlinear control of linear discrete-time systems with state and control constraints.* PhD thesis, University of Michigan, 1991.

[10] D. Q. Mayne and W. R. Schroeder. Robust time-optimal of constrained linear systems. Technical Report UCD-ECE-SCR-96/1, College of Engineering, University of California, Davis, July 1996.

[11] S.S. Keerthi and E.G. Gilbert. Computation of minimum-time feedback control laws for systems with state-control constraints. *IEEE Transactions on Automatic Control*, AC-32:432–435, 1987.

[12] J. D. Glover and F. C. Schweppe. Control of linear dynamic systems with set constrained disturbances. *IEEE Transactions on Automatic Control*, 16:411–423, 1971.

[13] D. P. Bertsekas and I. B. Rhodes. On the minimax reachability of target sets and target tubes. *Automatica*, 7:233–247, 1971.

[14] R. Schneider. *Convex bodies: The Brunn-Minkowski theory*, volume 44 of *Encyclopedia of Mathematics and its Applications*. Cambridge University press, Cambridge, England, 1993.

Hybrid Control for Global Stabilization of Nonlinear Systems[*]

N. Harris McClamroch[†] and Ilya Kolmanovsky[‡]
Department of Aerospace Engineering
The University of Michigan
Ann Arbor, Michigan 48109-2118.

Abstract

Stabilization of an equilibrium solution of nonlinear control systems has been extensively studied. Many elegant and important results are now available. Most of these results make use of smooth feedback to provide stabilization only within a local region containing the equilibrium. In this paper, we consider a class of nonlinear control systems, that includes nonholonomic control systems, and hence is known to present stabilization difficulties using smooth feedback. There are good theoretical and practical reasons for using hybrid feedback controllers to achieve closed loop stabilization for this class of systems. This paper presents a summary of one such hybrid feedback control approach, that guarantees global asymptotic stability. The development is a step towards a synthesis of classical nonlinear control methodology and intelligent control methodology.

1 BACKGROUND

Due to important technological advances, it is essential that control theorists continue to ask fundamental questions about the formulation and use of control theoretic models. Considerable attention has recently been given to control models formulated as hybrid systems. Classical smooth nonlinear control systems are included at one extreme, while at the other extreme are included finite automata. More typically, hybrid systems include both smooth nonlinear dynamical systems features as well as automata or discrete event features. Hybrid systems often arise as a consequence of logic-based switching in a supervisory control architecture.

Many important control systems are not smooth. The most common class arises from the inclusion of digital (or logical) components and interfaces [1].

---

[*]Edited version of paper presented at the Block Island Workshop on Control using Logic-Based Switching, September 29-October 2, 1995.

[†]Supported by NSF grant MSS-9114630.

[‡]Supported by a François-Xavier Bagnoud Fellowship.

There are many important nonlinear control systems that are smooth only in certain regions; from a global perspective they are smooth systems suitably patched together. If global stabilization problems are posed for such hybrid nonlinear control systems [2], it is presumably the case that hybrid feedback is required.

Although smooth feedback can be used to achieve local stabilization of smooth nonlinear control systems in many cases, there are good reasons for using hybrid feedback to obtain improved nonlocal properties; see, e.g., [3,4]. It is also the case that there are classes of smooth nonlinear control systems that cannot be stabilized even locally using smooth feedback [5], but they can be stabilized using hybrid feedback. Such classes of systems do arise in important applications. One class of such systems are nonholonomic control systems [6]; a survey of recent results in this area appears in [7].

For both of the above classes of nonlinear control systems, some form of hybrid feedback seems to be required. Experience suggests that simple discontinuous feedback controllers with simple switching surfaces do not suffice to stabilize these classes of systems. Rather, more general types of feedback controllers that require description using logical relations (if-then statements for example) seem to be required. In this paper, we refer to such logic based, discrete event, feedback controllers as hybrid controllers [8] since one can often express them in a hierarchical form consisting of classical feedback controllers with control parameters that are adjusted by a hybrid mechanism. Such control structures have been most commonly studied in the context of intelligent control [9,10]. The resulting closed loop defined by an interconnection of a smooth nonlinear system, a discrete event or logic based controller, and suitable interfaces, is referred to as a hybrid system.

2   A CLASS OF CASCADE NONLINEAR CONTROL SYSTEMS

We address a feedback stabilization problem for a class of globally defined nonlinear control systems which after preliminary state and control transformations take the form

$$\dot{\theta} = f(y, t), \tag{1}$$

$$\dot{x} = Ax + Bu, \tag{2}$$

$$y = Cx + Du, \tag{3}$$

where system (2)-(3) is referred to as the base subsystem and system (1) is referred to as the fiber subsystem. Here θ is a p-vector fiber variable, x is an n-vector base variable, y is an m-vector base output, u is an r-vector control variable, and $t \geq 0$ denotes time. In the fiber subsystem, the function $f(y, t)$ is globally defined, smooth, and time-periodic with a period $T > 0$. It is further assumed that $f(0, t) = 0$ for all $t \geq 0$ so that the origin is an equilibrium of (1)-(3) with $u = 0$. In the base subsystem, A is an $n \times n$ matrix, B is an $n \times r$ matrix, C is an $m \times n$ matrix and D is an $m \times r$ matrix. The base subsystem is uncoupled from the fiber subsystem, and the time rate of change of the fiber variables depend only on the base subsystem output and time. System (1)-(3) can be viewed as a cascade connection of a linear time-invariant subsystem, followed by a nonlinear time-periodic static subsystem and an integrator.

Control systems described by equations (1)-(3) are very general and can arise in many control problems, most notably as models for a large class of underactuated control systems. If $f(y, t)$ is linear and time invariant then a linear control system, in cascade form, is obtained. Equations (1)-(3) can also be used to describe bilinear control systems in a cascade form.

The formulation of equations (1)-(3) includes, as special cases, the normal forms for kinematic and dynamic models of nonholonomic Chaplygin systems [7]. To make this connection clear, consider the normal form for kinematic nonholonomic systems in Chaplygin form:

$$\begin{aligned}
\dot{\theta} &= g_1(v)\dot{v}_1 + \ldots + g_r(v)\dot{v}_r, \\
\dot{v}_i &= u_i, i = 1, \ldots, r,
\end{aligned} \qquad (4)$$

where (θ, v) is a $p + r$-vector state, $\theta = (\theta_1, \ldots, \theta_p)$, $v = (v_1, \ldots, v_r)$, and $u = (u_1, \ldots, u_r)$ is an r-vector control input. By defining $y = (v, \dot{v})$, $x = v$, $f(y, t) = \sum_{i=1}^{r} g_i(v)\dot{v}_i$, $A = 0_{r \times r}$, $B = I_{r \times r}$, $C = [I_{r \times r} 0_{r \times r}]^T$, $D = [0_{r \times r} I_{r \times r}]^T$, this system can be converted into the form (1)-(3). Examples of nonholonomic control systems which can be converted to the form (4) include multibody spacecraft conserving angular momentum [11] and vehicles with multiple trailers [12].

Cascade interconnections of linear and nonlinear systems of the form (1)-(3) can arise in study of partial feedback linearization [13,14], two-time-scale systems [15], and peaking phenomenon [16]. Under appropriate assumptions the cascade systems studied in these references may fit the formulation of the present paper. Many specific examples of nonlinear control problems studied in the literature also fit within the general assumptions of equations (1)-(3).

We describe a specific feedback construction procedure for nonlinear control systems described by equations (1)-(3). This construction procedure provides controllers that, under appropriate assumptions, globally asymptotically stabilize (1)-(3). These controllers operate by switching between various time-periodic control functions at discrete time instants. Hence this class of feedback controllers are examples of hybrid controllers. This feedback requires measurements of the fiber variables at the discrete time instants and continuous measurements of the base state variables.

In [17,18], a related approach for stabilization of drift-free, affine, time-invariant nonlinear controllable systems is proposed. Use is made of a family of periodic inputs that are universal nonsingular controls and result in periodic trajectories. Linearization about each of these trajectories is controllable. Consequently, a perturbation of a periodic input can be constructed to bring the state closer to the origin at the end of each cycle. Related ideas for non-holonomic control systems have also been proposed in [19]. Our approach also relies on a family of periodic control inputs but in a different way. For example, it is critical for our construction procedure that these periodic inputs do not result in periodic trajectories. Furthermore, formulation (1)-(3) includes systems which are not drift-free or time-invariant, to which the approach of [17,18] is not applicable.

3 CONSTRUCTION OF CONTROLLERS

Several assumptions are introduced. Then the form of the feedback controllers is identified. These feedback controllers are expressed in terms of certain control parameters whose values are adjusted according to an algorithmic procedure.

Assumption 1. The matrix A is Hurwitz.

Clearly, the assumption about A is equivalent to assuming that the pair (A, B) is stabilizable.

Assumption 2. The function $f(y, t)$ is everywhere continuously differentiable in the variable y and is continuous and T-periodic in the variable t.

The controller construction relies on a parameter dependent family of continuous T-periodic control functions of the form $U(\alpha, t)$, where $U(\alpha, t + T) = U(\alpha, t)$ for all $t \geq 0$ and α is a q-vector parameter.

Assumption 3. The family $U(\alpha, t)$ satisfies the following properties

$$U(0, t) = 0 \text{ for all } t \geq 0, \text{ and } \lim_{\alpha \to 0} \sup_{t \in [0, T]} \| U(\alpha, t) \| = 0. \tag{5}$$

One example of a control family satisfying (5) is provided by $U(\alpha, t) = F(t)\alpha$ where F is a continuous T-periodic $r \times q$ matrix function and α is a q-vector parameter. Many other choices of $U(\alpha, t)$ are also possible.

For the subsequent development, define the fiber controllability map

$$G(\alpha) = \int_0^T f(y^{ss}(\alpha, t), t) dt, \tag{6}$$

where $y^{ss}(\alpha, t)$ denotes the steady-state T-periodic solution of equations (2) and (3) with $u(t) = U(\alpha, t)$. The value of $G(\alpha)$ can be interpreted as the change in the fiber variable induced by one cycle of the steady-state base output $y^{ss}(\alpha, t)$. Since (2), (3) define a linear system and A is Hurwitz, there exists a steady state T-periodic base output for any T-periodic input.

Assumption 4. The fiber controllability map $G : R^q \to R^p$ is open at $\alpha = 0$, i.e., for any neighborhood V of 0 in R^q, $G(V)$ is a neighborhood of 0 in R^p.

Satisfaction of Assumption 4 requires that the fiber subsystem must be "controllable by the base output" and the control family $U(\alpha, t)$ must be "sufficiently rich in terms of the class of input functions allowed by the parameterization".

Let $t^k = kT$ and $\theta^k = \theta(t^k)$, where k is an integer. We now define a class of feedback controllers for system (1)-(3) as follows:

$$u(t) = U(\alpha^k, t), t^k \leq t < t^{k+1}. \qquad (7)$$

The stabilization problem, as formulated here, is to select a control parameter update rule that generates the sequence $\{\alpha^k\}$ so that the closed loop is globally asymptotically stable.

To illustrate that such update rules exist, consider the control parameter update rule developed in [11,20].

Algorithm 3.1. Suppose $0 < \gamma < 1$;

1. For k=0:   If $\theta(0) = 0$ and $x(0) = 0$, set $\alpha^0 = 0$;
              else select any $\alpha^0 \neq 0$ such that
              $\|\alpha^0\| < (\|x(0)\|^2 + \|\theta(0)\|^2)^{1/2}$.

2. For $k > 0$: If $\theta^k = 0$, set $\alpha^k = \alpha^{k-1}$.
              If $G^T(\alpha^{k-1})\theta^k < 0$, set $\alpha^k = \alpha^{k-1}$.
              If $G^T(\alpha^{k-1})\theta^k \geq 0$, $\theta^k \neq 0$, select α^k, $\|\alpha^k\| < \gamma\|\alpha^{k-1}\|$,
              such that $G^T(\alpha^k)\theta^k = -\rho^k\theta^k$ for some $\rho^k > 0$.

The feedback algorithm operates as follows. If initially the system is at the origin, then $\alpha^0 = 0$. Otherwise a nonzero value is assigned to α^0. For each $k > 1$ the selection of α^k is based θ^k, the observed value of the fiber variables at the time instant t^k. If $\theta^k = 0$ then $\alpha^k = \alpha^{k-1}$. Suppose $\theta^k \neq 0$. If, based on steady-state prediction provided by G, the value of α^{k-1} is expected to yield a nonzero net decay in $\|\theta\|$ over the kth period then $\alpha^k = \alpha^{k-1}$. Otherwise, α^k is selected so that the inner product of the vectors $G(\alpha^k)$ and θ^k is negative and so that the magnitude of α^k is a fraction of the magnitude of α^{k-1}. This selection procedure for α^k guarantees that if the base subsystem is in steady-state over the kth period, then $\theta^{k+1} = \theta^k + G(\alpha^k) = (1 - \rho^k)\theta^k$. Note that because of the transients, the actual net change in θ over the kth period may be different from the steady-state prediction provided by $G(\alpha^k)$. Assumption 4 guarantees that the required selection of α^k can be always made.

Our main result is provided by the following theorem. The proof is given in [11]. Similar results, restricted to a class of nonholonomic control systems, are available in [20].

Theorem 3.1.
Consider a feedback controller defined according to (7) and Algorithm 3.1, and suppose Assumptions 1, 2, 3 and 4 hold. Then, the origin of the closed loop system is globally asymptotically stable.

Remark 3.1.
Considerable flexibility is allowed in selecting the parameterized family of control functions; the only requirement is that specifications given by Assumptions

1, 2, 3, 4 be satisfied. If the system (1)-(3) is time invariant, there are no a priori specifications about the period of the parameterized family $U(\alpha, t)$. In fact a family of parameterized constant functions, corresponding to piecewise constant control inputs, may suffice.

Remark 3.2.
While in many cases evaluation of $G(\alpha)$ can be made directly using either analytical or numerical approaches, it turns out that Algorithm 3.1 can be based on an approximation of $G(\alpha)$. Suppose that $G(\alpha)$ can be represented as a sum of two functions, $G_1(\alpha)$ and $G_2(\alpha)$, where $G_1(\alpha)$ is open at the origin and $G_2(\alpha)$ is higher order at the origin, i.e., $\| G_2(\alpha) \| = o(\| G_1(\alpha) \|)$ as $\alpha \to 0$. Then, the closed loop can be shown to be globally asymptotically stable even when $G(\alpha)$ is replaced in Algorithm 3.1 by its approximation $G_1(\alpha)$ and α^0 is selected to be sufficiently small. In deciding if $G(\alpha)$, or its approximation $G_1(\alpha)$, are open at the origin, higher order local surjectivity criteria can be used, see [21]. In the case when $p = 1, \theta$ is a scalar, the only requirement is that $G(\alpha)$ and $G_1(\alpha)$ have the same sign for $\|\alpha\|$ small.

Remark 3.3.
Additional conditions can be given on the parameter update rule in Algorithm 3.1 that provide exponential convergence rate for each initial state.

Remark 3.4.
Algorithm 3.1 requires calculation of α^k at a time-instant t^k, based on the observed value of θ^k at the same time instant. The value of θ^k can be replaced in Algorithm 3.1 by a delayed value of the fiber variables and the conclusion of Theorem 3.1 remain valid. This property facilitates real time controller implementation.

Remark 3.5.
A preliminary investigation suggests that the above control construction results in an exceedingly robust closed loop. It is clear that the specific choice of matrices A, B, C, D, and the vector function $f(y, t)$ are relevant only to the extent that they influence the fiber controllability map. The nature of this robustness to model uncertainty has yet to be fully explored. However, it is clear that large uncertainty in the function $f(y, t)$ can be accommodated outside a small neighborhood of the origin by selection of initial control parameters with small norm. The penalty for obtaining such robustness is decreased convergence rate. Consequently, there is a trade off between increased robustness (at least in terms of uncertainty in the function $f(y, t)$) and decreased speed of response.

Remark 3.6.
If there is substantial model uncertainty so that the fiber controllability map $G(\alpha)$ cannot be accurately computed based on the model, neural network or other learning schemes can be employed to identify and approximate the map $G(\alpha)$, and this approximation can be used in construction of the hybrid feed-

back control.

Remark 3.7.
The linearity of the base subsystem is used in the proof of Theorem 3.1 to ensure that for all α sufficiently small in magnitude all trajectories of the base subsystem forced with $u(t) = U(\alpha, t), t \geq 0$, converge exponentially to the periodic trajectory $y^{ss}(\alpha, t)$. This property may hold even if the base subsystem is nonlinear or has delays; in this case the result in Theorem 3.1 is still valid.

4 STABILIZATION OF NONHOLONOMIC SYSTEMS IN POWER FORM

Consider a kinematic nonholonomic system in power form with two inputs [22]:

$$\dot{\theta}_1 = y_1 y_2,$$
$$\vdots$$
$$\dot{\theta}_p = \frac{1}{p!}(y_1)^p y_2,$$
$$\dot{x}_1 = u_1, \dot{x}_2 = u_2,$$
$$y_1 = x_1, y_2 = u_2.$$

$$(8)$$

Although the base dynamics is not asymptotically stable, it is, clearly, stabilizable. Let $q = p + 1$ and $\alpha = (\alpha_1, \ldots, \alpha_{p+1})$ be the q-vector parameter. The controller for the power form, defined with $T = 2\pi$ according to a specific parameterization that satisfies the required specifications, is

$$u_1(t) = -rx_1 + r\alpha_1^k(\cos rt - \sin rt), \ 2k\frac{\pi}{r} \leq t < 2(k+1)\frac{\pi}{r},$$

$$u_2(t) = -rx_2 + r\sum_{j=2}^{p+1} \alpha_j^k \left(\frac{\sin r(j-1)t}{(j-1)}\right.$$
$$+ \left. \cos r(j-1)t\right), \ 2k\frac{\pi}{r} \leq t < 2(k+1)\frac{\pi}{r}, \quad (9)$$

where the feedback gain $r > 0$ determines the speed of response. After some calculations we obtain an explicit expression for

$$G(\alpha) = H(\alpha_1) \begin{bmatrix} \alpha_2 \\ \vdots \\ \alpha_{p+1} \end{bmatrix}, \quad (10)$$

where $H(\alpha_1)$ is a $p \times p$ matrix given by

$$H(\alpha_1) = \pi \begin{bmatrix} a_{11}\alpha_1 & 0 & \ldots & 0 \\ a_{12}(\alpha_1)^2/2 & a_{22}(\alpha_1)^2/2 & \ldots & 0 \\ \vdots & \vdots & \ldots & 0 \\ a_{1p}(\alpha_1)^p/p! & a_{2p}(\alpha_1)^p/p! & \ldots & a_{pp}(\alpha_1)^p/p! \end{bmatrix}$$

If $\alpha_1 \neq 0$, the matrix $H(\alpha_1)$ is invertible. Hence, G is open at the origin. It is easy to express Algorithm 3.1 for this case in a more explicit form:

Algorithm 4.1. Suppose $0 < \gamma < 1$;

1. For k=0:   If $\theta(0) = 0$ and $x(0) = 0$, set $\alpha^0 = 0$;
   else select any $\alpha^0 \neq 0$ such that $||\alpha^0|| < (||x(0)||^2 + ||\theta(0)||^2)^{1/2}$

2. For $k > 0$:   If $\theta^k = 0$, set $\alpha^k = \alpha^{k-1}$.
   If $G^T(\alpha^{k-1})\theta^k < 0$, set $\alpha^k = \alpha^{k-1}$.
   If $G^T(\alpha^{k-1})\theta^k \geq 0$, $\theta^k \neq 0$, select α^k with $||\alpha^k|| < \gamma||\alpha^{k-1}||$,
   solve the linear equation $H(\alpha^k)v = -\theta^k$ for a p-vector v and se

$$
\begin{bmatrix} \alpha_2^k \\ \vdots \\ \alpha_{p+1}^k \end{bmatrix} = \frac{\gamma}{\sqrt{p+1}}||\alpha^{k-1}||\frac{v}{||v||}.
$$

This completes the feedback controller construction for nonholonomic systems in power form.

Remark 4.1:
Various other feedback controllers for the power form (and the related chained form) have been proposed in the literature, see e.g. [12,23,24,25,26,27].

Remark 4.2:
Extensions of the proposed controllers to the case of nonholonomic systems in power form with more than two inputs [27] and to the case of nonholonomic systems in dynamic (or extended) power form [11] are possible.

Remark 4.3.
In our earlier work [28], a formally different hybrid feedback structure that guarantees global asymptotic stability of the origin for nonholonomic control systems in power form has been proposed. Actually, the hybrid structure in [28] also results in amplitude modulated periodic base responses that are selected to force the fiber variables and the base variables simultaneously to the origin. There are many hybrid control structures that can achieve global asymptotic stability.

5   EXAMPLES

We consider three examples. The first example is a nonholonomic system of the form (1)-(3). The second example is a linear control system of the form (1)-(3). The same hybrid controller constructed to stabilize the nonholonomic system in the first example is shown to stabilize this linear system. The third example is a nonholonomic system that cannot be transformed into the power form. In all cases, hybrid controllers are constructed that provide global stabilization to the origin.

Example 5.1:
Consider a nonholonomic control system given by

$$\dot{\theta} = -y_1 \dot{y}_2,$$
$$\dot{x}_1 = x_2, \dot{x}_2 = u_1,$$
$$\dot{x}_3 = x_4, \ \dot{x}_4 = u_2,$$
$$y_1 = x_1, \ y_2 = x_4.$$

One controller, defined to meet the required specifications, is

$$u_1(t) = -x_1 - x_2 + \alpha^k \cos t + \alpha^k, \ 2k\pi \leq t < 2(k+1)\pi,$$
$$u_2(t) = -x_3 - x_4 + | \alpha^k | \sin t, \ 2k\pi \leq t < 2(k+1)\pi.$$

It is easy to show that $G(\alpha) = -0.5\alpha \mid \alpha \mid$. Thus $G(\alpha)$ is open and Algorithm 3.1 can be expressed as follows: If $\alpha^{k-1}\theta^k > 0$ or $\theta^k = 0$, then select $\alpha^k = \alpha^{k-1}$; If $\alpha^{k-1}\theta^k \leq 0$ or $\theta^k \neq 0$, then select $\alpha^k = \gamma|\alpha^{k-1}|\text{sign}(\theta^k)$.

Thus the resulting closed loop is globally asymptotically stable. The controller was implemented with $\gamma = 0.8$. The time histories of some of the states for a typical closed loop response are shown in Figure 1.

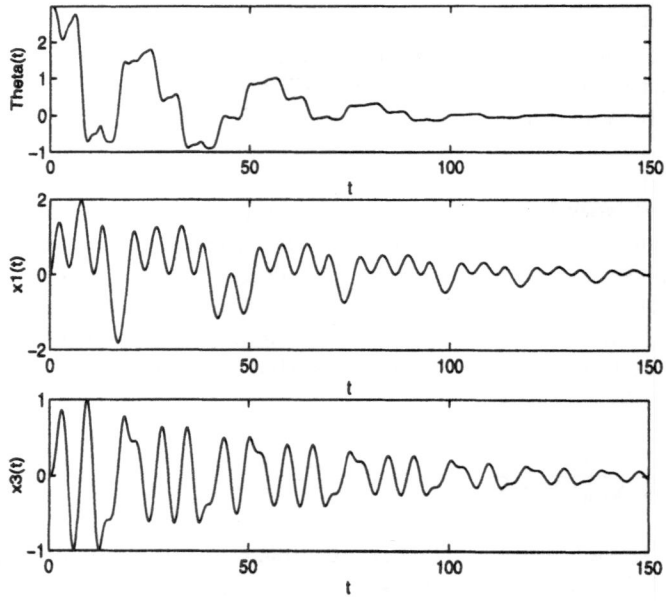

Figure 1.

Example 5.2:

Now consider a linear control system of the form (1)-(3) given by

$$\dot{\theta} = y_1 - \dot{y}_2,$$
$$\dot{x}_1 = x_2, \ \dot{x}_2 = u_1,$$
$$\dot{x}_3 = x_4, \ \dot{x}_4 = u_2,$$
$$y_1 = x_1, \ y_2 = x_4,$$

and the controller

$$u_1(t) = -x_1 - x_2 + \alpha^k \cos t + \alpha^k, \ 2k\pi \leq t < 2(k+1)\pi,$$
$$u_2(t) = -x_3 - x_4 + \mid \alpha^k \mid \sin t, \ 2k\pi \leq t < 2(k+1)\pi.$$

An easy computation shows that $G(\alpha) = -\alpha$. Thus $G(\alpha)$ is open at the origin and Algorithm 3.1 can be expressed as follows: If $\alpha^{k-1}\theta^k > 0$ or $\theta^k = 0$, then select $\alpha^k = \alpha^{k-1}$; If $\alpha^{k-1}\theta^k \leq 0$ or $\theta^k \neq 0$, then select $\alpha^k = \gamma|\alpha^{k-1}|\text{sign}(\theta^k$

Thus the resulting closed loop is globally asymptotically stable. The controller was implemented with $\gamma = 0.8$. The time histories of some of the states for a typical closed loop response are shown in Figure 2.

Note that this is exactly the same controller, with the same parameter update algorithm, as in Example 5.1. Thus the same hybrid controller simultaneously globally stabilizes the origin of the two control systems, one a nonholonomic system and the other a linear system. The key fact is that for the given controller form, the two fiber controllability maps satisfy $G(\alpha) > 0$ for $\alpha < 0, G(0) = 0$, and $G(\alpha) < 0$ for $\alpha > 0$; hence the parameter update algorithms are identical for the two cases. This common controller is observed to provide extremely robust closed loop properties, simultaneously stabilizing a class of plants that include nonholonomic and linear plants.

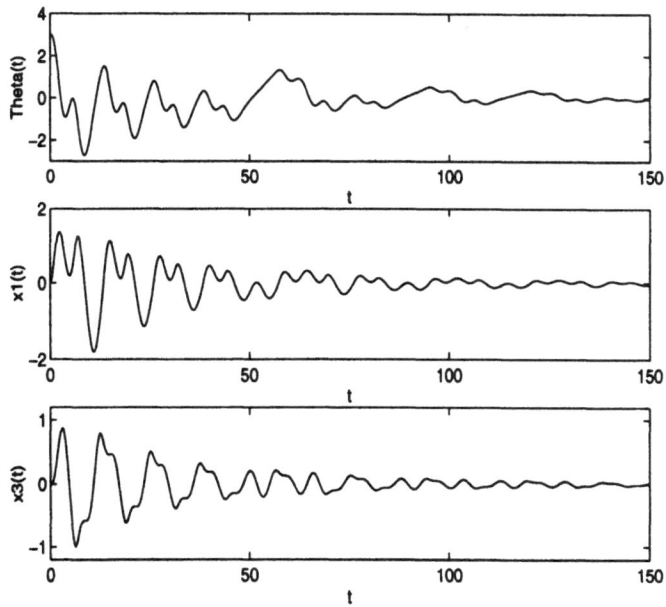

Figure 2.

Example 5.3:

The third example is the nonholonomic system

$$\dot{\theta} = \frac{y_3 + y_4}{1 + y_1^2 + y_2^2},$$
$$\dot{x}_1 = u_1, \quad \dot{x}_2 = u_2,$$
$$y_1 = x_1, \quad y_2 = x_2, \quad y_3 = u_1, \quad y_4 = u_2.$$

In any neighborhood of the origin this system violates conditions for conversion into the power (or chained) form [22]. Consequently, stabilization techniques developed for systems in power (or chained) form cannot be used. Nevertheless, feedback controllers can be constructed using the procedure described above.

A controller, defined according to the required specifications, is

$$u_1(t) = -x_1 + \alpha^k \sin t, \quad 2k\pi \le t < 2(k+1)\pi,$$
$$u_2(t) = -x_2 + \alpha^k + \alpha^k \cos t, \quad 2k\pi \le t < 2(k+1)\pi.$$

Values of $G(\alpha)$ can be computed numerically and it can be verified that $G(\alpha) > 0$ for $\alpha < 0, G(0) = 0$, and $G(\alpha) < 0$ for $\alpha > 0$. If the control parameters are based on fiber variable measurements that are delayed by one half period and are updated according to Algorithm 3.1, the closed loop is guaranteed to be globally asymptotically stable. The controller, based on the delayed measurements, was implemented with $\gamma = 0.8$. The time histories of some of the states for a typical closed loop response are shown in Figure 3.

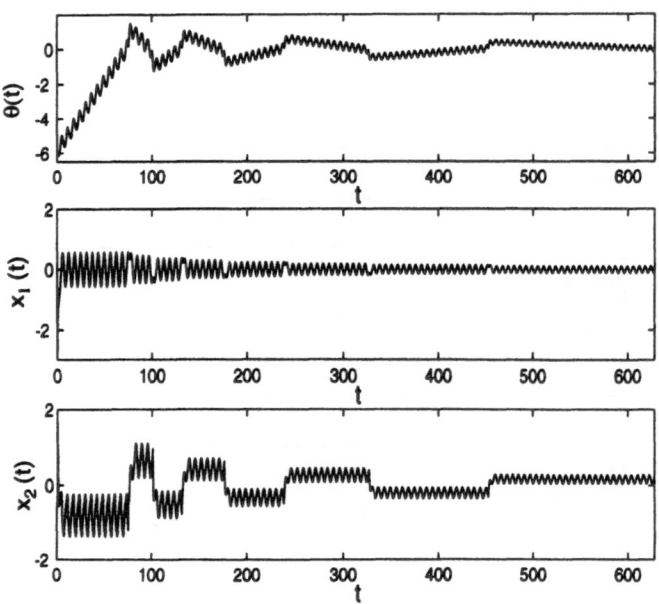

Figure 3.

6   CONCLUDING REMARKS

We have introduced feedback controllers, in a hybrid form, for globally stabilizing a class of nonlinear control systems to the origin. The controllers excite the base dynamics periodically and adjust the excitation at discrete time instants by switching from one periodic input to another, thereby achieving stabilization of the system. Although exponential stabilization is not guaranteed, the major benefits of the methodology are that global stabilization is provided and the required assumptions are extremely weak. These benefits are obtained by using a carefully developed hybrid feedback controller.

References

[1] Stiver J A, Antsaklis P J and Lemmon M D. Digital control from a hybrid perspective. Proceedings of 33rd Conference on Decision and Control 1994; 33:4241-4246

[2] Caines P E and Ortega R. The semi-lattice of piecewise constant controls for non-linear systems: A possible foundation for fuzzy control. Proceedings of IFAC Symposium on Nonlinear Control System Design 1995

[3] Guckenheimer J. A robust hybrid stabilization strategy for equilibria. IEEE Transactions on Automatic Control 1995; 40:321-326

[4] Pait F M and Morse A S. A cyclic switching strategy for parameter-adaptive control. IEEE Transactions on Automatic Control 1994; 39:1172-1183

[5] Brockett R W. Asymptotic stability and feedback stabilization. In: Differential Geometric Control Theory. Birkhauser, Boston, 1983, pp 181-191

[6] Bloch A M, Reyhanoglu M and McClamroch N H. Control and stabilization of nonholonomic dynamic systems. IEEE Transactions on Automatic Control 1992; 37:1746-1757

[7] Kolmanovsky I and McClamroch N H. Developments in nonholonomic control problems. IEEE Control Magazine1995; 15:20-36

[8] Lemmon M D and Antsaklis P J. Event identification in hybrid control systems. Proceedings of 32nd Conference on Decision and Control 1993; 32:2323-2328

[9] Passino K M. Intelligent control for autonomous systems. IEEE Spectrum 1995; 55-62

[10] Passino K M. Bridging the gap between conventional and intelligent control. IEEE Control Systems Magazine 1993; 13:12-18

[11] Kolmanovsky I. On motion planning and feedback control of nonholonomic dynamic systems with applications to attitude control of multibody spacecraft. Ph.D. dissertation, The University of Michigan, 1995

[12] Sordalen O J and Wichlund K Y. Exponential stabilization of a car with n trailers. Proceedings of the 32nd IEEE Conference on Decision and Control 1993; 32:978-983

[13] Byrnes C I and Isidori A. Global feedback stabilization of nonlinear minimum phase systems. Proceedings of the 24th IEEE Conference on Decision and Control 1985; 24:1031-1037

[14] Marino R. Feedback stabilization of single-input nonlinear systems. Systems & Control Letters 1988; 10:201-206

[15] Saksena V R, O'Reilly J and Kokotovic P V. Singular perturbations and two-scale methods in control theory: Survey 1976-1983. Automatica 1984; 20:273-293

[16] Sussmann H and Kokotovic P V. The peaking phenomenon and the global stabilization of nonlinear systems. IEEE Transactions on Automatic Control 1991; 36:424-440

[17] Sontag E D. Universal nonsingular controls. Systems & Controls Letters 1992; 19:221-224

[18] Sontag E D. Control of systems without drift via generic loops. IEEE Transactions on Automatic Control 1995; 40:1210-1219

[19] Tilbury D, Sastry S, Murray R, and Laumond J P. Stabilization of trajectories for systems with nonholonomic constraints. IEEE Transactions on Automatic Control 1994; 39:216-222

[20] Kolmanovsky I and McClamroch N H. Stabilization of nonholonomic Chaplygin systems with linear base space dynamics. Proceedings of 34th Conference on Decision and Control 1995. 35:27-32

[21] Grasse K A. A higher-order sufficient condition for local surjectivity. Nonlinear Analysis: TMA 1986; 10:87-96

[22] Murray R M and Sastry S S. Nonholonomic motion planning: steering using sinusoids. IEEE Transactions on Automatic Control 1993; 38:700-716

[23] M'Closkey R T and Murray R. Convergence rates for nonholonomic systems in power form. Proceedings of the American Control Conference 1993; 32:2967-2972

[24] Pomet J B. Explicit design of time-varying stabilizing control laws for a class of controllable systems without drift. Systems & Control Letters 1992; 18:147-158

[25] Sordalen O J and Egeland O. Exponential stabilization of nonholonomic chained systems. IEEE Transactions on Automatic Control 1995; 40:35-49

[26] M'Closkey R T and Murray R M. Exponential stabilization of driftless nonlinear control systems via time-varying, homogeneous feedback. Proceedings of 33rd IEEE Conference on Decision and Control 1994; 33:1317-1322

[27] Walsh G C and Bushnell L G. Stabilization of multiple input chained form control systems. Proceedings of the 32nd Conference on Decision and Control 1993; 1:959-964

[28] Kolmanovsky I, Reyhanoglu M and McClamroch N H. Discontinuous feedback stabilization of nonholonomic systems in extended power form. Systems and Control Letters 1996; 27:29-36

A Minimum Energy Approach to Switching Control for Mechanical Systems

Carla A. Schwartz

University of Florida Dept of Electrical and Computer Eng
405 CSE PO Box 116130
Gainesville, FL 32611-6130 USA

Egbert Maben

Dept. Aerospace Engineering, Mechanics and Engineering Science
PO Box 116250
Gainesville, FL 32611-6250 USA

Abstract

A minimum energy control method for on-off decentralized control of mechanical systems is introduced. Energy consumption is minimized by turning on an actuator when the structure does work on it, and turning it off when the actuator would impart energy on the structure to improve overall system response.

1 Introduction

In this work, a method for minimum energy on-off switching control of mechanical systems is presented. By switching the controller on when the work done on the controller is positive, and off otherwise, the overall damping of the system response may be improved without the use of external energy.

With the advent of the space shuttle, there is considerable interest in large space structures. As the size of spacecraft increases, the need to suppress the vibrations introduced into the spacecraft body, must also be incorporated into the controller design.

Flight duration for spacecraft is limited in many cases by the amount of on-board fuel. On-board fuel is needed to fire the thrusters and apogee/perigee motors on satellites. Thruster firings are also needed for station-keeping and to correct orbital errors.

Since fuel consumption is critical to spacecraft flight, and controlling spacecraft requires fuel, it seems natural to use fuel consumption (as measured by energy) as an optimization criterion for control system design.

The work presented here is based on the intuition that using a fuel minimization criterion that works in conjunction with conservation principles will produce energy efficient methods for control of vibrating systems. The computational process by which the controller scheme is determined would introduce delays in practice. These delays are ignored in this analysis, but their effect can be easily incorporated into the analytical results presented here.

Several authors have used or proposed the use of on-off control for attitude and shape control of large flexible space structures [1, 2, 3, 4, 7, 8, 10, 11, 12]. Of that body of work [2, 3, 4, 5, 8, 10, 11] used the principle of fuel minimization as a criterion for the implementation of on-off switching control. In [4], Foster and Silverberg introduced a method which involved the construction of an observer to estimate the times of occurrence of, as well as the values, of the relative minima and maxima of the states (nodal positions and velocities). On-off controllers have also been applied in suppressing the structural vibrations introduced by wind and earthquakes [9] in civil structures.

The work described in this paper differs from previous work on on-off control in that in this work, energy is used as a measure of fuel consumption. It is assumed throughout this work that the actuator is of the spring type, i.e., it generates a force proportional to displacement in a mechanical structure. Conservation laws which drive the actuator on-off scheme constrain fuel consumption to be minimized.

This paper is organized as follows. In Section 2, a method for increasing the overall damping in a system by using on-off control is explained. Conclusions follow in Section 3.

2 Main Results: Damping Vibrations Using On-Off Control

2.1 Introduction

When a structure with an actuator attached to it undergoes vibrations, the structure does work on that actuator and vice-versa. In this section a method is described in which work done on the actuators will be taken advantage of for fuel minimization. The ideas for this method were inspired by previous work on on-off control [4, 5].

Consider a structure in which some of the vibrating modes are well-damped while some are poorly damped. The use of on-off control in a conservative system makes possible the distribution of energy throughout the structure in order to achieve a desired overall damping in the system. In the following analysis, it is assumed that before control is applied, the system has one highly damped mode while all others are lightly damped. The actuator location is assumed to be such that the resulting system is controllable from that location.

The analysis will be carried out for one actuator but can be extended to the case of several actuators. The proposed switching laws are such that the actuator is on when the structure is doing work on the actuator, and off otherwise. After the actuator is turned off it remains off until the structure once again does work on the actuator. There are two possibilities for treating the energy absorbed by the actuator after the actuator is switched off. Either it can be dumped into some sort of energy dissipating device, such as a resistor, or stored for future use in a storage device, such as a gyroscope or a battery.

2.2   Energy Analysis

Consider a second order dynamical system given by

$$M\ddot{x} + D\dot{x} + Kx = F \tag{1}$$

where $M^T = M > 0$ is the mass matrix, $K^T = K \geq 0$ is the stiffness matrix, $D^T = D \geq 0$ is the damping matrix, F is the $n \times n$ forcing term, and x is the $n \times 1$ vector of nodal displacements.

Under control that is applied through state feedback, $F = (\Delta Kx)$, the dynamics become

$$M\ddot{x} + D\dot{x} + (K + \Delta K)x = 0, \tag{2}$$

where $(K + \Delta K)^T = (K + \Delta K) \geq 0$. For the case of only one actuator, assume that ΔK has only one nonzero entry on the diagonal, $\Delta K_a > 0$, associated with the displacement, x_a, which corresponds to the actuator location. Even though the controller is switched on and off, the values of displacements and velocities are continuous at all times.

The controller will be switched on according to a scheme described here. If the actuator is of the spring type and were on continuously, it would periodically absorb energy from and dissipate energy to the vibrating structure. If the actuator is switched off when it would be releasing energy and switched on only when it would be absorbing energy from the structure, the objective of increasing the damping in the structure using a minimum of external energy may be achieved. This concept comprises the crux of the proposed switching scheme and results in a minimization of fuel use.

When the controller is off, the energy in the system is given by

$$E_{off} = \frac{1}{2}[\dot{x}^T M \dot{x} + x^T K x - \dot{x}^T D x]. \tag{3}$$

Since the system is dissipative, the rate of change of energy is given by

$$\dot{E}_{off} = \dot{x}^T [M\ddot{x} + (K)x] - \dot{x}^T D\ddot{x}. \tag{4}$$

While the controller is on, the rate of change of energy is given by

$$\dot{E}_{on} = \dot{x}^T [M\ddot{x} + (K + \Delta K)x] - \dot{x}^T D\ddot{x}. \tag{5}$$

The signal $\dot{E}_{on} - \dot{E}_{off}$ gives a measure of the rate of work done on all actuators if they were on continuously. In the case of one actuator, $\dot{E}_{on} - \dot{E}_{off}$ is positive (work is done on the actuator by the structure), if and only if $\Delta K_a x_a > 0$. In the case of multiple actuators, if i denotes the location of the ith actuator, $[\Delta K_i x_i]$ is monitored and the actuators are switched on when the force signal acting on them is positive. This concept is summarized in the following switching procedure.

Procedure for switching with one actuator:

1. Between switching times, monitor the signal $\Delta K_a x_a$, where ΔK_a is the nonzero entry on the ΔK matrix.

2. When $\Delta K_a x_a$ is positive, switch the controller on. Switch the controller off at the end of a predetermined time which is related to the system time constant as determined by the lowest period of oscillation in the structure,[1] and release (or store) the energy absorbed by the spring when it is switched off.

2.3 Example:

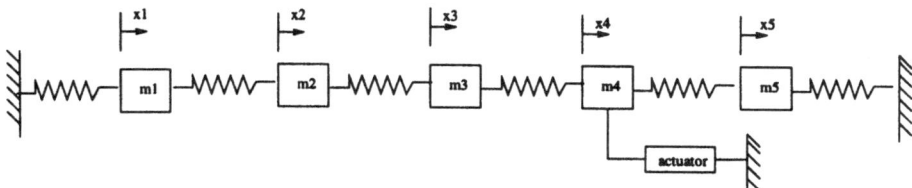

Figure 1: A 5DOF Mechanical System, with actuator attached to fourth mass

Consider the five degree of freedom spring mass system shown in figure (1) described by the second order system,

$$M\ddot{x} + D\dot{x} + Kx = f.$$

The mass, damping, and stiffness matrices are respectively chosen as

$$M = \begin{bmatrix} 1 & 0 & 0 & 0 & 0 \\ 0 & 1 & 0 & 0 & 0 \\ 0 & 0 & 1 & 0 & 0 \\ 0 & 0 & 0 & 1 & 0 \\ 0 & 0 & 0 & 0 & 1 \end{bmatrix}$$

$$D = \begin{bmatrix} 0.055 & 0.065 & 0.022 & -0.039 & -0.026 \\ 0.065 & 0.161 & 0.063 & -0.088 & -0.059 \\ 0.022 & 0.063 & 0.084 & -0.028 & -0.025 \\ -0.039 & -0.088 & -0.028 & 0.217 & 0.102 \\ -0.026 & -0.059 & -0.025 & 0.102 & 0.117 \end{bmatrix}$$

---

[1]In the example in the next subsection, this time is set to be one tenth of lowest period of oscillation.

$$K = \begin{bmatrix} 3 & -1 & 0 & 0 & 0 \\ -1 & 2 & -1 & 0 & 0 \\ 0 & -1 & 2 & -1 & 0 \\ 0 & 0 & -1 & 2 & -1 \\ 0 & 0 & 0 & -1 & 3 \end{bmatrix}$$

In this example $x_a = x_4$. All but the fourth mode have very little damping (equivalent damping factor $\simeq 0.002$), while the fourth mode is highly damped (equivalent damping factor $\simeq 0.4$). The simulations of the control procedure of the previous section show this method to be very effective in damping out the vibrations of the measured response even when the initial conditions were very close to the mode shapes of a lesser damped mode of the uncontrolled structure, a worst case scenario.

Figures 2 through 6 show the nodal (position) response of the uncontrolled and controlled system when the initial conditions were set to the mode shape of one of the lesser damped mode of the uncontrolled structure. The simulation was carried out for each of the mode shapes. However, the results are only shown for one set of initial conditions. From these plots it is clear that the proposed method is effective in increasing the damping.

3   Conclusions

A new minimum energy method for decentralized on-off switching control of mechanical structures is introduced. It is demonstrated that energy consumption is minimized by turning on the actuator when a structure is doing work on the actuator, and turning it off after prescribed on-times. This method is shown to be effective in distributing the energy throughout an otherwise lightly damped system. It can also be shown that this proposed methodology can be used to achieve a desired average damping rate when each mode may be uniquely controlled. In practice, some energy input for signal monitoring, as well as energy to drive an electro-mechanically switched controller using a computer or a microprocessor, would be required when implementing this method.

References

[1] W.E. Vander Velde, J. He, "Design of Space structure control systems using on-off thrusters," *Journal of Guidance, Control and Dynamics*, Vol 6,No 1, 1983, pp. 53–60.

[2] M.E. Brown, "Rapid slewing maneuvers of a flexible spacecraft using on-off thrusters," Charles Stark Draper Laboratory Inc, Cambridge,Ma, Internal Report CSDL- T-825, 1983.

[3] L.Silverberg, "Uniform Damping Control of Spacecraft," *Journal of Guidance,Control and Dynamics*, Vol 9, No 2, 1986, pp. 221–227.

[4] L.Silverberg, "On Off Decentralized Control of Flexible structures," *ASME Journal of Dynamic Systems Measurement and Control*, 1991 i, Vol 113, pp. 41–47.

[5] S.F. Masri, G.A. Bekey, T.K. Caughey, "Optimum Pulse Control of Flexible Structures," *Journal of Applied Mechanics*, Vol 48, No 9, 1981, pp. 619–626.

[6] D.J.Inman, *Vibrations, Control, Measurements*, Prentice Hall, 1991.

[7] A. Arbel, N.K. Gupta, "Robust Collocated control for large flexible space structures," *Journal of Guidance and Control*, vol. 4, Sep-oct 1981, pp. 480–486.

[8] S.F. Masri, G.A. Bekey, T.K. Caughey, "Optimum Pulse Control of Flexible Structures," *ASME Journal of Applied Mechanics*, Vol. 48, Sept. 1981, pp. 619–626.

[9] M. Abdel Rohman, H.H.E. Leipholz, "Structural Control By Pole Assignment method," *Journal of Engineering Mechanics Division ASCE*, Vol 104, No EM5, Oct 1978, pp 1159–1175.

[10] Seywald H., Kumar R. R., Deshpande S.S., Heck M., "Minimum fuel spacecraft reorientation," *Journal of Guidance ,Control and Dynamics*, Vol 17, No 1, Jan-Feb 1994, pp. 21–29.

[11] Redmond J., Silverberg L., "Fuel Consumption in Optimal Control," *Journal of Guidance Control and Dynamics*, Vol 13, No 2, Mar-Apr, 1992, pp. 424–430.

[12] Neustadt L. , "Synthesizing time optimal control systems," *Journal of Mathematical Analysis and Applications*, No 1, 1960, pp. 484–493.

[13] Peter Khan , *Mathematical methods for scientists and engineers*, John Wiley and Sons. 1990.

[14] Caughey T.K., O'Kelly, M.E.J., "Classical Normal Modes in Damped Linear Dynamic Systems," *ASME Journal of Applied Mechanics*, No 32, 1965, pp. 583–588.

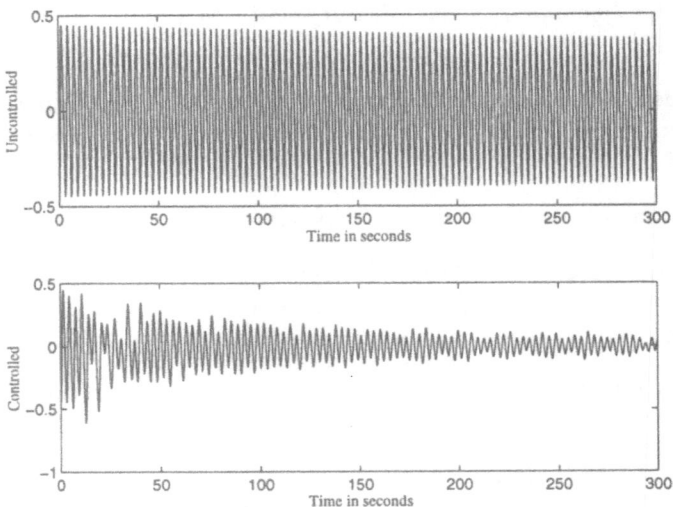

Figure 2: Response at Node 1

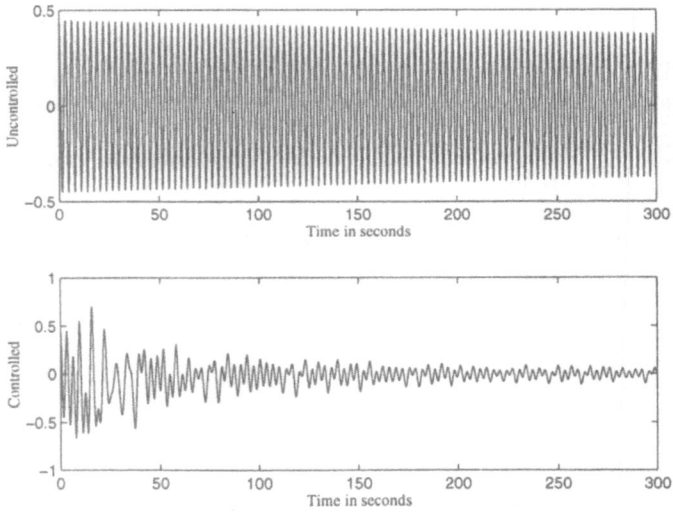

Figure 3: Response at Node 2

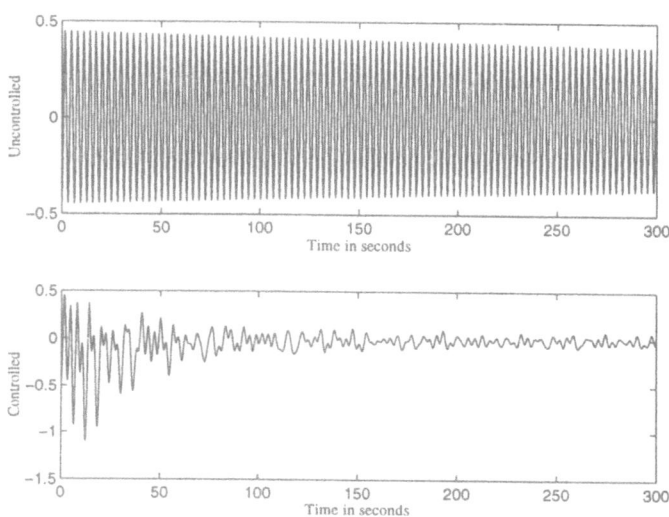

Figure 4: Response at Node 3

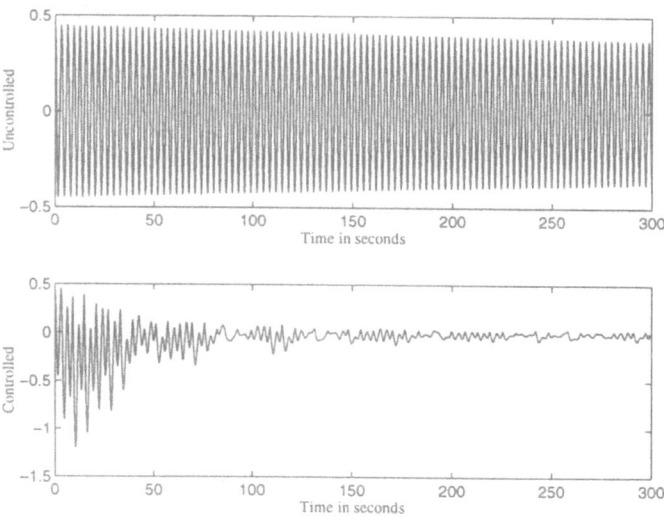

Figure 5: Response at Node 4

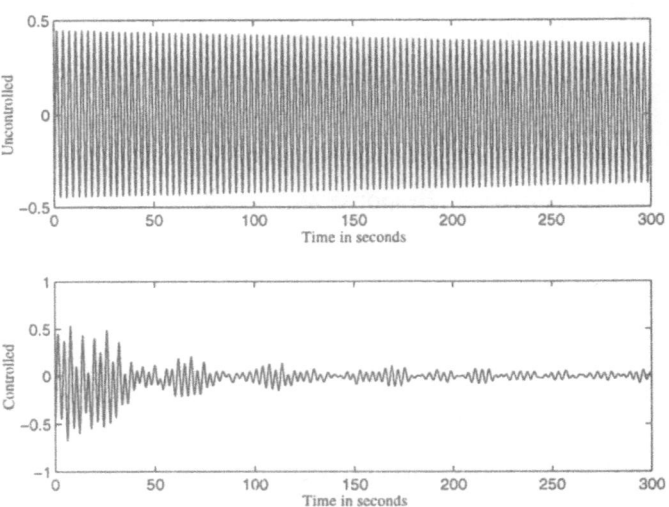

Figure 6: Response at Node 5

Lagrangian Modeling and Control of Switch Regulated DC-to-DC Power Converters *

Romeo Ortega
HEUDIASYC, URA CNRS 817
Université de Technologie de Compiegne
Centre de Recherches de Royallieu BP 649
Compiegne Cedex, France
e-mail: rortega@hds.univ-compiegne.fr

Hebertt Sira-Ramírez
Departamento Sistemas de Control
Escuela de Ingeniería de Sistemas
Universidad de Los Andes
Mérida, Edo. Mérida, Venezuela
e-mail: isira@ing.ula.ve

Keywords : DC–to–DC Power Converters, Lagrangian Dynamics, Passivity-Based Sliding Control.

1 Introduction

Modeling and regulation of switched dc-to-dc power converters was initiated by the pioneering work of Middlebrook and Ćuk [1] in the mid seventies. The area has undergone a wealth of practical and theoretical development as evidenced by the growing list of research monographs, and textbooks, devoted to the subject (see Severns and Bloom [2], and Kassakian *et al* [3]).

In this article, a Lagrangian dynamics approach is used for deriving a physically motivated model of the dc–to–dc power converters of the "Boost" type. The approach, however, is suitable to be applied to any kind of dc-to-dc power converter. The proposed modeling technique is based on a suitable parametriza-

---

*This work was supported by the Consejo de Desarrollo Científico, Humanístico and Tecnológico of the Universidad de Los Andes, under Research Grant I-456-94, by the French Embassy in Caracas, Venezuela, through the Programme de Cooperation Postgradué (PCP), and by the Consejo Nacional de Desarrollo Científico, Humanístico and Tecnológico (CONICIT) of Venezuela

tion, in terms of the switch position parameter, of the Euler-Lagrange functions describing each intervening system and subsequent application of the Lagrangian formalism. The resulting system is also shown to be a Lagrangian system, hence, passivity based regulation is proposed as a natural controller design technique (see Ortega *et al*, [4] for details). The second part of the article designs and compares the performance of traditional (static) sliding mode controller and that of a dynamical passivity-based sliding mode controller. The comparison is carried out in terms of evaluating a physically motivated scalar performance index involving the total stored energy. It is shown that, depending on the choice of the controller's initial state, the passivity based controller might render a finite performance index while the traditional controller results in an unbounded index.

2   Modeling of Switched Euler–Lagrange Systems

An Euler–Lagrange system is classically characterized by the following set of nonlinear differential equations, known as *Lagrange equations*,

$$\frac{\mathrm{d}}{\mathrm{d}t}\left(\frac{\partial \mathcal{L}}{\partial \dot{q}}\right) - \frac{\partial \mathcal{L}}{\partial q} = -\frac{\partial \mathcal{D}}{\partial \dot{q}} + \mathcal{F}_q \tag{2.1}$$

where q is the vector of *generalized positions*, assumed to have n components, represented by $q_1, \ldots, q_n$, and $\dot{q}$ is the vector of *generalized velocities*. The scalar function $\mathcal{L}$ is the *Lagrangian* of the system, defined as the difference between the *kinetic energy* of the system, denoted by $\mathcal{T}(\dot{q}, q)$, and the *potential energy* of the system, denoted by $\mathcal{V}(q)$, i.e,

$$\mathcal{L}(\dot{q}, q) = \mathcal{T}(\dot{q}, q) - \mathcal{V}(q) \tag{2.2}$$

The function $\mathcal{D}(\dot{q})$ is the *Rayleigh dissipation cofunction* of the system. The vector $\mathcal{F}_q = (\mathcal{F}_{q_1}, \ldots, \mathcal{F}_{q_n})$ represents the ordered components of the set of *generalized forcing functions* associated with each generalized coordinate.

We refer to the set of functions $(\mathcal{T}, \mathcal{V}, \mathcal{D}, \mathcal{F})$ as the *Euler–Lagrange functions* of the system and symply express a system Σ by the ordered cuadruple

$$\Sigma = (\mathcal{T}, \mathcal{V}, \mathcal{D}, \mathcal{F}) \tag{2.3}$$

We are particularly interested in dynamical systems containing a single *switch*, regarded as the only *control function* of the system. The switch position, denoted by the scalar u, is assumed to take values on a discrete set of the form $\{0, 1\}$. We assume that for each one of the switch position values, the resulting system is an Euler-Lagrange system (EL system for short) characterized by its corresponding EL parameters.

Definition 2.1 *We define a switched EL function $\mathcal{M}_u$, associated with the EL functions $\mathcal{M}_0$ and $\mathcal{M}_1$, as a function, parametrized by the switch position u, which is consistent with $\mathcal{M}_0$ and $\mathcal{M}_1$, for the corresponding values of the switch position parameter, $u \in \{0, 1\}$, i.e.,*

$$\mathcal{M}_u|_{u=0} = \mathcal{M}_0 \quad ; \quad \mathcal{M}_u|_{u=1} = \mathcal{M}_1 \qquad (2.4)$$

A system arsing from switchings among the EL systems Σ_0 and Σ_1 is said to be a switched EL system, whenever it is completely characterized by the set of consistent switched EL functions

$$\Sigma_u = (\mathcal{T}_u, \mathcal{V}_u, \mathcal{D}_u, \mathcal{F}_u) \qquad (2.5)$$

Assume we are given two EL system models, Σ_0 and Σ_1, characterized by EL parameters, $(\mathcal{T}_0, \mathcal{V}_0, \mathcal{D}_0, \mathcal{F}_0)$ and $(\mathcal{T}_1, \mathcal{V}_1, \mathcal{D}_1, \mathcal{F}_1)$, respectively. Our basic modeling problem consists, generally speaking, in determining a consistent parametrization of the EL functions, $(\mathcal{T}_u, \mathcal{V}_u, \mathcal{D}_u, \mathcal{F}_u)$ in terms of the switch position parameter, u, with corresponding switched Lagrangian $\mathcal{L}_u$, such that the system model obtained by direct application of the EL equations (2.5) on $\mathcal{L}_u$, results in a parametrized model, Σ_u, which is consistent,in the sense described above, with the models Σ_0 and Σ_1 .

2.1 A Lagrangian Approach to the Modeling of a "Boost" Converter

Consider the switch–regulated "Boost" converter circuit of Figure 1. We consider, separately, the Lagrange dynamics formulation of the two circuits associated with each one of the two possible positions of the regulating switch. In order to use standard notation we refer to the input current x_1 in terms of the derivative of the circulating charge q_L, as $\dot{q}_L$. Also the capacitor voltage x_2 will be written as q_C/C where q_C is the electrical charge stored in the output capacitor. The switch position paramerter, u, is assumed to take values in the discrete set $\{0, 1\}$

Consider then $u = 1$. In this case, two separate, or decoupled, circuits are clearly obtained and the corresponding Lagrange dynamics formulation can be carried out as follows.

Define $\mathcal{T}_1(\dot{q}_L)$ and $\mathcal{V}_1(q_C)$ as the kinetic and potential energies of the circuit, respectively. We denote by $\mathcal{D}_1(\dot{q}_C)$ the Rayleigh dissipation cofunction of the circuit. These quantities are readily found to be

$$\mathcal{T}_1(\dot{q}_L) = \frac{1}{2} L (\dot{q}_L)^2$$

$$\mathcal{V}_1(q_C) = \frac{1}{2C} q_C^2$$

$$\mathcal{D}_1(\dot{q}_C) = \frac{1}{2} R (\dot{q}_C)^2$$

$$\mathcal{F}_{q_L}^1 = E \quad ; \quad \mathcal{F}_{q_C}^1 = 0 \qquad (2.6)$$

where $\mathcal{F}_{q_L}^1$ and $\mathcal{F}_{q_C}^1$ are the *generalized forcing* functions associated with the coordinates q_L and q_C, respectively.

Consider now the case $u = 0$. The corresponding Lagrange dynamics formulation is carried out in the next paragraphs.

Define $\mathcal{T}_0(\dot{q}_L)$ and $\mathcal{V}_0(q_C)$ as the kinetic and potential energies of the circuit, respectively. We denote by $\mathcal{D}_0(\dot{q}_L, \dot{q}_C)$ the Rayleigh dissipation function of the circuit. These quantities are readily found to be,

$$
\begin{aligned}
\mathcal{T}_0(\dot{q}_L) &= \frac{1}{2} L \left(\dot{q}_L \right)^2 \\
\mathcal{V}_0(q_C) &= \frac{1}{2C} q_C^2 \\
\mathcal{D}_0(\dot{q}_L, \dot{q}_C) &= \frac{1}{2} R \left(\dot{q}_L - \dot{q}_C \right)^2 \\
\mathcal{F}_{q_L}^0 &= E \quad ; \quad \mathcal{F}_{q_C}^0 = 0
\end{aligned}
\tag{2.7}
$$

where, $\mathcal{F}_{q_L}^0$ and $\mathcal{F}_{q_C}^0$ are the *generalized forcing* functions associated with the coordinates q_L and q_C, respectively.

The EL parameters of the two situations, generated by the different switch position values, result in identical kinetic and potential energies. The switching action merely changes the Rayleigh dissipation cofunction between the values $\mathcal{D}_0(\dot{q}_C)$ and $\mathcal{D}_1(\dot{q}_L, \dot{q}_C)$. Therefore, the *dissipation structure* of the system is the only one affected by the switch position.

$$
\begin{aligned}
\mathcal{T}_u(\dot{q}_L) &= \frac{1}{2} L \left(\dot{q}_L \right)^2 \\
\mathcal{V}_u(q_C) &= \frac{1}{2C} q_C^2 \\
\mathcal{D}_u(\dot{q}_L, \dot{q}_C) &= \frac{1}{2} R [(1 - u)\dot{q}_L - \dot{q}_C]^2 \\
\mathcal{F}_{q_L}^u &= E \quad ; \quad \mathcal{F}_{q_C}^u = 0
\end{aligned}
\tag{2.8}
$$

The switched lagrangian function associated with the above defined EL parameters is given by

$$
\mathcal{L}_u = \mathcal{T}_u(\dot{q}_L) - \mathcal{V}_u(q_C) = \frac{1}{2} L \left(\dot{q}_L \right)^2 - \frac{1}{2C} q_C^2
\tag{2.9}
$$

One then proceeds, using the Lagrange equations (2.1), to formally obtain the parametrized differential equations defining the switch regulated system corresponding to (2.8). Such equations are given by,

$$
\begin{aligned}
\frac{\mathrm{d}}{\mathrm{d}t} \left(\frac{\partial \mathcal{L}_u}{\partial \dot{q}_L} \right) - \frac{\partial \mathcal{L}_u}{\partial q_L} &= -\frac{\partial \mathcal{D}_u}{\partial \dot{q}_L} + \mathcal{F}_{q_L} \\
\frac{\mathrm{d}}{\mathrm{d}t} \left(\frac{\partial \mathcal{L}_u}{\partial \dot{q}_C} \right) - \frac{\partial \mathcal{L}_u}{\partial q_C} &= -\frac{\partial \mathcal{D}_u}{\partial \dot{q}_C} + \mathcal{F}_{q_C}
\end{aligned}
\tag{2.10}
$$

Use of (2.9) on (2.10) results in the following set of differential equations

$$L\ddot{q}_L = -(1-u)R[(1-u)\dot{q}_L - \dot{q}_C] + E$$
$$\frac{q_C}{C} = R[(1-u)\dot{q}_L - \dot{q}_C] \qquad (2.11)$$

which can be rewritten, after substitution of the second equation into the first, as

$$\ddot{q}_L = -(1-u)\frac{q_C}{LC} + \frac{E}{L}$$
$$\dot{q}_C = -\frac{1}{RC}q_C + (1-u)\dot{q}_L \qquad (2.12)$$

Using $x_1 = \dot{q}_L$ and $x_2 = q_C/C$ one obtains

$$\dot{x}_1 = -(1-u)\frac{1}{L}x_2 + \frac{E}{L}$$
$$\dot{x}_2 = (1-u)\frac{1}{C}x_1 - \frac{1}{RC}x_2 \qquad (2.13)$$

The proposed switched dynamics (2.13) coincides with the classical state model developed in [1]. The fact that switched circuits, such as the above presented "Boost" converter, can be modeled using the Lagrangian formalism, implies that a *passivity-based approach* can be naturally attempted in the design of stabilizing feedback control policies. In order to establish suitable comparisons we first revisit the traditional static sliding mode controller and its main properties.

3 Regulation of the "Boost" Converter

3.1 Traditional Sliding Mode Control of the "Boost" Converter

Consider the "Boost" converter circuit, shown in Figure 1, described by the set of differential equations (2.13).

We shall denote by $\overline{x}_1$ and $\overline{x}_2$ the state variables of the system under *ideal sliding mode* conditions. In other words, $\overline{x}_1$ and $\overline{x}_2$, represent the "average" values of the state variables under sliding mode operation. The "equivalent control", denoted by u_{eq}, represents an ideal, i.e., a *virtual* feedback control action that smoothly keeps the controlled state trajectories of the system evolving on the sliding surface, provided motions are started, precisely, at the sliding surface itself (see Utkin [5] for definitions).

In order to avoid a well-knwon unstable closed loop behavior due to the non-minimum phase character of the converter we proceed, instead of regulating the output capacitor votage, x_2, to *indirectly* regulate such a variable as indicated in the following proposition.

Proposition 3.1 *Consider the switching line* $s = x_1 - V_d^2/RE$, *where* $V_d > 0$
is a desired constant capacitor voltage value. The switching policy, given by

$$u = 0.5\,[\,1 - \text{sign}\,(s)\,] = 0.5\,\big[\,1 - \text{sign}\,(x_1 - V_d^2/RE)\,\big] \qquad (3.1)$$

locally creates a stable sliding regime on the line $s = 0$ *with ideal sliding dynamics characterized by*

$$\bar{x}_1 = \frac{V_d^2}{RE} \;\; ; \;\; \dot{\bar{x}}_2 = -\frac{1}{RC}\left[\bar{x}_2 - \frac{V_d^2}{\bar{x}_2}\right] \;\; ; \;\; u_{eq} = 1 - \frac{E}{\bar{x}_2} \qquad (3.2)$$

Moreover, the ideal sliding dynamics behaviour of the capacitor voltage variable, described by (3.2), can be explicitly computed as

$$\bar{x}_2(t) = \sqrt{V_d^2 + \big[\bar{x}_2^2(t_h) - V_d^2\big]\,e^{-\frac{2}{RC}(t - t_h)}} \qquad (3.3)$$

where t_h *stands for the reaching instant of the sliding line* $s = 0$ *and* $\bar{x}_2(t_h)$ *is the capacitor voltage at time* t_h.

Figure 2 depicts a typical "start up" state variables evolution, from zero initial conditions, in a current–mode controlled "Boost" converter such as that of Proposition 3.1.

A measure of the performance of the sliding mode controlled system, described above, is obtained by using the integral of the stored stabilization error energy. This quantity is given by

$$\mathcal{I}_B = \int_0^\infty H(\tau)\mathrm{d}\,\tau = \int_0^\infty \frac{1}{2}\left[L\left(x_1(\tau) - \frac{V_d^2}{RE}\right)^2 + C\left(x_2(\tau) - V_d\right)^2\right]\mathrm{d}\,\tau$$
$$(3.4)$$

Such a performance criterion can also be regarded as a weighted integral square state stabilization error for the state vector. We simply address such an index as the "WISSSE" index.

Proposition 3.2 *The WISSSE index, computed along the sliding mode controlled trajectories of the "Boost" converter, is unbounded, independently of the initial conditions of the converter.*

3.2   Passivity-Based Sliding "Current-Mode" Controller for the "Boost" Converter

In the following developments we introduce an auxiliary state vector, denoted by x_d. The basic idea is to take x_d as a "desired" vector trajectory for the converter state vector x. This auxiliary vector variable will be determined on the basis of energy shape considerations, and passivity, imposed on the evolution of the error vector $x - x_d$. The feedback regulation of the auxiliary state x_d, towards the desired constant equilibrium value of the state x, will in fact result in the specification of a *dynamical output feedback controller* for the original converter state. We will be using a sliding mode control viewpoint for the regulation of x_d towards the desired equilibrium value of x.

We rewrite the "Boost" converter equations (2.13) in matrix-vector form as

$$\mathcal{D}_B \dot{x} + (1 - \mu)\mathcal{J}_B x + \mathcal{R}_B x = \mathcal{E}_B \tag{3.5}$$

where

$$\mathcal{D}_B = \begin{bmatrix} L & 0 \\ 0 & C \end{bmatrix} \; ; \; \mathcal{J}_B = \begin{bmatrix} 0 & 1 \\ -1 & 0 \end{bmatrix} \; ; \; \mathcal{R}_B = \begin{bmatrix} 0 & 0 \\ 0 & 1/R \end{bmatrix} \; ; \; \mathcal{E}_B = \begin{bmatrix} E \\ 0 \end{bmatrix} \tag{3.6}$$

Consider the stored stabilization error energy, H_d, of the state x with respect to the auxiliary state variable x_d,

$$H_d = \frac{1}{2}(x - x_d)^T \mathcal{D}_B (x - x_d) \tag{3.7}$$

Suppose x_d satisfies the following controlled differential equation

$$\mathcal{D}_B \dot{\tilde{x}} + (1 - u)\mathcal{J}_B \tilde{x} + \mathcal{R}_{Bd}\tilde{x} = 0 \tag{3.8}$$

where $\mathcal{R}_{Bd} = \mathcal{R} + \mathcal{R}_B$, with $\mathcal{R} = \text{diag}[R_1 \; 0]$. Then, the following proposition holds valid.

Proposition 3.3 *Given a desired state vector trajectory $x_d(t)$ for the converter state vector $x(t)$, the error vector $\tilde{x} = x - x_d$ asymptotically decays to zero, from any arbitrary initial condition $\tilde{x}(0)$, whenever $x_d(t)$ is obtained as a solution of the controlled differential equation:*

$$\mathcal{D}_B \dot{x}_d + (1 - u)\mathcal{J}_B x_d + R_{Bd}x_d = \mathcal{E}_B$$

for any given control policy u which is equally applied to both the plant and the auxiliary system. Moreover, for some positive constants α, and β, which may be, respectively, taken as $\alpha = \min\{R_1, 1/R\}$ and $\beta = \max\{L, C\}$, the time derivative of the total stored error energy H_d satisfies

$$\dot{H}_d = -(x - x_d)^T \mathcal{R}_{Bd}(x - x_d) \leq -\frac{\alpha}{\beta} H_d \leq 0$$

Consider now the auxiliary system defining x_d, explicitly written as

$$L\dot{x}_{1d} + (1 - u)x_{2d} - R_1(x_1 - x_{1d}) = E$$
$$C\dot{x}_{2d} - (1 - u)x_{1d} + \frac{1}{R}x_{2d} = 0 \tag{3.9}$$

The following proposition depicts the most important features of a sliding current-mode regulation policy of the auxiliary system (3.9) towards the desired constant equilibrium state $(x_{1d}(\infty), x_{2d}(\infty)) = (V_d^2/RE, V_d)$ of the "Boost" converter.

Proposition 3.4 *Consider the switching line* $s = x_{1d} - V_d^2/RE$, *where* $V_d > 0$ *is a desired constant capacitor voltage value for the auxiliary variable* x_{2d} *and for the converter's capacitor voltage* x_2. *The switching policy, given by*

$$u = 0.5[1 - \text{sign}(s)] = 0.5[1 - \text{sign}(x_{1d} - V_d^2/RE)] \qquad (3.10)$$

locally creates a sliding regime on the line $s = 0$. *Moreover, if the sliding-mode switching policy (3.10) is applied to* both *the converter and the auxiliary system, the converter state trajectory* $x(t)$ *converges towards the auxiliary state trajectory* $x_d(t)$ *and, in turn,* $x_d(t)$ *converges towards the desired equilibrium state. i.e.,*

$$(x_1, x_2) \rightarrow (x_{1d}, x_{2d}) \rightarrow \left(\frac{V_d^2}{RE}, V_d\right)$$

The ideal sliding dynamics is then characterized by

$$\bar{x}_{1d} = \frac{V_d^2}{RE} \; ; \; \dot{\bar{x}}_{2d} = -\frac{1}{RC}\left[\bar{x}_{2d} - \left(\frac{V_d^2}{E}\right)\frac{E + R_1(\bar{x}_1 - V_d^2/RE)}{\bar{x}_{2d}}\right]$$

$$u_{eq} = 1 - \frac{E + R_1(\bar{x}_1 - V_d^2/RE)}{\bar{x}_{2d}} \qquad (3.11)$$

where $\bar{x}_1$ *is the converter's inductor current under sliding mode conditions, primarily occurring in the controller's state space and induced, through the control input* u *on the controlled system state space.*

Figure 3 depicts simulations of a typical closed loop state behaviour for a passivity-based regulated converter system whose converter's initial state starte from several initial conditions.

We can prove the following result.

Theorem 3.5 *Consider the WISSSE performance index (3.4). The passivity-based sliding current-mode controller, described in Proposition 3.4, yields identically unbounded WISSSE index behaviour as the traditional sliding current-mode static controller of Proposition 3.1, provided initial conditions for the controller and the plant are chosen to be identical. If, on the other hand, initial conditions for the dynamical controller are chosen to be precisely at the required constant state equilibrium vector for the controlled system, then the WISSSE index is* finite. *This property holds true when the initial states of the converter are taken to be different from those of the controlled converter, provided it is guaranteed that* $\bar{x}_2(t) < V_d$ *for all* $t > t_h$.

References

[1] Middlebrook R D, Ćuk S. A General Unified Approach to Modeling Switching–Converter Power Stages. IEEE Power Electronics Specialists Conference (PESC) 1976; 18-34.

[2] Severns R P, Bloom G E. Modern DC–to–DC Switchmode Power Converter Circuits, Van Nostrand-Reinhold, New York 1983.

[3] Kassakian J G, Schlecht M, Verghese G C. Principles of Power Electronics, Addison-Wesley, Reading, Mass. 1991.

[4] Ortega R, Loría A, Kelly R, Praly L. On Passivity–Based Output Feedback Global Stabilization of Euler–Lagrange Systems. Int J of Robust and Nonl Contr 1995; 5: 313-324, .

[5] Utkin V I. Sliding Modes and Their Applications in Variable Structure Systems. MIR Editors, Moscow 1978.

FIGURES

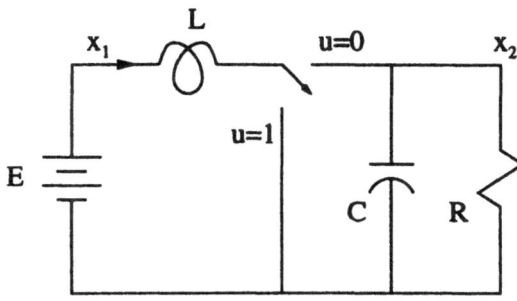

Figure 1: The "Boost" Converter Circuit.

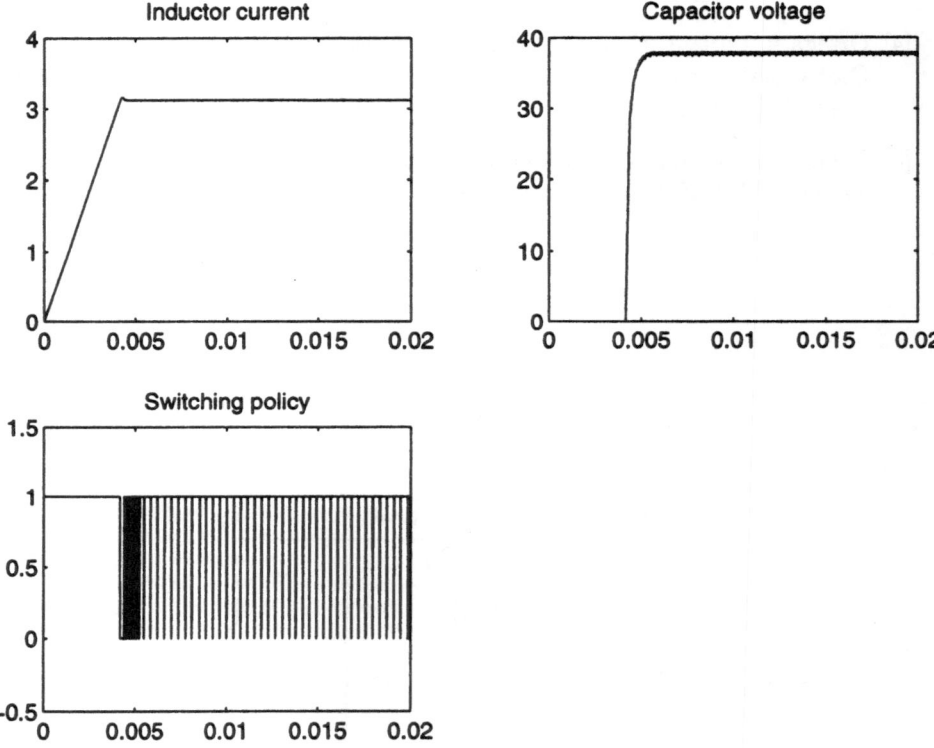

Figure 2: Typical Sliding "Current-Mode" Controlled State Responses for the "Boost" Converter.

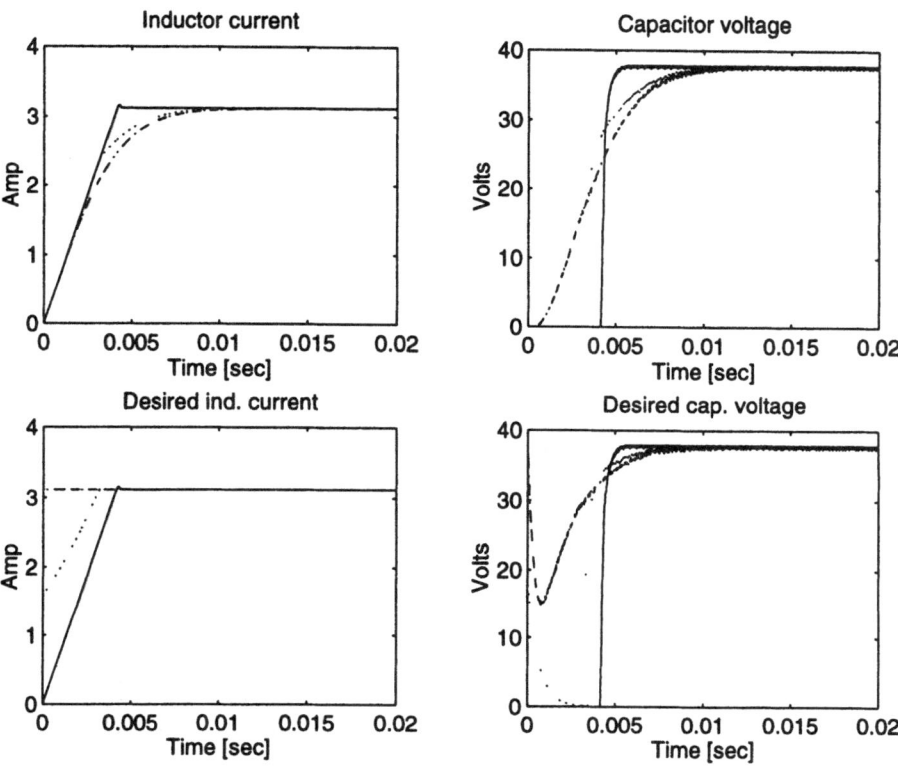

Figure 3: Controller and Plant State Resposes for Different Controller Initial Conditions, (—) $(x_{1d}(0), x_{2d}(0)) = (0,0)$; $(\cdots)$ $(x_{1d}(0), x_{2d}(0)) = (1.6, 19)$ (---) $(x_{1d}(0), x_{2d}(0)) = (3.125, 37.5)$.

Control of Underactuated Mechanical Systems Using Switching and Saturation*

Mark W. Spong
Coordinated Science
Laboratory
University of Illinois
1308 W. Main St.
Urbana, Ilinois 61801
USA

Laurent Praly
Centre Automatique
et Systèmes
École des Mines de Paris
35 rue St. Honoré
77305 Fontainebleau cédex
FRANCE

Abstract

In this paper we present some ideas on the control of underactuated mechanical systems using switching and saturation. We focus on the swingup control problem for a class of "gymnast" robots and also for the classical cart–pole system. The design methodology is based on partial feedback linearization in a first stage to linearize the actuated degrees of freedom followed by the control of the transfer of energy from the actuated to the unactuated degrees of freedom in a second stage. In a typical swingup control the desired equilibrium is unstable in the closed loop system as a consequence of the non-minimum phase behavior of the system. For this reason it is necessary to switch controllers at the appropriate time to a controller which renders the equilibrium stable. The successful implementation of the switching control has proved to be nontrivial, both in simulation and in experiment. We discuss both local and global design methods and present some simulation results.

1 Introduction

Underactuated mechanical systems are mechanical systems with fewer actuators than degrees–of–freedom and arise in several ways, from intentional design as in the brachiation robot of Fukuda [1] or the Acrobot [2], in mobile robot systems when a manipulator arm is attached to a mobile platform, a space platform, or an undersea vehicle, or because of the mathematical model used for

---

*The research of the first author is partially supported by the National Science Foundation under grants CMS-9402229, and INT-9415757

control design as when joint flexibility is included in the model [3]. In the latter sense, then, all mechanical systems are underactuated if one wishes to control flexible modes that are not directly actuated (the noncollocation problem), or even to include such things as actuator dynamics in the model description.

We consider an n–degree–of–freedom system with generalized coordinates $q^1, \ldots, q^n$, and $m < n$ actuators, each of which directly actuates a single degree of freedom. We partition the vector $q \in R^n$ of generalized coordinates as $q_1 \in R^\ell$ and $q_2 \in R^m$, where $q_1 \in R^\ell$ represents the unactuated (passive) joints and $q_2 \in R^m$ represents the actuated (active) joints. The Euler–Lagrange equations of motion of such a system are then given by [4]

$$M_{11}\ddot{q}_1 + M_{12}\ddot{q}_2 + h_1 + \phi_1 \;=\; 0 \qquad (1)$$
$$M_{21}\ddot{q}_1 + M_{22}\ddot{q}_2 + h_2 + \phi_2 \;=\; \tau \qquad (2)$$

where

$$M(q) = \begin{bmatrix} M_{11} & M_{12} \\ M_{21} & M_{22} \end{bmatrix} \qquad (3)$$

is the symmetric, positive definite inertia matrix, the vector functions $h_1(q, \dot{q}) \in R^\ell$ and $h_2(q, \dot{q}) \in R^m$ contain Coriolis and centrifugal terms, the vector functions $\phi_1(q) \in R^\ell$ and $\phi_2(q) \in R^m$ contain gravitational terms, and $\tau \in R^m$ represents the input generalized force produced by the m actuators at the active joints. For notational simplicity we will henceforth not write the explicit dependence on q of these coefficients.

2 Partial Feedback Linearization

Unlike fully actuated systems, which are always feedback linearizable, the system (1)–(2) is not linearizable in the q–coordinates, although in some cases, the system is linearizable after a nonlinear coordinate transformation. However, we may still linearize a portion of the system in the original q–coordinates. To see this, consider the first equation (1)

$$M_{11}\ddot{q}_1 + M_{12}\ddot{q}_2 + h_1 + \phi_1 = 0 \qquad (4)$$

The term M_{11} is an invertible $\ell \times \ell$ matrix as a consequence of the uniform positive definiteness of the robot inertia matrix M in (3). Therefore we may solve for $\ddot{q}_1$ in equation (4) as

$$\ddot{q}_1 = -M_{11}^{-1}(M_{12}\ddot{q}_2 + h_1 + \phi_1) \qquad (5)$$

and substitute the resulting expression (5) into (2) to obtain

$$\bar{M}_{22}\ddot{q}_2 + \bar{h}_2 + \bar{\phi}_2 = \tau \qquad (6)$$

where the terms $\bar{M}_{22}, \bar{h}_2, \bar{\phi}_2$ are given by

$$\bar{M}_{22} \;=\; M_{22} - M_{21}M_{11}^{-1}M_{12}$$
$$\bar{h}_2 \;=\; h_2 - M_{21}M_{11}^{-1}h_1$$
$$\bar{\phi}_2 \;=\; \phi_2 - M_{21}M_{11}^{-1}\phi_1$$

It is easily shown that the $m \times m$ matrix $\bar{M}_{22}$ is itself symmetric and positive definite. A partial feedback linearizing controller can therefore be defined for equation (6) according to

$$\tau = \bar{M}_{22}u + \bar{h}_2 + \bar{\phi}_2 \tag{7}$$

where $u \in R^m$ is an additional control input yet to be defined. The complete system up to this point may be written as

$$M_{11}\ddot{q}_1 + h_1 + \phi_1 = -M_{12}u \tag{8}$$
$$\ddot{q}_2 = u \tag{9}$$

Setting

$$u = -k_1 q_2 - k_2 \dot{q}_2 + \bar{u} \tag{10}$$

and defining state variables

$$\begin{array}{ll} z_1 = q_2 & z_2 = \dot{q}_2 \\ \eta_1 = q_1 & \eta_2 = \dot{q}_1 \end{array} \tag{11}$$

we may write the system in state space as

$$\dot{z} = Az + B\bar{u} \tag{12}$$
$$\dot{\eta} = w(z, \eta, \bar{u}) \tag{13}$$

where $z^T = (z_1^T, z_2^T)$, $\eta^T = (\eta_1^T, \eta_2^T)$, and the matrix A is Hurwitz. We see from (12) and (13) that $z = 0$, $\bar{u} = 0$ defines an invariant manifold in state space. Since A is Hurwitz for positive values of gains in the matrices k_p and k_d this manifold is attractive. The dynamics on the manifold are given by

$$\dot{\eta} = w(0, \eta) \tag{14}$$

We now take as our starting point the problem of designing the control input u to stabilize the system (12)–(13). This class of systems falls into the class of feedforward systems considered by Teel [5], Mazenc and Praly [6], and Janković, et. al. [7].

3 Swingup Control: Case Studies

3.1 The Acrobot

The Acrobot is a two–link planar robot with a single actuator at the elbow. The equations of motion are given by [4]

$$m_{11}\ddot{q}_1 + m_{12}\ddot{q}_2 + h_1 + \phi_1 = 0 \tag{15}$$
$$m_{21}\ddot{q}_1 + m_{22}\ddot{q}_2 + h_2 + \phi_2 = \tau \tag{16}$$

where

$$
\begin{aligned}
m_{11} &= m_1 \ell_{c1}^2 + m_2(\ell_1^2 + \ell_{c2}^2 + 2\ell_1 \ell_{c2} \cos(q_2)) + I_1 + I_2 \\
m_{22} &= m_2 \ell_{c2}^2 + I_2 \\
m_{12} &= m_{21} = m_2(\ell_{c2}^2 + \ell_1 \ell_{c2} \cos(q_2)) + I_2 \\
h_1 &= -m_2 \ell_1 \ell_{c2} \sin(q_2)\dot{q}_2^2 - 2m_2 \ell_1 \ell_{c2} \sin(q_2)\dot{q}_2 \dot{q}_1 \\
h_2 &= m_2 \ell_1 \ell_{c2} \sin(q_2)\dot{q}_1^2 \\
\phi_1 &= (m_1 \ell_{c1} + m_2 \ell_1)g \cos(q_1) + m_2 \ell_{c2} g \cos(q_1 + q_2) \\
\phi_2 &= m_2 \ell_{c2} g \cos(q_1 + q_2)
\end{aligned}
$$

The parameters m_i, ℓ_i, ℓ_{ci}, and I_i are masses, link lengths, centers of masses, and moments of inertia, respectively. The zero configuration, $q_i = 0$, in this model corresponds to the arm extended horizontally. Therefore, the swing up task is move the robot from the vertically downward configuration $q_1 = -\pi/2$, $q_2 = 0$ to the inverted configuration $q_1 = +\pi/2$, $q_2 = 0$. Our strategy is as follows: We first apply the partial feedback linearization control (7) with the outer loop term given by (10). The resulting system can be written as

$$
m_{11}\ddot{q}_1 + h_1 + \phi_1 = -m_{12}(\bar{u} - k_2 \dot{q}_2 - k_1 q_2) \tag{17}
$$

$$
\ddot{q}_2 + k_2 \dot{q}_2 + k_1 q_2 = \bar{u} \tag{18}
$$

We then choose the additional control $\bar{u}$ to swing the second link "in phase" with the motion of the first link in such a way that the amplitude of the swing of the first link increases with each swing. This can be accomplished with either a switching control or a saturating control. It is shown in [8] that the simple choice of $\bar{u}$ given by

$$
\bar{u} = k_3 \, \text{sat}(\dot{q}_1) \tag{19}
$$

where sat() is the saturation function, increases the energy and swings up the Acrobot. In the closed loop system, the open loop stable equilibrium configuration, $q_1 = -\pi/2$, $q_2 = 0$, becomes unstable and the trajectory is driven towards the inverted configuration. The final step is to switch to a "balancing controller" when the "swingup controller" $\bar{u}$ brings the state into the basin of attraction of the balancing controller. We have investigated various methods for designing balancing controllers, chiefly pseudo–linearization [2] and Linear–Quadratic methods. Figure 1 shows a swing up motion using this approach.

The difficult part of this strategy is to design the gains k_i above so that the state enters into the basin of attraction of the balancing controller. This involves searching for robust balancing controllers with large basins of attraction and proper tuning of the gains, both of which are nontrivial problems.

3.2 Three–Link Gymnast Robot

Next we apply the partial feedback linearization control to execute a so–called *giant swing* maneuver and balance for a three–link planar gymnast robot with

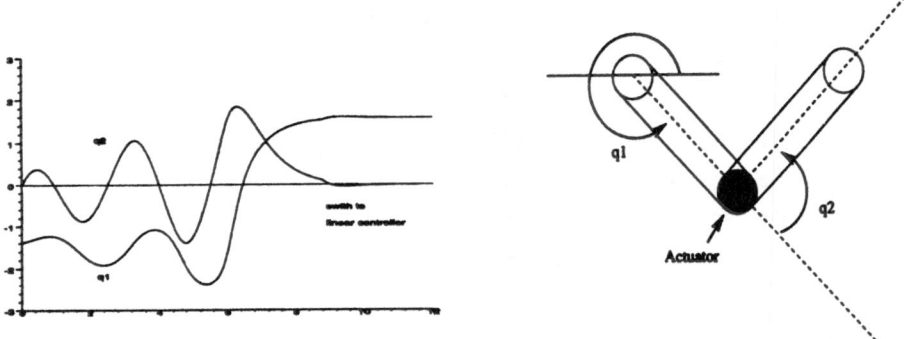

Figure 1: Swingup and Balance of The Acrobot

two actuators. The equations of motion are of the form

$$m_{11}\ddot{q}_1 + m_{12}\ddot{q}_2 + m_{13}\ddot{q}_3 + h_1 + \phi_1 \;=\; 0 \qquad\qquad (20)$$
$$m_{21}\ddot{q}_1 + m_{22}\ddot{q}_2 + m_{23}\ddot{q}_3 + h_2 + \phi_2 \;=\; \tau_2 \qquad\qquad (21)$$
$$m_{31}\ddot{q}_1 + m_{32}\ddot{q}_2 + m_{33}\ddot{q}_3 + h_3 + \phi_3 \;=\; \tau_3 \qquad\qquad (22)$$

In this case we can linearize two of the three degrees of freedom to obtain

$$m_{11}\ddot{q}_1 + h_1 + \phi_1 \;=\; -m_{12}u_2 + m_{13}u_3 \qquad\qquad (23)$$
$$\ddot{q}_2 \;=\; u_2 \qquad\qquad (24)$$
$$\ddot{q}_3 \;=\; u_3 \qquad\qquad (25)$$

Using our insight gained from the Acrobot, we specify the outer loop controls according to

$$u_2 \;=\; -k_{21}q_2 - k_{22}\dot{q}_2 + \bar{u}_2 \qquad\qquad (26)$$
$$u_3 \;=\; -k_{32}q_3 - k_{32}\dot{q}_3 + \bar{u}_3 \qquad\qquad (27)$$

with

$$\bar{u}_2 \;=\; k_{23}\,\mathrm{sat}(\dot{q}_1) \qquad\qquad (28)$$
$$\bar{u}_3 \;=\; k_{33}\,\mathrm{sat}(\dot{q}_2) \qquad\qquad (29)$$

Figure 2 shows a plot of the resulting giant swing maneuver including a switch to a linear controller at the end to balance the robot in the inverted position.

3.3 The Cart–Pole System

In this section we treat the familiar cart–pole system and show how the same design ideas as above can be applied. The Euler-Lagrange equations for this

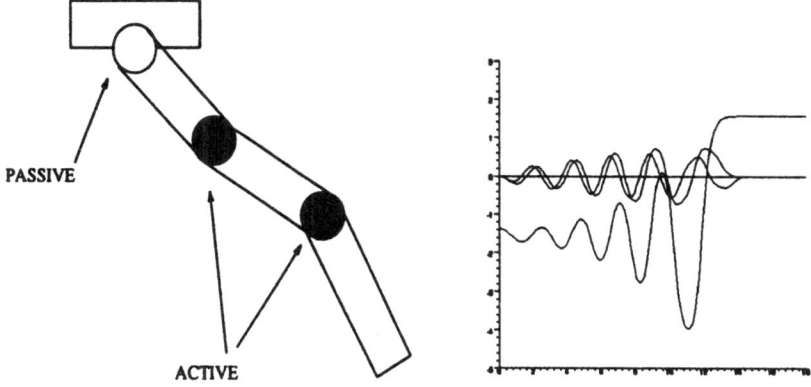

Figure 2: Three Link Gymnast Robot: Swingup and Balance

system are

$$(M + m)\ddot{x} + ml\cos(\theta)\ddot{\theta} = ml\dot{\theta}^2\sin(\theta) + F \tag{30}$$
$$\ddot{x}\cos(\theta) + l\ddot{\theta} = g\sin(\theta) \tag{31}$$

where x is the cart position and θ is the pendulum angle measured from the vertical. Applying the partial feedback linearization control, it is easy to show that the cart–pole system may be written as:

$$\dot{x} = v \tag{32}$$
$$\dot{v} = u \tag{33}$$
$$\dot{\theta} = \omega \tag{34}$$
$$\dot{\omega} = \sin(\theta) - \cos(\theta)u \tag{35}$$

where we have used the more descriptive notation, x, v, θ and ω instead of z_i and η_i to represent the cart position and velocity, and pendulum angle and angular velocity, respectively, and have normalized the parameter values. Using our above strategy, the simple control

$$u = -k_1 x - k_2 v + k_3 \operatorname{sat}(\omega) \tag{36}$$

can be used to swing up the pendulum and regulate the cart position for suitable choices of the gain parameters. However, we can improve the above controller by borrowing from section 3.4 as follows. We note that the total energy of the pendulum is given as

$$E = \frac{1}{2}\omega^2 + \cos(\theta) \tag{37}$$

and that $E = 1$ corresponds to the upright position of the pendulum. Then it can be shown that the control

$$u = -k_1 x - k_2 v + k_3(E - 1)\cos(\theta)\omega \tag{38}$$

locally guarantees that the energy converges to unity and that the cart position
and velocity are regulated to zero, as shown by Chung and Hauser in [9]. We
conjecture here that the control

$$\bar{u} = k_3 \, \text{sat}((E - 1)\cos(\theta)\omega - k_1 x - k_2 v) \tag{39}$$

renders the above result semi–global. Figure (3) shows the response of the
cart–pole using the control (39).

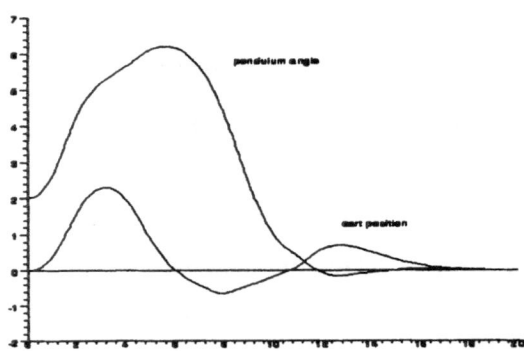

Figure 3: Cart–Pole Response Using (39)

3.4   An Almost Globally Stabilizing Controller

Whereas we expect only semi-global stabilization by using (39), an almost
globally stabilizing controller can be obtained as follows : we design a first
controller whose objective is to lead the pendulum to its homoclinic orbit and
the cart to its desired position. As a consequence the state of the cart-pole
system reaches, in finite time, a neighborhood of its desired equilibrium. Then,
as in the previous section, another controller is used to stabilize this point.

The first design is achieved by applying the recursive technique of adding
integration as introduced in [6]. For the cart-pole system, two stages are needed.

Consider, in a first stage, the system

$$\dot{v} = u_{s1}, \quad \dot{\theta} = \omega, \quad \dot{\omega} = \sin(\theta) - \cos(\theta)\, u_{s1}. \tag{40}$$

Defining the energy E as above and noting that

$$\dot{E} = -\cos(\theta)\omega u_{s1} \tag{41}$$

we can write, partially, the dynamics of this system (40) in terms of E and v
as

$$\dot{v} = u_{s1}, \quad \dot{E} = -\cos(\theta)\,\omega\, u_{s1}, \tag{42}$$

implying that the Jurdjevic–Quinn technique applies to the (v, E)–system [11, 10]. This leads us to introduce :

$$V_{s1}(v, \theta, \omega) = \Phi_1(E - 1) + \frac{k_v}{2} v^2 , \qquad (43)$$

where k_v is a strictly positive real number and Φ_1 is a positive definite and proper C^2 function, defined on $[-2, +\infty)$ which satisfies :

$$\max \{|\Phi_1'(s)|, |\Phi_1''(s)|\} \leq \frac{k_E}{2\sqrt{s + 2}} , \quad \forall s \in (-2, +\infty) , \qquad (44)$$

with k_E a real number in $(0, 1]$, e.g.

$$\Phi_1(s) = \frac{k_E}{8} \log(1 + s^2) . \qquad (45)$$

We get :

$$\dot{V}_{s1} = [-\Phi_1'(E - 1) \omega \cos(\theta) + k_v v] u_{s1} . \qquad (46)$$

It follows that $\dot{V}_{s1}$ can be made non positive by the following feedback law :

$$u_{s1}(v, \theta, \omega) = \frac{\Phi_1'(E - 1) \omega \cos(\theta) - k_v v}{1 + |v| + \Phi_1''(E - 1) \omega \cos(\theta) \sin(\theta)} . \qquad (47)$$

Note that, since :

$$|\omega| \leq \sqrt{2} \sqrt{E + 1} , \quad \cos(\theta) \sin(\theta) \leq \frac{1}{2} , \qquad (48)$$

we have :

$$|u_{s1}(v, \theta, \omega)| \leq \frac{k_E}{1 - k_E/2} + k_v , \qquad (49)$$

implying that $|u_{s1}|$ can be made arbitrarily small with an appropriate choice of k_E and k_v.

In the second stage, we consider the complete cart-pole system. To be able to apply once again Jurdjevic and Quinn technique, we follow the suggestion of [6] and introduce the following change of variables :

$$y = k_v x + v + \frac{v|v|}{2} - \Phi_1'(E - 1) \sin(\theta) , \quad u = u_{s1}(v, \theta, \omega) + u_{s2} . \quad (50)$$

This allows us to rewrite the dynamics of the cart-pole system as

$$\dot{y} = [1 + |v| + \Phi_1'' \omega \cos(\theta) \sin(\theta)] u_{s2} \qquad (51)$$
$$\dot{v} = u_{s1}(v, \theta, \omega) + u_{s2} \qquad (52)$$
$$\dot{\theta} = \omega \qquad (53)$$
$$\dot{\omega} = \sin(\theta) - \cos(\theta) (u_{s1}(v, \theta, \omega) + u_{s2}) \qquad (54)$$

From the result of the first stage and by applying the Jurdjevic and Quinn technique, we introduce :

$$V_1(x, v, \theta, \omega) = V_{s2}(x, v, \theta, \omega) = \Phi_1(E - 1) + \frac{k_v}{2} v^2 + \Phi_2(y) , \qquad (55)$$

where Φ_2 is a positive definite and proper C^1 function. We have :

$$\dot{V}_1 = [1 + |v| + \Phi_1'' \omega \cos(\theta) \sin(\theta)] [(\Phi_2' - u_{s1}) u_{s2} - u_{s1}^2] . \quad (56)$$

It follows that the following bounded feedback law makes $\dot{V}_2$ non positive :

$$u_1(x, v, \theta, \omega) = u_{s1}(v, \theta, \omega) - \text{sat} (\Phi_2'(y) - u_{s1}(v, \theta, \omega)) . \quad (57)$$

This feedback law makes the solution given by the cart staying at its desired position and the homoclinic orbit of the pendulum asymptotically stable with basin of attraction the whole state space minus a set of measure zero. This set is the stable manifold of the equilibrium point corresponding to the downward vertical position of the pole.

To complete the design, we consider a balancing controller of the form :

$$u_2(x, v, \omega, \theta) = a_1 x + a_2 v + a_3 \cos(\theta) \omega + a_4 \sin(\theta) , \quad (58)$$

where the real numbers a_i's are chosen so that the closed-loop linearized at the desired equilibrium admits this point has a locally asymptotically stable point. Corresponding to this control law, given some threshold $u_{\max}$, we can find a positive definite matrix P and a positive real number p such that by letting :

$$V_2(x, v, \theta, \omega) = (x \ v \ \omega \ \sin(\tfrac{1}{2}\theta)) P \begin{pmatrix} x \\ v \\ \omega \\ \sin(\tfrac{1}{2}\theta)) \end{pmatrix} , \quad (59)$$

we have, when $u = u_2(x, v, \omega, \theta)$,

$$0 < V_2(x, v, \omega, \theta) \le p \quad \Longrightarrow \quad \begin{cases} \dot{V}_2(x, v, \omega, \theta) < 0 , \\ |u_2(x, v, \omega, \theta)| \le u_{\max} . \end{cases} \quad (60)$$

The feedback law is then :

$$u = u_\sigma(x, v, \theta, \omega) , \quad (61)$$

where σ is a new state variable with values in $\{1, 2\}$ and whose dynamics are :

$$\sigma(t) = 1 \quad \text{if} \quad \{p \le V_2\} \quad \text{or} \quad \{\sigma^-(t) = 1 \text{ and } k_V \, p < V_2\} ,$$
$$= 2 \quad \text{if} \quad \{V_2 \le k_V \, p\} \quad \text{or} \quad \{\sigma^-(t) = 2 \text{ and } V_2 < p\} ,$$

where k_V is a real number in $(0, 1)$ and $\sigma^-(t)$ denotes the limit of $\sigma(\tau)$ when τ tends from below to t. Figure (4) shows the response of the cart–pole to the above controller (61).

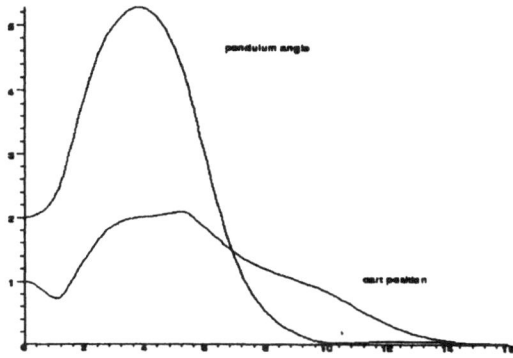

Figure 4: Cart–Pole Response Using (61)

4   Discussion

The class of underactuated mechanical systems considered here is an example of nonlinear systems in feedforward form. The design methodology presented in Section 2, takes advantage of physical insight about the dynamics of underactuated systems, and produces controllers that are simple to implement. These controllers are difficult to tune, however, and guarantee only local stability, in general. Thus, these systems are good vehicles for investigating tuning algorithms based on repetitive learning and logic based switching.

The more general procedure outline in Section 3.4, on the other hand, is more difficult to design but guarantees (almost) global stability. The design difficulty arises from the necessity to construct certain changes of variables defined as the solution of a PDE. Thus more research is needed to identify classes of systems, such as the cart–pole, for which such changes of coordinates can be readily computed.

Acknowledgments: The first author would like to thank Rogelio Lozano and Bernard Brogliato for several recent discussions on the control of underactuated mechanical systems.

References

[1] Saito, F., Fukuda, T., and Arai, F., "Swing and Locomotion Control for Two–Link Brachiation Robot," *Proc. 1993 IEEE Int. Conf. on Robotics and Automation*, pp. 719-724, Atlanta, GA, 1993.

[2] Bortoff, S.A., *Pseudolinearization using Spline Functions with Application to the Acrobot*, Ph.D. Thesis, Dept. of Electrical and Computer Engineering, University of Illinois at Urbana–Champaign, 1992.

[3] Spong, M.W., "Modeling and Control of Elastic Joint Robots", *ASME J. Dyn. Systems, Meas. and Control*, Vol. 109, pp. 310-319, December, 1987.

[4] Spong, M.W., and Vidyasagar, M., *Robot Dynamics and Control*, John Wiley & Sons, Inc., New York, 1989.

[5] Teel, A.R., "Using saturation to stabilize a class of single-input partially linear composite systems," IFAC NOLCOS'92 Symposium, Bordeaux, June 1992.
See also : "A Nonlinear Small Gain Theorem for the Analysis of Control Systems with Saturation," *IEEE Trans. on Automatic Control*, submitted, 1994.

[6] Mazenc, F., and Praly, L., "Adding an Integration and Global Asymptotic Stabilization of Feedforward Systems," *IEEE Trans. on Automatic Control*, submitted, 1994. See also : *Proceed. 33rd IEEE Conference on Decision and Control*, December 1994.

[7] Janković, M., Sepulchre, R., and Kokotović, P.V., "Global Stabilization of an Enlarged Class of Cascade Nonlinear Systems," preprint, 1995.

[8] Spong, M.W., "The Swing Up Control Problem for the Acrobot," *IEEE Control Systems Magazine*, Vol. 15, No. 1, pp. 49–55, Feb. 1995.

[9] Chung, C.C. and Hauser, J., "Nonlinear Control of a Swinging Pendulum," *Automatica*, Vol. 31, No. 6, pp. 851–862, 1995.

[10] Jurdjevic, V., and Quinn, J.P., "Controllability and Stability," *Journal of Differential Equations*, Vol. 4, pp. 381–389, 1978.

[11] Bacciotti, A., "Local Stabilizability of Nonlinear Control Systems," *Series on Advances in Mathematics for Applied Sciences*, Vol. 8, World Scientific, 1992.

[12] Wiklund, M., Kristenson, A., and Astrom, K.J., "A New Strategy for Swinging up an Inverted Pendulum," *Proc. IFAC Symposium*, Sydney, Australia, 1993.

Hybrid Systems in Automotive Control Applications

Ken Butts, Ilya Kolmanovsky, N. Sivashankar, Jing Sun
Scientific Research Laboratories
Ford Motor Company
20000 Rotunda Drive, SRL MD 2036
Dearborn, MI 48121 – 2053.

Abstract

Engineers and researchers in the automotive industry encounter numerous hybrid control problems on a daily basis, some stem from hardware design or the hybrid nature of physical processes, others result from the implementation of logic based controllers. In this paper, we present several standard automotive control problems from the hybrid system perspective, and discuss the industrial needs in terms of analytical methods and tools to address them. We hope to stimulate interest in hybrid systems by describing some concrete examples in automotive control applications, and communicate the lack of practical engineering tools particularly suitable for representing and designing hybrid systems.

1   Introduction

We consider a system to be *hybrid* when it is convenient to use two or more representational frameworks to describe the system's dynamic behavior. Given this view, we routinely encounter hybrid systems of various types in automotive systems development. In this paper we hope to familiarize the reader with these systems by providing examples that are organized in four loosely defined categories:

1. Subsystems with discrete operational settings;

2. Plant models with switching dynamics;

3. Switching structure controllers; and

4. Plant models that evolve in mixed domains.

In addition, we provide an in-depth discussion of the problems encountered by hybrid system developers by viewing the idle speed control problem (an automotive standard) as an example of Category 4. The need for improved hybrid system modeling, analysis and design tools is a recurring theme throughout the paper and is the subject of the remainder of this section.

In some sense, logic-based switching has been used to control automobiles since the introduction of the multi-speed transmission. In this hybrid system, the driver manipulates the continuously variable throttle opening and the discretely variable transmission gear ratio to achieve the desired vehicle acceleration. More recently, the introduction of the electronic control module has resulted in the realization of untold logic-based switching controllers. Some significant uses of logic are to 1) identify and manage the various operational modes, 2) account for significant plant nonlinearities and 3) mitigate the effects of sensor or actuator failure.

Hybrid behavior is not limited to the controller subsystems as exemplified by the engine processes that evolve in both the time and engine crankshaft position domains. Some processes, such as a clutch engagement whose equations of motion change (switch) at the transition from slipping to locked clutch, are event driven. And many systems have discrete operational settings. Examples include the previously mentioned transmission, intake tuning systems, engine cylinder de-activation systems and variable geometry turbocharging systems.

In order to be viable in intensely competitive global markets, automotive manufacturers are redefining their product development processes (Figure 1) based on systems engineering principles [12]. The new processes are driven by structured systems requirements that are used to define and validate the work product (i.e. plant models, control laws and software implementation) at each stage of development. It is clear that the use of dynamic systems tools during the initial stages of development would reduce the required development resources. Unfortunately, the theoretical foundations for hybrid systems development are still emerging and engineering tools for modeling, analysis, synthesis and rapid prototyping of such systems are not yet available. Consequently, hybrid automotive systems have traditionally been developed in the heuristic manner depicted in Figure 1. Of course, the desired transition from the heuristic to the systematic development processes cannot be completed until hybrid system methods and tools are mature.

We hope to stimulate interest in hybrid systems by describing and categorizing some automotive examples in Section 2. In Section 3, we formulate an interesting hybrid system analysis and controller design problem that is based on the engine idle speed control system. Finally in Section 4, we present a method for modeling hybrid systems that partially satisfies the needs of the systems engineering development process.

2   Examples of Hybrid Systems

Controlled automotive processes can often be modeled as hybrid systems. A hybrid system is described by analog and discrete state variables that can evolve either in continuous or discrete domain. If a variable evolves in discrete domain, it can be either updated periodically or triggered by events whose

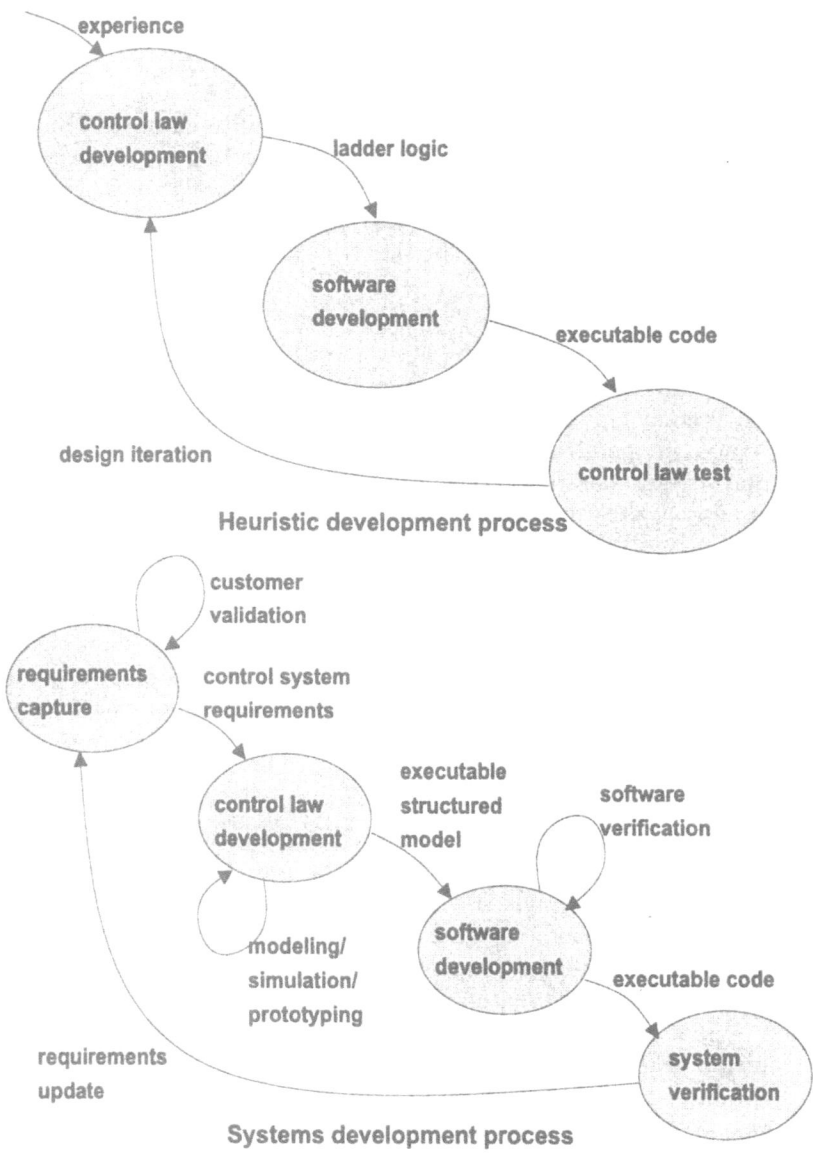

Figure 1: Control system development processes

occurrence signals certain behavior of the process. In a typical scenario, a high level logic-based supervisor monitors the behavior of the process and switches low level controllers as required by various events.

The purpose of this section is to illustrate four major ways in which models of controlled processes with hybrid features can arise:

1. Subsystems (hardware pieces) often have a finite number of discrete operational settings (e.g. an on/off switch). Selecting a specific setting can be viewed as a discontinuous control action since it often modifies the behavior of the subsystem in a discontinuous fashion. During controlled operation, the settings are updated by a logic-based supervisor which monitors the process.

2. The behavior of a plant or a plant model may change rather drastically (and often in a discontinuous fashion) when the plant variables enter or exit certain regions or operating modes. During controlled operation, changes in operating modes are detected by a logic-based supervisor. In many cases, operating mode changes require the supervisor to switch to an appropriate set of low level controllers for the new operating mode.

3. Even in those cases where the behavior of the plant can be adequately represented without hybrid features, these features may be introduced by a switching controller structure. In this scenario, separate controllers are designed to cover different regions (or modes) of plant operation. In this case the modes of operation arise not because of easily identifiable switches in plant dynamics but as regions where various controllers can be used.

4. The nature of engine operation is such that the updates of many relevant control variables take place each time the crankshaft undergoes a rotation by an appropriate angle. Because the engine speed varies with time these updates are aperiodic in time and can be viewed as triggered by events which occur each time the crankshaft traverses an incremental rotational angle. These event based processes evolve simultaneously with other processes, such as intake manifold dynamics, which are better represented in time domain.

It should be emphasized that distinguishing between these four cases is somewhat subjective. In fact, a system which belongs to one class can often be viewed in a different way that makes it belong to another class. In many cases, all these four hybrid characteristics can be encompassed in one system which may consist of subsystems belonging to different categories.

We now provide specific examples which illustrate the first three categories. An interesting example which illustrates the fourth category is provided in a separate section.

2.1   Subsystems with Discrete Operational Settings

As a first example, consider a turbocharged diesel engine [20], where the purpose of the turbocharger is to increase the torque and power output of the engine. An automotive turbocharger consists of a centrifugal compressor that is driven by a gas turbine, which in turn is driven by engine exhaust gases. To obtain better performance over a wide range of operating speeds, variable geometry turbochargers (see e.g. [20, 24, 26, 25]) can be used. A typical variable geometry turbocharger uses nozzle vane drive for the turbine which allows to set nozzle vanes in at least two discrete open positions: "large" or "small". Each setting corresponds to a certain angle of gas flow past nozzle vane. For low engine speeds, when there is a lack of intake air, the control system selects "small" setting to increase the boost pressure and torque, and improve combustion quality and exhaust smoke emissions. For high engine speeds, there is, typically, no lack of intake air, and the control system selects "large" setting to decrease exhaust manifold pressure and thus reduce pumping losses and improve fuel economy.

Variable (or dual) displacement engines [19] provide a further example where a discrete control action can discontinuously modify the operation of an engine in a fundamental way. In such an engine, the number of active cylinders is changed dynamically to provide a balanced trade-off among competing requirements on fuel economy, emission, and performance.

Another familiar example is the automobile transmission. The transmission can be viewed as having a finite number of discrete settings (speeds or gear ratios). The selection of one of the settings can be viewed as a discontinuous control action. Logic-based approaches are natural for controlling the transmission, see [23] for an example. This paper uses a fuzzy-logic approach for scheduling transmission shifts. In a companion paper [27] in these proceedings, a novel approach to controller design for powertrain systems is developed. The reader is referred to this paper for more information. Actually, the approach in [27] is quite general and may be useful in controlling other systems with discrete operational settings, including the examples described above.

2.2   Plant Models with Switching Dynamics

Transmission clutch components are typically represented by plant models which exhibit switching dynamics. The clutch can operate in three different modes: unlocked and slipping, unlocked and not slipping, and locked. Not only the dynamic behavior of the clutch (and thus of the automobile) changes when the clutch goes from the unlocked mode to the locked mode but also the number of degrees of freedom changes. A simplified Simulink/Matlab model of a clutch is shown in Figure 2 [16]. When the clutch is unlocked and slipping the engine speed (w_e) is not matched to the vehicle speed (w_v) and the torque which is transmitted across the clutch is determined by the pressure plate loading (normal force), controlled by the driver's clutch pedal. After the lockup, the torque transmitted across the clutch is different than during the engagement process.

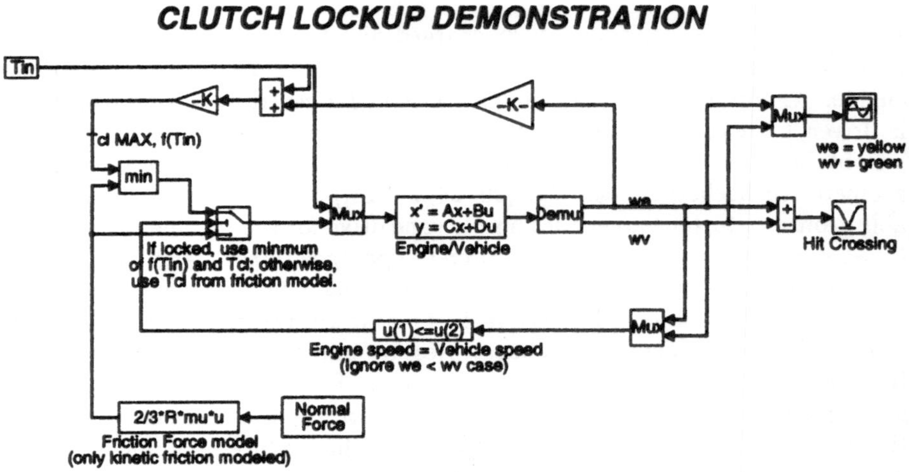

CLUTCH LOCKUP DEMONSTRATION

Figure 2: Transmission clutch model

If the clutch torque capacity is not exceeded, the torque produced by the engine, T_{in}, drives both the engine and the vehicle inertia. If the torque capacity of the clutch is exceeded (e.g. when T_{in} is large), the clutch unlocks and the torque which drives the vehicle inertia is no longer determined by T_{in}. Controlling the clutch engagement process using normal force and throttle inputs is a hybrid control problem [22].

Another example where switching dynamics can be clearly identified is provided by the throttle operation. The air flow across the throttle can be in one of the following different modes: choked (when the flow is only a function of the opening), unchoked (the flow is a function of the opening and the pressure difference across the valve), and wide open (the flow does not depend on the opening and pressure, but is solely a function of engine speed).

2.3   Switching Structure Controllers

In designing controllers for complex plants it is typical to use logic to qualitatively assess and classify the behavior of the plant. This assessment results in a specification of a family of operating modes for the process. Low level controller design is accomplished separately for each mode and logic-based switchings between various controllers are used to control the process across several modes.

For spark ignited engines, examples of modes include [6]: cranking (engine speed less than threshold value), closed throttle (throttle angle less than a threshold value), part throttle (throttle angle is between two threshold values) and wide open throttle (throttle angle is greater than a threshold value). If the engine coolant temperature is below normal the engine is said to be in a cold running mode. Further subdivisions of modes are possible: e.g. closed throttle mode can be subdivided into idling and deceleration modes. Further examples of operating modes include altitude compensation mode, clear-flood mode, battery voltage correction mode, and various failure modes, see e.g. [28]. The modes are not necessarily disjoint, e.g. the engine may be running part throttle and in a cold running mode. The change in an operating mode of an engine is an event which may require the control system to switch to a new set of controllers appropriate for the new mode. In some cases, the new mode of operation is undesirable, and the control system ensures that the engine exits it. For example, in part or closed throttle mode and when the engine is not in a cold running mode, a feedback controller is used to adjust fuel supply to keep the air-fuel ratio near the optimal operating point [6]. The feedback controller uses the information provided by an oxygen (EGO) sensor in the exhaust. In other modes of operation the feedback controller is turned off and various open loop controllers are used. This happens, for example, in the cold running mode, when the EGO sensor does not operate properly [6]. The cold running mode is undesirable and the control system provides extra exhaust heat by retarding the spark to exit this mode quickly [1]. Once the engine is normalized, the feedback controller is turned on. Actually, for cost reasons, the EGO sensor in production vehicles is, typically, a binary switching sensor which can be thought of as having two discrete states, indicating lean or rich combustion. See [17] for control approaches applicable to engines with binary discrete-event EGO sensors.

Another example is provided by an antilock braking system (ABS) (see e.g. [28]) which prevents the wheels from going into a skid during braking thereby allowing the vehicle to stop faster with improved directional stability. The ABS periodically interrupts the force applied to the brakes. The duration and the frequency of these interruptions are determined by a control logic. In the mode where a lockup is anticipated (the lockup is defined, in some cases, as the condition when one wheel speed decelerates significantly faster than other wheel speeds), hydraulic pressure to the brake of that particular wheel is diverted. This logic or mode-based operation of the ABS results in a hybrid control system.

Failure modes add further complexity to control system design. A detection of a failure mode requires appropriate adjustment of the control action. For example, if the nozzle vane of a variable geometry turbocharger sticks at "small" setting, large increase of boost pressure and turbocharger rotating speed would result when the engine speed increases. In this case, it is necessary to decrease fuel flow to protect the engine. On the other hand, if the turbine is stuck at a "large" setting severe black smoke emissions may result when the engine speed decreases unless the fuel flow is adjusted [29].

3 Engine Idle Speed Control

As mentioned earlier, we view any system that can be conveniently represented in more than one framework as a hybrid system. Because of its reciprocating nature, an internal combustion engine is a typical example whose dynamics can be better captured by a mixture of time-based representation (such as throttle body, intake manifold) and crankangle or event-based representation (such as intake-to-power delay, air charge into the cylinder and spark timing). In this section, we use the idle speed control example to elaborate on this particular hybrid problem. Even though the example is specific to idle speed control, we believe that the underlying problem and the related issues are very representative of automotive control applications.

The primary objective of engine idle speed control (ISC) is to maintain engine speed and reject torque disturbance from accessory loads while the primary throttle is closed. Since ISC encompasses almost all important aspects of vehicle performance including fuel economy, noise-vibration-harshness (NVH) quality, combustion stability, accessory performance, etc., the problem has been investigated in many different frameworks by control researchers and practitioners in the automotive industry and academic community (see the survey paper [11] and references therein). In this section, however, we focus on the issues particularly relevant to the very hybrid nature of the problem, and discuss their implication on required analysis and implementation tools for analyzing and developing hybrid control systems such as ISC.

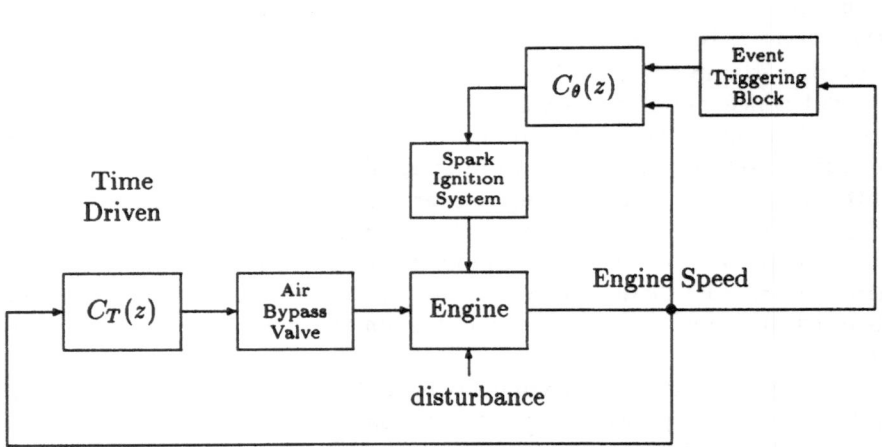

Figure 3: Engine idle speed control system

The general configuration of an ISC system is shown in Figure 3. Typically, two actuators are available for regulating engine speed while idling: the air bypass valve and spark timing. On the surface, Figure 3 represents a multi-input-single-output problem which may lead some readers to speculate that one of the controls might be redundant. In reality, however, the air bypass valve and spark controls are both used for many practical reasons. The bypass valve has large control authority, but it is relatively slow because of the manifold dynamics and intake-to-power stroke time delay. On the other hand, the spark control responds very fast while it has limited control authority [18]. Due to other considerations such as catalyst overheating, emission, engine combustion quality, it is also desirable to return the spark to its nominal operating point without a bias at the steady state. The constraints on the two control actuators dictate that the complementary strengths of the spark (fast control) and air (slow with integration) controls be used.

The example described above is a classical case of one of the hybrid systems (Category 4) described in Section 2. This is a non–conventional control system with two different control loops:

1. the control loop with the air–bypass–valve (ABV) actuator, henceforth called the air control loop, operating at a fixed time–based sampling rate and

2. the spark control loop being updated every engine event, i.e., operating at a time–varying sampling rate.

Due to many practical operational requirements and on-vehicle computational considerations, the air bypass valve and spark control loops are designed and operated in two different domains. The spark is synchronized with engine events and has to be fired during every compression stroke for each cylinder. This dictates that the sampling scheme as well as the design model for spark control be event based and aligned with the crank angle. For spark ignition engine with n cylinders, there is a spark firing event for every $720/n$ crank angle degrees. This translates to $\left(\frac{120}{nN}\right)$ seconds in time–domain where N is the engine speed in revolutions per minute (RPM). The sampling rate in the spark loop becomes irregular in the real time domain because it is engine speed dependent. For the air control, on the other hand, the actuator is a solenoid valve whose dynamics do not change with the engine's natural events, and it can be treated conveniently in the fixed sampling framework. In fact, because of limited on-board computation resources, only the most crucial control tasks such as spark can be updated every engine event. The mixing of time–periodic and event–periodic control updates leads to many interesting issues. However, there is a lack of analytical and software tools to address these issues precisely. Some of these issues are discussed below.

Analysis Issues: As mentioned earlier, the engine, as a dynamic system, is a hybrid process. Incorporating the spark control loop and the air control loop operating in two different domains introduces another dimension to this hybrid control problem. Given a spark loop and an air loop controller, one would like to analyze the performance and robustness of the closed loop hybrid system. A preliminary analysis of this hybrid control system could be performed as follows. For a fixed engine idle speed, the event–periodic spark loop becomes time–periodic with a time period of $\left(\frac{120}{nN}\right)$ seconds. In this scenario, the hybrid control system may be analyzed as a multi–rate digital control system. This approach can be used to analyze the steady state properties of the control system such as levels of disturbance rejection and tracking. However, the transient behavior of the system cannot be analyzed with this approach. Under dynamic conditions, this system may be viewed as a multi–rate system at frozen time instants. But, the multi–rate periodicity changes dynamically. We believe that there is sufficient structure to this problem which may be exploited to obtain insightful analysis methods.

Design Issues: To understand the performance tradeoff and use the complementary strengths of the two control loops, it is important to treat the idle speed control problem as one multivariable control problem. The two control loops operating in two different domains make it difficult to formulate this control problem in one unified framework. As a result, the two control loops are designed independently. This could result in a potential conflict situation between the spark loop and the air loop. We believe that a design approach for hybrid systems will help to resolve such issues.

Simulation Issues: In order to reduce the time between design iterations, it is important to use simulation tools to assess the performance of a given design. The simulation environment needs to provide standard templates that are relevant to hybrid systems. Currently, the tools that can simulate such systems require a lot of engineering effort on the part of the user and they are computationally time consuming.

It is clear from the above discussion that effective analysis methods and commercial tools are needed to address typical industrial hybrid control problems such as the ISC problem. Researchers seemed to have realized this need in the last few years as evidenced by increasing interest in this subject. This has resulted in the development of methods which have the potential to become practical engineering tools to treat hybrid systems in the future. We describe one such method in the next section.

4   An Engineering Representation for Hybrid Systems Development

In this section we present a modeling environment that is consistent with the examples described in Section 2. The developers can model, simulate and code-generate (for rapid-prototyping purposes) the system's hybrid behavior within this environment. Moreover, care has been taken to ensure that the environment is based on a formal foundation that can serve as a bridge between the engineering and research communities to advance the development of hybrid system analysis and controller design methodologies.

Recently, automotive system developers have used structured methods (such as Hatley-Pirbhai [10]) to describe systems in functional (what the system does) and architectural (how the system is implemented) models. Because control designers are typically concerned with the dynamic behavior of the system, they primarily deal with the functional model. This model is comprised of 1) transformational process models such as data flow diagrams, control system block diagrams or logic pseudo code, 2) control models in which combinational logic and finite-state-machines act as supervisory controllers to enable / disable the associated transformational processes and 3) a data dictionary in which information entities are catalogued and described in a structured form.

We have found that the methods based on the Harel statechart [9] are particularly effective in capturing the mode-based nature of the typical embedded control law. The statechart allows the control designer to capture complex finite-state-machine behavior without succumbing to the complexity explosion exhibited by traditional state-transition diagrams. Figure 4 demonstrates this characteristic for two equivalent system representations. Also, note that the natural functional hierarchy of a mode-based system (idle, stop and normal modes in Figure 4) is apparent in the statechart representation. Concurrency, hierarchy, history (not shown) and the compound transition (not shown) are powerful statechart semantics that improve the graphical representation. Supporting tools have evolved so that it is possible to animate, simulate and code generate statechart based structured models. In addition, the statechart's formal underpinnings enable the system representation to be analyzed for state reachability, deadlock, nondeterminism, and racing conditions [21].

The structured representation has a secondary benefit as well. It is common practice to use this representation as the communication medium between the control law design and software implementation phases of systems development. As such, there is minimal modifications required to transfer information between these two critical development phases. The data dictionary helps to ensure that all data and control signals are defined, typed and consistent throughout the model hierarchy. Because the standard automotive embedded microcontroller uses fixed-point arithmetic [8], an improper selection of data types can degrade controller performance [4]. The control designer needs to be conscious of the downstream fixed-point implementation details such as signal wordlength and binary point placement.

Unfortunately, the existing structured methods tools are not designed to model continuous-time systems. Consequently, it is not natural to model the plant or feedback controller dynamics in these structured methods tools. In [5] we have proposed that the integration of structured methods (CASE) and computer-aided-control-system-design (CACSD) tools would yield a more suitable engineering environment for automotive systems development. For example, the hybrid engine idle speed control system presented in Section 3 can be recast in this integrated CASE / CACSD method. Figure 5 shows a data flow diagram representing the major idle speed system components and the flow of information throughout the system. We have chosen to identify those processes evolving in the time domain by affixing *TIME* to the process name. Similarly, those processes that evolve in the crank angle domain are identified by the *EVNT* marker. Though not included in this paper, the form of the detailed representation of the sub-processes is given according on the nature of the particular process. For example, the *BREATHING DYNAMICS TIME* process is naturally described by a conventional CACSD block diagram while the *CALCULATE SPEED EVNT* is algorithmic in nature and best captured by CASE pseudo-code. In this model the *CALCULATE SPEED EVNT* statechart

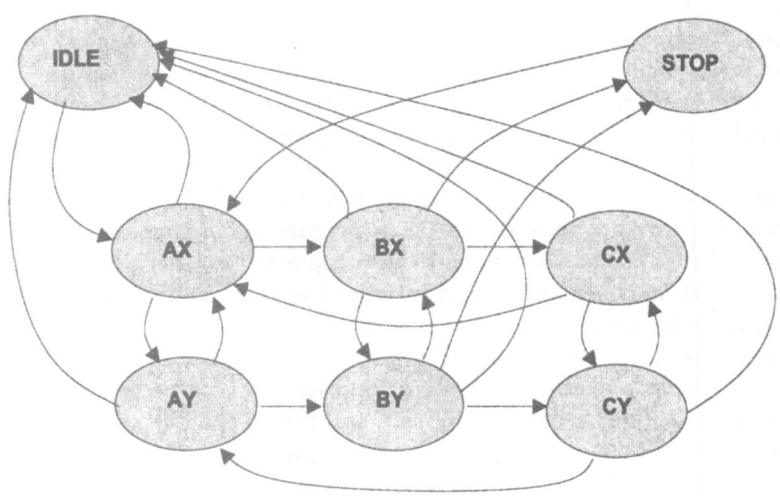

State transition diagram

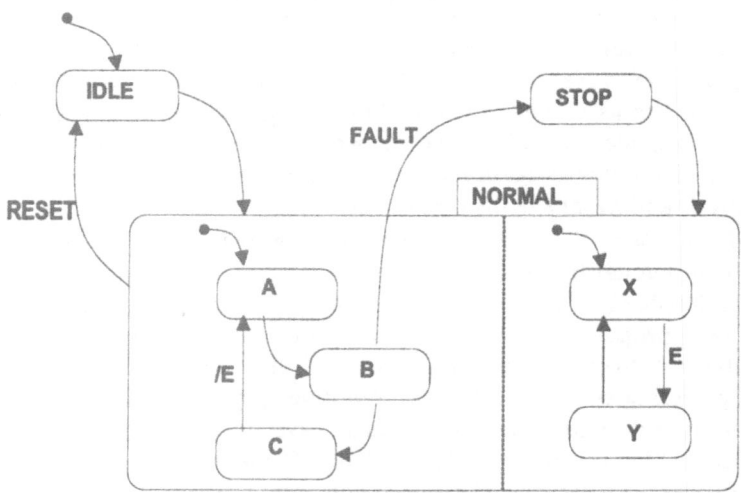

Harel statechart

Figure 4: Comparison of finite-state-machine representations

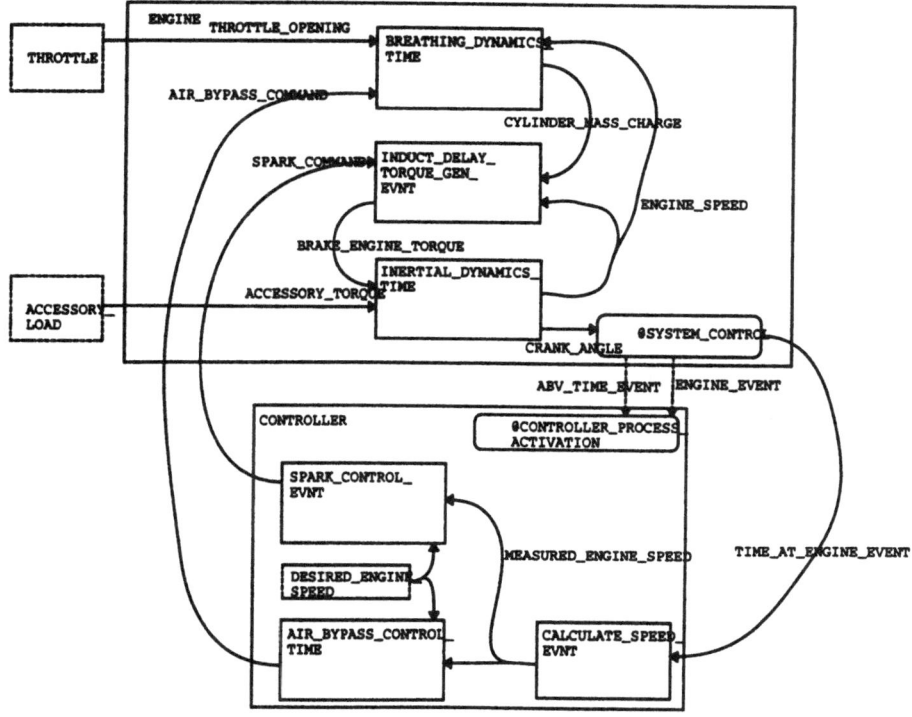

Figure 5: Structured representation of the idle speed control system

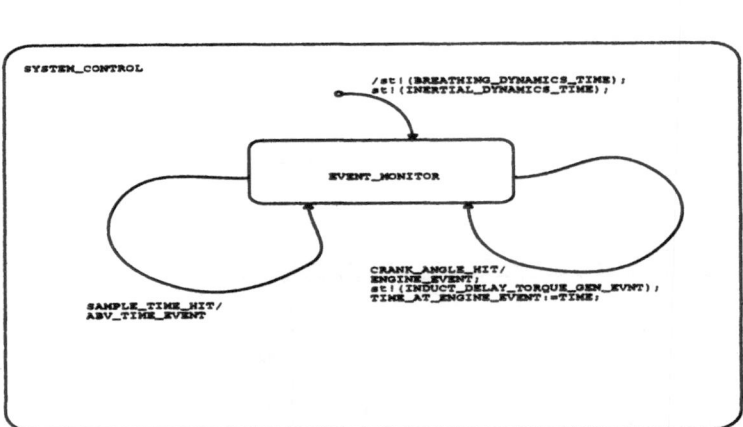

Figure 6: Plant model statechart

(Figure 6) monitors the evolution of crankshaft position and time to generate their corresponding controller sampling events, *ENGINE EVENT* and *SAMPLE TIME EVENT*. Also, *CALCULATE SPEED EVNT* triggers the event based induction-to-power stroke delay process and captures the free-running time for engine speed calculation when *ENGINE EVENT* is generated. The *CONTROLLER PROCESS ACTIVATION* statechart (Figure 7) responds to the sampling events by controlling the execution of the controller sub-processes *CALCULATE SPEED EVNT*, *SPARK CONTROL EVNT* and *AIR BYPASS CONTROL TIME*. Given this model we could study the closed-loop system performance through simulation and then quickly transition to experimental testing via rapid-prototyping. Of course this environment only partially fulfills our needs as we are not providing the tools to perform rigorous analysis or to design the spark and air-bypass controllers in a systematic manner.

This new hybrid systems development environment has been well received by CASE and CACSD tool vendors as other automotive and aerospace companies have expressed similar requirements. In fact, some tool vendors are currently developing integration interfaces and product enhancements and we expect hybrid system modeling, simulation and code generation capabilities to become available in the near term. Because the Harel statecharts have a formal structure, it is hoped that the proposed hybrid systems representation will give way to systematic analysis and controller synthesis techniques. We are encouraged that some statechart based research is already underway. Examples include [2, 3, 7, 14] and [15].

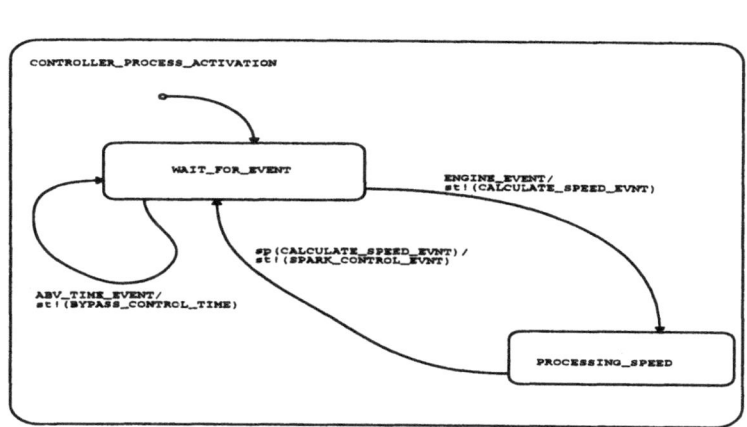

Figure 7: Controller statechart

5 Final Remarks

We have provided a sample of the various types of hybrid systems encountered in automotive product development, and we have outlined an engineering environment that will enable the systems to be modeled, simulated and tested. The need for the development of the associated hybrid systems analysis and design techniques has also been demonstrated.

References

[1] Achleitner, E., Hosp, W., Koch, A., and Schurz, W., "Electronic engine control system for gasoline engines for LEV and ULEV standards," SAE Technical Paper No. 950479, 1995.

[2] Boissier, R., Dima, B., Razafindramary, D., and Soriano, T., "Hybrid systems modeling and validating using Statecharts and Grafcet," *Proceedings of Real Time Programming 1994 Workshop*, IFAC, pp. 73-79, June 1994.

[3] Brave, Y., and Heymann, M., "Control of Discrete Event Systems Modeled as Hierarchical State Machines," *IEEE Transactions on Automatic Control, vol. 38, no. 12*, pp. 1803–1819, December, 1993.

[4] Butts, K. R., *Issues in the Implementation of Automotive Control Systems*, doctoral dissertation, The University of Michigan, 1993.

[5] Butts, K. R., Stetson, D. A., and Cook, J. A., "Computer Aided Engineering for Automotive Powertrain Controller Development," *DSC-Vol. 56/DE-Vol.86, Advanced Automotive Technologies, Proceedings of the 1995 ASME IMECE*, ASME, pp. 3403-3405, 1995.

[6] Czadzeck, G.H., and Reid, R.A., "Ford's 1980 Central Fuel Injection System," SAE Technical Paper No. 790742, 1979.

[7] Eckrich, M. J., Kempf, G. G., and Rumpf, O.J., "Testing Automotive Systems Modeled by Finite State Machines," SAE technical paper No. 940136, 1994.

[8] Hanselmann, H., "Implementation of Digital Controllers - A Survey," *Automatica, vol. 23, no. 1*, pp. 7–32, 1987.

[9] Harel, D., and Pnueli, A., "Statecharts: A Visual Formalism for Complex Systems," *Science of Computer Programming 8*, pp. 231–274, 1987.

[10] Hatley, D.J., and Pirbhai, I.A., *Strategies for Real-Time System Specification*, Dorset House, New York, 1987.

[11] Hrovat, D. and J. Sun, "Models and Control Methodologies for IC Engine Idle Speed Control," Proceedings of the 13th IFAC, 1996.

[12] *IEEE Trial-Use Standard for Application and Management of the Systems Engineering Process (IEEE Std 1220)*, IEEE, 1994.

[13] Kang, K.C., Kwang-Il Ko, "PARTS - a temporal logic-based real-time software specification method supporting multiple-viewpoints," *Proceedings of the First Asia-Pacific Software Engineering Conference*, IEEE Computer Society Press, pp. 328–335, 1994.

[14] Malec, J., "On formal analysis of emergent properties," *Proceedings of Current Trends in AI Planning. EWSp 1993 - 2nd European Workshop on Planning*, IOS Press, Amsterdam, pp. 213–225, 1994.

[15] Manna, Z. and Pnueli, A., "Verifying Hybrid Systems," *Hybrid Systems Lecture Notes in Computer Science 736*, Springer-Verlag, pp. 4–35, 1993.

[16] MathWorks, The, "Automotive examples using SIMULINK," distribution diskette, The MathWorks, Inc., October, 1995.

[17] Moraal, P., "Adaptive compensation of fuel dynamics in as SI engine using a switching EGO sensor," *Proceedings of the 34th IEEE CDC*, pp. 661–666, 1995.

[18] Morris, R.L., Warlick, M.V., and Borcherts, R.H., "Engine idle dynamics and control: A 5.8L application," SAE technical paper No. 820778, 1982.

[19] Pouliot, H. N., W. R. Delameter and C. W. Robinson, "A Variable-Displacement Spark-Ignition Engine," SAE technical paper 770114.

[20] Siuru Jr., W.D., "Automotive superchargers and turbochargers," in *Handbook of Turbomachinery*, E. Logan, Jr. editor, Marcel Dekker, 1994.

[21] *Statemate - Analyzer Reference Manual Version 5.0*, i-Logix, Inc., 1992.

[22] Szadkowski, A., and McNerney, G.J., "Clutch engagement simulation: Engagement with throttle," SAE technical paper 922483.

[23] Yamaguchi, H., Narita, Y., Takahashi, H., and Katou, Y., "Automatic transmission shift schedule control using fuzzy logic", SAE technical paper 930674.

[24] Wallace, F.J., Howard, D., and Roberts, E.W., "Variable geometry turbocharging - optimization and control under steady state conditions," *Proceedings of 3rd International Conference on Turbocharging and Turbochargers*, IMechE Conference Publications, 1986-4, pp. 215-226.

[25] Wallace, F.J., Howard, D., and Anderson, U., "Variable geometry turbocharging - control under transient conditions," *Proceedings of 3rd International Conference on Turbocharging and Turbochargers*, IMechE Conference Publications, 1986-4, pp. 227-240.

[26] Watson, N., K. Banisoleiman, "A variable-geometry turbocharge control system for high output diesel engines," SAE technical paper 880118.

[27] Wang, L.Y, "Optimal hybrid control with applications to automotive powertrain systems," these proceedings.

[28] Weathers, T., and Hunter, C., *Computerized Engine Control and Diagnostics*, Prentice Hall, 1990.

[29] Yokota, K. and H. Hattori, "A high BMEP diesel engine with variable geometry turbocharger," *Proceedings of 3rd International Conference on Turbocharging and Turbochargers*, IMechE Conference Publications, 1986.

Optimal Hybrid Control with Applications to Automotive Powertrain Systems

Le Yi Wang,* Ali Beydoun,†
Jeffrey Cook,† Jing Sun†, Ilya Kolmanovsky‡

1   Introduction

In this paper, the problem of optimal control for a class of hybrid systems is investigated. The class of hybrid systems is characterized by discrete-time state space models where control inputs include both analog control variables and discrete actions. The problem is strongly motivated by automotive powertrain control problems in which transmission gear shift and other new engine technology introduce naturally discrete control actions, on top of other conventional continuous control variables.

An optimal control problem is formulated in which classical performance indices, such as quadratic ones, are modified to reflect penalties on discrete actions. Control strategies are discussed for solving such optimal hybrid control problems. Several fundamental issues have emerged in this investigation, including design complexity, implementation complexity, stability constraints, as well as achievable performance. Some potential approaches for resolving these issues are discussed and tested via simulation on a simplified engine-transmission model.

Hybrid control problems are now very actively pursued in the control field (see, e.g., [1] [2] [4] [5] [8] [9] [10]). The development of control methods for hybrid systems is still at its infancy. The approaches followed by researchers are vastly diversified, ranging from scheduling or switching control, numerical fuzzy control, to conceptual intelligent control frameworks. However, there are still significant gaps between good intuition and rigorous problem formulations, between industry applications and mathematically tractable problems, and between trial-and-error control methods and rigorous analysis. We hope that research discussed here will eventually provide some benchmark hybrid control problems to stimulate and facilitate future research in this very promising area.

---

*Department of Electrical Engineering, Wayne State University, Detroit, Michigan 48202, Tel. 313-577-4715, Fax 313-577-1101, Email lywang@ece.eng.wayne.edu

†Advanced Vehicle Technology, Ford Motor Company, ETC Building, Rm. C360, 17000 Rotunda Dr., Dearborn, Michigan 48121

‡Scientific Research Laboratories, Ford Motor Company, P.O. Box 2053, MD1170 SRL, Dearborn, Michigan 48121-2053

2   Problem Formulation

We are considering in this paper discrete-time hybrid systems expressed in the state-space form

$$x(t+1) = f_{k(t)}(x(t), u(t)), \tag{1}$$

where $x(t) \in \mathbb{R}^n$ is the state, $u(t) \in \mathbb{R}^m$ the continuous (analog) control input, and $k(t) \in \Phi = \{1, 2, \cdots, d\}$ the discrete control action. $f_{k(t)}(x(t), u(t))$ is assumed to be continuous in x and u. The terms "continuous" and "discrete" are used to distinguish value sets of control signals, and should not be confused with their traditional applications in terms such as "continuous-time" and "discrete-time".

This type of hybrid systems are strongly motivated by automotive powertrain control problems.

2.1   Powertrain Hybrid Systems

Consider a powertrain system shown in Figure 1. Control signals may contain throttle angle θ, spark advance δ, exhaust gas recirculation EGR, air/fuel ratio AFR, transmission gear position G. The outputs of the system may include vehicle acceleration, emission pollutants (such as HC, CO, and NO_x), fuel consumption, etc. The design objective is to achieve: (1) Fast and smooth acceleration response to the driver's commands; (2) Low fuel consumption; and (3) Low levels of pollutant emissions; among others. These performance measures are to be achieved in the presence of modeling errors and disturbances such as load changes from air conditioning engagement and road conditions.

Powertrain systems are inherently hybrid. Gear selections are clearly discrete actions which can only assume a finite number of values. In contrast, θ, δ, EGR, fuel mass charge are analog signals.

The essential features of the hybrid control problems we encounter in powertrain systems include the following: (1) Discrete actions, such as gear selections, are control variables, rather than uncontrollable events; (2) While discrete actions change system dynamics, due to inertia in engine and transmission systems they do not result in jumps in system states such as engine speed, vehicle speed, manifold pressure, engine torque. As a result, system states are still analog variables. The hybrid systems (1) capture these essential features. In this research, we are seeking control strategies for such systems and their applications to powertrain control problems.

Example 1: Simplified Joint Engine-Transmission Models

We will derive a simplified engine-transmission model for testing and verification of design strategies.

(a) Engine Models [11]:

$$
\begin{aligned}
N &= \quad \text{engine speed (RPM)} \\
p &= \quad \text{manifold pressure (in. Hg)} \\
T_e &= \quad \text{engine torque (lbf-ft)}
\end{aligned}
$$

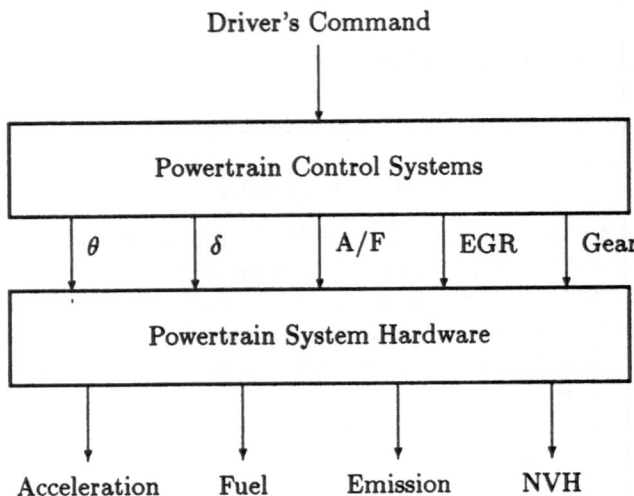

Figure 1: Powertrain Systems

$$
\begin{aligned}
\delta &= \text{spark advance (degree)} \\
\theta &= \text{throttle angle (degree)} \\
AFR &= \text{air/fuel ratio (unitless)} \\
EGR &= \text{exhaust gas recirculation (\%)}
\end{aligned}
$$

<u>Manifold dynamics:</u>
$$\dot{p} = a_1 N + a_2 N p^2 + u \tag{2}$$

where $a_1 = -0.7791; a_2 = -0.0278$. u is a nonlinear function of the throttle angle θ and manifold pressure p. Although the actual analog control variable is θ, we will simply regard u as the analog control input since the nonlinear function is invertible in the operating range of interest.

<u>Engine torque expression:</u>
$$
\begin{aligned}
T_e = 1.3558(&-115 + 0.9061\dot{M} + 22AFR - 0.82AFR^2 + 0.927\delta \\
&-0.0227\delta^2 + 0.00092\delta N - 0.0179N - 0.000029N^2 - 0.779EGR)
\end{aligned}
$$

where
$$\dot{M} = 0.01925N + 0.0006875Np^2.$$

For fixed AFR, EGR, and spark advance, the engine torque expression is simplified to
$$T_e = b_1 + b_2 N + b_3 N^2 + b_4 N p^2. \tag{3}$$

(b) Simplified Transmission

$$
\begin{aligned}
v &= \text{vehicle speed (mph)} \\
\beta(i) &= \text{N/v ratio at gear } i, i = 1, 2, 3, 4 \text{ (RPM/mph)}
\end{aligned}
$$

$$N = \beta(i)v, \qquad \text{at gear } i = 1, 2, 3, 4 \tag{4}$$

A typical case used in simulation here is

$$\beta(1) = 120; \quad \beta(2) = 67; \quad \beta(3) = 44; \quad \beta(4) = 31.$$

(c) Basic Vehicle Dynamics

$$
\begin{aligned}
P_e &= \quad \text{engine power (hp)} \\
ac &= \quad \text{vehicle acceleration (mph/s)} \\
F &= \quad \text{force applied on vehicle (lbf)} \\
P_L &= \quad \text{vehicle load power (hp)} \\
W &= \quad \text{vehicle weight (lbf)} \\
g &= \quad \text{gravity acceleration } (32.174 \ ft/s^2)
\end{aligned}
$$

Basic relations: A typical case of the vehicle load power is

$$P_L = c_1 + c_2 v^2$$

with $c_1 = 1.5; c_2 = 0.0056$. And

$$P_e = \frac{2\pi T_e N}{550 \times 60}, \quad F = \frac{(P_e - P_L)550}{1.467 v}, \quad ac = \frac{F}{1.467\frac{W}{g}}$$

For simulation here, $W = 3500(lbf)$.

(d) Joint Models
Combining the previous expressions with fixed AFR, EGR and δ, we can obtain a simplified joint engine-transmission model in the form of

$$
\begin{aligned}
\dot{p} &= \quad a_1 N + a_2 N p^2 + u \\
\dot{v} &= \quad d_1\beta(i) + d_2\beta(i)^2 v + d_3\beta(i)^3 v^2 + d_4\beta(i)^2 vp^2 + \frac{d_5}{v} + d_6 v \tag{5}
\end{aligned}
$$

For $AFR = 14.6; EGR = 0; \delta = 15$, we have $d_1 = 0.0244, d_2 = 8.0916E - 006, d_3 = -1.7587E - 008, d_4 = 3.7779E - 007, d_5 = -3.5240, d_6 = -0.0132$.

Apparently, the system (5) is a hybrid system with analog control u and discrete action i. After the standard sampling process with sampling interval T, the system (5) is tranformed into a discrete-time hybrid system

$$
\begin{aligned}
p(t+1) &= \quad f_1(p(t), v(t), u(t), i(t)) \\
v(t+1) &= \quad f_2(p(t), v(t), u(t), i(t)) \tag{6}
\end{aligned}
$$

2.2 Optimal Hybrid Control Problems

In this paper, we formulate a generic hybrid control problem as follows. The performance index is expressed as

$$J(x, u, k) = \sum_{t=0}^{N} F_1(x(t), u(t), t) + F_2(k)$$

where $F_1(x, u, t) \geq 0$ is a penalty function on state and analog control variables, and $F_2(k) \geq 0$ a penalty on discrete actions.

Problem *For systems (1) under initial conditions $x(0)$ and $k(-1)$, design an analog control sequence $u(t), t = 0, ..., N-1$, and a discrete action sequence $k(t), t = 0, ..., N-1$ such that J is minimized,*

$$J_{opt}(x(0), k(-1))$$
$$= \inf_{\substack{k(t) \in \{1,\cdots,d\}, \\ t=0,1,\cdots,N-1}} \inf_{\substack{u(t) \in R^m, \\ t=0,1,\cdots,N-1}} J(x, u, k) \qquad (7)$$

While it has been well established in classical control problems that performance indices shall generically penalize deviations of x and u from 0, a reasonable penalty on the discrete action k is often its switching rather than its absolute value. For instance, to avoid busy shifting of transmission gears, the shifting, rather than gear positions, should be penalized. For simplicity, in this project we will start with quadratic measures, modified to reflect certain penalty terms on discrete actions. Quadratic measures are relatively simple to handle but still reasonable for practical control problems. It should be pointed out here that other performance measures can certainly be employed without changing the conceptual framework. However, the issues of computational complexity may differ significantly from the development here.

More precisely, we will employ the following index for optimization: For the given initial condition $x(0)$ and $k(-1)$, the performance index is in the form of

$$J(x, u, k) = \sum_{t=0}^{N-1} [x^T(t)Q(t)x(t) + u^T(t)R(t)u(t)]$$
$$+ x^T(N)Sx(N) + \sum_{t=0}^{N-1} SP(k(t), k(t-1)) \qquad (8)$$

where all weighting matrices are positive definite, and $SP(i, j)$ is the penalty for discrete actions, i.e., switching from i to j. Note that while usually $SP(i, i) = 0$, it is not a necessary assumption here.

3   Optimal Hybrid Control

Generically speaking, the optimization problem can be solved as follows: For a given discrete action sequence $k = \{k(0), k(1), ..., k(N-1)\} \in \Phi^N$, denote

$$J(k, x(0)) = \inf_{\substack{u(t) \in R^m, \\ t=0,1,\cdots,N-1}} J(u, x, k) \qquad (9)$$

subject to the system equation (1). Note that the optimization problem (9) is a classical nonlinear optimal control problem.

Then, the optimal hybrid control is achieved by

$$J_{opt}(x(0), k(-1)) = \inf_{k \in \Phi^N} J(k, x(0)) \qquad (10)$$

The natural decomposition of the optimization problem (7) into a classical optimal control (9) and a finite optimization (10) shows that the optimal controller assumes a hierarchical structure. It is noted, however, that this hierarchical structure differs from the common two-level structure proposed for intelligent control systems in which the lower level is of classical dynamic control, and the higher level consists of a discrete-event system. In fact, the interaction between the optimization problems (9) and (10) demonstrates that the common two-level structure may inherently miss optimal solutions.

This design procedure encounters, however, the problem of computational complexity which will be addressed next.

We shall start with simple cases and tackle more complex and generic problems progressively afterwards.

3.1 Hybrid Linear Quadratic Control Problems

When the underlying systems are linear, except the switching effects, the hybrid system becomes

$$x(t+1) = \sum_{i=1}^{d} C_i(k(t))(A_i x(t) + B_i u(t)) \tag{11}$$

where $C_i(k)$ are the selection functions defined by $C_i(k) = 1$, for $k = i$; $= 0$, for $k \neq i$, and each subsystem is assumed to be controllable. Observe that for a given discrete action sequence $k = k(0), ..., k(N-1)$, the optimization problem (9), subject to the system (11) and index (8), is an ordinary LQ problem for time-varying systems. As a result, the optimal u, as a function of k and x, can be obtained by using Riccati equation methods. Naturally, the optimal discrete action can be obtained by solving (10).

¿From this procedure, it can be seen that the optimal u will still be in a state feedback form. While this design procedure is conceptually easy to understand, the issues of stability and computational complexity, which is exponential to the data window length N, impose great challenge for practical implementations. To make the design procedure feasible and to accommodate the practical requirements of small window lengths in implementation, N must be reasonably small. However, small N may result in the loss of stability when this design is repeated over moving data windows during on-line implementation. This intuitive trade-off between design complexity and stability (also performance) leads to the discussions of some essential issues in the subsequent sections.

3.2 Stability and Performance

For design implementation, the control interval N must be sufficiently small so that tracking can be achieved without significant delays (current control strategies are mostly static mappings, and hence require only small data windows). When the optimization procedure is applied on-line, a fundamental issue arises: Will the optimal control guarantee stability?

For a given length N, we are seeking conditions under which the following inequality holds: ($\|\cdot\|$ is the Euclidean norm)

$$\|x(N)\| \leq \|x(0)\|, \qquad \forall x(0) \in \Omega \tag{12}$$

where Ω is the set of all permissible initial states. This invariance property (or contraction property) will imply stability of the design procedure when it is implemented on-line. If (12) is enhanced by

$$\|x(N)\| \leq \alpha \|x(0)\|, \qquad \forall x(0) \in \Omega \tag{13}$$

for some $0 < \alpha < 1$, then we will have the asymptotic property

$$x(iN) \to 0, \qquad i \to \infty$$

which will imply (under mild conditions on the system and weightings) that if the design procedure is implemented on every data window of length N: $\{iN, ..., (i+1)N - 1\}, i = 0, \cdots$, the resulting (hybrid) closed-loop system will be asymptotically stable.

Under the constraint (13), the optimization problem (9) becomes:

For a given $x(0)$ and discrete sequence k, solve

$$J(k, x(0)) = \inf_{\substack{u(t) \in R^m, \\ t=0,1,\cdots,N-1}} J(u, x, k)$$

subject to (1) and (13).

The complete solution to this constrained optimization problem is currently unknown. While the problem can be solved numerically by employing some standard nonlinear programing algorithms or convex optimization methods, it is of importance to obtain characterizations of the sulotions to the problem.

For linear systems, the problem can be simplified. A potential candidate for the stability requirements is the following sufficient condition. Observe that in the case of linear systems, the optimal u and k result in a closed-loop system which relates $x(0)$ to $x(N)$ by a mapping

$$x(N) = M(u, k)x(0).$$

Then, a sufficient condition for (12) is

$$\sigma_{max}(M) \leq 1;$$

and for (13)

$$\sigma_{max}(M) < 1$$

where σ_{max} denotes the largest singular value.

3.3 Complexity

The optimization (10) is a minimization over a finite set of size d^N. The solution will be a function of the initial condition $x(0)$ and $k(-1)$. Note that the complexity of this optimization problem grows exponentially with N. Also, the dependence of this optimization problem on the initial state $x(0)$ implies that it must be implemented on-line. This raises serious issues of computational feasibility of optimal controllers.

To resolve this problem, we observe that many discrete sequences k are always non-optimal, no matter what initial states are. Off-line elimination of such inactive sequences will dramatically reduce the complexity of on-line implementation of the optimization (10). One possibility is to use the Monte Carlo method to eliminate inactive sequences. The main ideas can be summarized as follows:

(a) First, we eliminate sequences which are not permitted due to practical considerations. For instance, gear shift from 1 to 4 is sometimes not desirable on existing hardware systems and transmission control design.

(b) We generate randomly a large set of permissible initial states $x(0)$. Optimal discrete sequences for this set will be retained as the active set of sequences. The remaining sequences will be eliminated. The probability of losing optimal sequences due to this elimination is a monotone decreasing function of the size of the set, and hence can be controlled.

(c) Furthermore, we will partition the set of initial states into equivalent classes by the relation:

$$x^1(0) \equiv x^2(0)$$

if they result in the same optimal discrete sequence. Suppose, there are l such equivalent classes. Then, we can construct, say, L dividing hyperplanes to separate these classes:

$$h_j(x(0)) = 0, \qquad j = 1, 2, \cdots, L$$

During on-line implementation, we only need to test the L conditions to arrive at the optimal sequence.

4 Case Studies: Automotive Powertrain Control

Consider a linearized hybrid system

$$x(t + 1) = A_i x(t) + B_i u(t), \qquad i = 1, \cdots, d$$

where $A_i \in \mathbb{R}^{n \times n}$, $B_i \in \mathbb{R}^{n \times m}$. Each pair (A_i, B_i) is assumed controllable. For a given window size N, there are d^N possible discrete sequences in the window. Given weighting matrices $Q(n \times n)$, $S(n \times n)$, $R(m \times m)$ and switching penalty matrix $SP(d \times d)$ ($SP(i, j)$ is the penalty for switching from i to j), the design procedure can be specified as follows:

1. Generate the admissible set Ξ of discrete action sequences.
2. For a given $k \in \Xi$, design the optimal LQ control u for the corresponding time-varying system.
3. Randomly generate a set Γ of the permissible initial states $x(0)$.
4. For each $x(0) \in \Gamma$ and $k(-1)$, obtain the corresponding optimal sequence k and optimal analog control u.
5. Verify the stability condition:

$$\|x(N)\| < \|x(0)\|$$

6. Eliminate inactive discrete sequences.
7. Partition Γ into equivalent classes based on optimal discrete sequences.
8. Construct dividing hyperplanes which characterize the equivalent classes.

Example 2:

Near a given vehicle speed v and for a given sampling interval T, the joint model of Example 1 can be locally linearized and discretized to obtain a linearized hybrid system

$$x(t+1) = A_i x(t) + B_i u(t), \qquad i = 1, 2, 3, 4.$$

For simulation here, $v = 50(mph)$ and $T = 0.2(s)$ are used. We obtain the linearized discrete-time systems ($x_1 = p, x_2 = v$):

Gear $i = 1$,

$$A_1 = \begin{bmatrix} -65.72 & -185.3929 \\ 1.7197 & 0.685 \end{bmatrix}, \qquad B_1 = \begin{bmatrix} 0.2 \\ 0 \end{bmatrix}$$

Gear $i = 2$,

$$A_2 = \begin{bmatrix} -36.252 & -56.5792 \\ 0.3775 & 0.9411 \end{bmatrix}, \qquad B_2 = \begin{bmatrix} 0.2 \\ 0 \end{bmatrix}$$

Gear $i = 3$,

$$A_3 = \begin{bmatrix} -23,464 & -24.3802 \\ 0.1238 & 0.9813 \end{bmatrix}, \qquad B_3 = \begin{bmatrix} 0.2 \\ 0 \end{bmatrix}$$

Gear $i = 4$,

$$A_4 = \begin{bmatrix} -16.236 & -13.3137 \\ 0.0509 & 0.9923 \end{bmatrix}, \qquad B_4 = \begin{bmatrix} 0.2 \\ 0 \end{bmatrix}$$

After defining the performance index (8) with specified Q, R, S matrices, as well as switching penalty matrix SP, we can compute off-line the optimal gear sequence $i(t)$ and analog control $u(t)$. Note that the optimal control will be a function of initial gear position $i(-1)$ and initial state $x(0)$. Simulation results are omitted here.

5   Discussions

Hybrid systems arise naturally in typical automotive control problems. The optimal hybrid control problems formulated in this note represent a preliminary attempt in extracting essential features from automotive hybrid systems to formulate reliable benchmark problems for rigorous studies of hybrid control systems. Many fundamental issues have emerged from this study, including stability, performance, robustness and complexity, which we do not have comprehensive answers to offer at present. It is believed, however, that efforts in this direction will be of importance in both theoretical studies and practical implementations of hybrid control systems.

References

[1] R.W. Brockett, Hybrid models for motion control systems, *Essays on Control: Perspectives in the Theory and its Applications*, Tentleman and Willems, eds, pp. 29-54, 1993.

[2] P.E. Caines and Y. Wei, Hierarchical hybrid control systems, *Notes from Workshop on Control Using Logic-Based Switching*, Block Island, RI, pp. 50-57, 1995.

[3] J.A. Cook, B.K. Powell and J.W. Grizzle, Modeling and analysis of an inherently multi-rate sampling fuel injected engine idle speed control loop, *J. Dynamic Systems, Measurement and Control*, V. 109, No. 4, pp. 405-411, 1987.

[4] S.J. Cusumano and K. Poolla, Adaptive control of uncertain systems: A new approach, *Proc. 1988 ACC Conference*, pp. 355-359, 1988.

[5] M. Fu and B.R. Barmish, Adaptive stabilization of linear systems via switching controls, *IEEE Trans. Auto. Cont.*, pp. 1079-1103, 1986.

[6] F.R. Gantmacher, The theory of matrices, Vol. I & II, Chelsea Pub., 1960.

[7] J.W. Grizzle, K.L. Dobbins and J.A. Cook, Individual cylinder air/fuel ratio control with a single EGO sensor, *IEEE Tran. Vehicular Technology* Vol. 40, pp. 280- 286, 1991.

[8] A.S. Morse, Supervisory control of families of linear set-point controllers, *Proc. the 32nd CDC Conference*, pp. 1055-1060, 1993.

[9] A.S. Morse, Dwell time switching, *Proc. the 2nd ECC Conference*, pp. 176-181, 1993.

[10] K. Poolla and J.S. Shamma, Optimal asymptotic robust performance through logic based switching, *Notes from Workshop on Control Using Logic-Based Switching*, Block Island, RI, pp. 44-49, 1995.

[11] B.K. Powell and J.A. Cook, Modeling of an internal combustion engine for control analysis, *Control Systems Magazine*, V. 8, pp. 20-26, 1988.

[12] L.Y. Wang, W. Follmer, J. Cook and J. Sun, Generic Powertrain Control, Ford Technical Report, 1995.

Optimal Asymptotic Robust Performance through Logic-Based Switching

Jeff S. Shamma

Department of Aerospace Engineering and Engineering Mechanics

The University of Texas at Austin

Austin, TX 78712

Kameshwar Poolla

Department of Mechanical Engineering

University of California at Berkeley

Berkeley, CA 94720

Abstract

In this paper we introduce a novel measure of asymptotic disturbance rejection for a feedback system. This notion of asymptotic performance is particularly well suited to the robust control of systems which exhibit both parametric and dynamic modeling uncertainty. We then derive a switching type controller that provides optimal asymptotic disturbance rejection properties. The particular notion of disturbance rejection we consider is rejection of persistent bounded disturbances (i.e., ℓ^1-optimal control). While asymptotic performance is guaranteed, we also provide bounds that quantify the transient response behavior of this particular control scheme.

1 Introduction

An important question in feedback control design is the potential benefit of using nonlinear controllers to control linear plants (i.e., plants for which a linearized description of the dynamics is adequate). In other words, given a family $\mathcal{F}$ of possible plant models, and given a desired performance objective, does there exist a controller (possibly nonlinear) which achieves the desired performance objective for every admissible plant in $\mathcal{F}$? This issue is at the heart of robust and adaptive control design (cf., [1, 3]).

The problem addressed here is described as follows. Let $J(P)$ denote the performance measure

$$J(P) \stackrel{\text{def}}{=} \inf_K \{ \|T(P, K)\| : K \text{ is any stabilizing compensator}\},$$

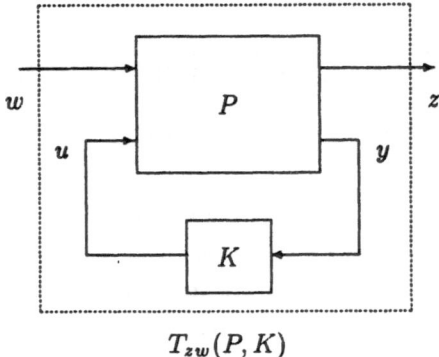

$$T_{zw}(P, K)$$

Figure 1: Block diagram for disturbance rejection

where $T(P, K)$ is a given operator depending on the plant P and controller K (e.g., $T(P, K) = (I + PK)^{-1}$), and $\|\cdot\|$ represents a level of disturbance rejection. The particular notion of performance used here is rejection of persistent bounded disturbances (i.e., ℓ^1-optimal control [2]). Then $\sup_{P \in \mathcal{F}} J(P)$ represents a *lower bound* on the achievable performance for the family $\mathcal{F}$. In [8], it was shown via counterexample that this lower bound need not be approachable. In this paper, we introduce a notion of *asymptotic* disturbance rejection. We then construct a nonlinear controller which approaches the lower bound $\sup_{P \in \mathcal{F}} J(P)$ asymptotically. While the asymptotic performance is guaranteed, we also provide a bound on the transient performance of this particular control scheme.

Similar objectives are considered in the general research area of robust adaptive control. (See [4] for an excellent overview as well as references therein.) These works develop adaptive control schemes which maintain stability in the presence of modeling errors and disturbances. In the present paper, the class of disturbances and modeling errors are stated beforehand as part of the design criteria. This is close in spirit to the approach of [6]. Consideration of transient responses in an adaptive control setting is also considered in [7].

Note: A full version of this paper, "Optimal asymptotic robust performance via nonlinear controllers" has appeared in *International Journal of Control*, vol. 62, no. 6, 1995, pp. 1367–1389.

2   Formulation of Performance Specifications

Consider the feedback diagram of Figure 1. In this figure, P denotes the plant, K denotes a compensator, and the signals w, z, y, and u are defined as follows: w, exogenous disturbances; z, signals to be regulated; y, measured plant outputs; and u, control inputs to the plant.

Let $T_{zw}(P, K)$ denote the resulting closed-loop dynamics from w to z.

A performance objective for robust nonlinear controllers in now discussed.

The use of robust nonlinear controllers arises in certain cases where the plant is known only to lie is some given *family* of plants. The goal is then to design a compensator so that the effect of exogenous disturbances on regulated outputs is small for all admissible plants. Given this plant uncertainty, one cannot expect the controller to achieve the desired performance immediately. The reason is that the control scheme typically must experience a period of identification[1] before the desired performance is achieved. In other words, the desired performance is achieved "asymptotically." In fact, one can construct simple examples for which no controller—nonlinear or not—can achieve immediately the desired performance [8].

Notation: Let $\mathcal{D}_\alpha$ denote a deadzone of level α.

Definition 2.1 *Let Ω be a known class of disturbances. A compensator K is said to achieve an **asymptotic Ω-signal weak disturbance rejection of** μ for the plant, P, if*

$$\mathcal{D}_\mu T_{zw}(P, K)w \in \mathcal{L}^1, \quad \forall w \in \Omega.$$

In other words, this definition states that the magnitudes of regulated outputs are allowed to exceed their desired bound, μ, as long as this excess (measured by the deadzone operator $\mathcal{D}_\mu$) is only a transient phenomenon.

A class of disturbances particularly suited to this notion of performance is the following.

Definition 2.2 *The **asymptotic ρ-ball in** $\mathcal{L}^\infty$, denoted $\mathcal{AB}_{\mathcal{L}^\infty}(\rho)$, is defined as*

$$\mathcal{AB}_{\mathcal{L}^\infty}(\rho) \stackrel{\text{def}}{=} \{w \in \mathcal{L}_e^\infty : \mathcal{D}_\rho w \in \mathcal{L}^1\}.$$

In case $\rho = 1$, the simplified notation $\mathcal{AB}_{\mathcal{L}^\infty}$ is used.

Now suppose that $\mathcal{AB}_{\mathcal{L}^\infty}$ represents the admissible class of disturbances. Thus, every $w \in \mathcal{AB}_{\mathcal{L}^\infty}$ admits a decomposition

$$w = w_{ss} + w_{tr}.$$

The signal w_{ss} represents a persistent (steady-state) part of the disturbances and satisfies $\|w_{ss}\|_{\mathcal{L}^\infty} \leq 1$. The signal $w_{tr} \in \mathcal{L}^1$ represents a transient part, perhaps due to initial conditions or various transient phenomena.

3 Asymptotic Performance via Switching Controllers

In this section, the problem of disturbance rejection with an uncertain plant is discussed. The approach taken here may be summarized as follows. It is assumed that the true plant, say P_o, belongs to a large family of plants, $\mathcal{F}$.

---

[1]In this context, the term "identification" loosely means identification of unknown plant parameters, identification of the correct controller, or both.

Now this family is too large to design a controller (using conventional linear robust design techniques, such as [2, 5]) which achieves the desired disturbance rejection. Thus, one divides the large family, $\mathcal{F}$, into a finite number of smaller families, $\mathcal{F}_i$, such that P_o is contained in the union of the $\mathcal{F}_i$. This decomposition is made such that one may design controllers, K_i, which achieve the desired performance for each family, $\mathcal{F}_i$. The key idea is then to *switch* between the K_i such that the desired performance is achieved asymptotically. That is, one continues to switch between the K_i's while monitoring the performance of each one. When it has been inferred that one of these controllers, say K_{n^*}, delivers the desired performance for the true plant, P_o, switching stops.

3.1 Problem Formulation and Main Results

The feedback configuration under consideration is a specialized form of the more general Figure 1 with the specialization being that **all regulated variables are also measured.**

The following assumptions are made.

Assumption 3.1 *The plant, P_o, satisfies*

$$P_o \in \bigcup_{i=1}^{N} \mathcal{F}_i,$$

where,

$$\mathcal{F}_i \subset \mathcal{A}(\sigma_i), \quad i = 1, \ldots, N.$$

Assumption 3.2 *For each family, $\mathcal{F}_i$, there exists a $K_i \in \mathcal{A}_-(0)$ such that*

1. *K_i finite-gain $\mathcal{L}^\infty$-stabilizes every $P \in \mathcal{F}_i$ uniformly.*

2. *K_i achieves a disturbance rejection of μ for every $P \in \mathcal{F}_i$.*

Essentially, these assumptions state that the true linear time-invariant plant lies in the union of a finite collection of linear time-invariant plants. Furthermore, for each family, one has a *stable* finite-gain stabilizing controller which achieves a disturbance rejection of μ. It is important to emphasize that each controller exhibits *robust performance* in that it achieves the desired disturbance rejection uniformly over the entire sub-family.

The main result is now stated.

Theorem 3.1 *Consider the feedback configuration of Figure 1 under Assumptions 3.1–3.2 and $z = y$. Given any $\rho \geq 0$ and $\varepsilon > 0$, there exists a (nonlinear) finite-gain $\mathcal{L}^\infty$-stable compensator, $K_o : \mathcal{L}_e^\infty \to \mathcal{L}_e^\infty$, which achieves an asymptotic $\mathcal{AB}_{\mathcal{L}^\infty}(\rho)$-signal weak disturbance rejection of $(\mu + \varepsilon)\rho$ for every plant in $\bigcup_{i=1}^{N} \mathcal{F}_i$, hence for the true plant, P_o. That is, for any $w \in \mathcal{L}_e^\infty$,*

$$\mathcal{D}_\rho w \in \mathcal{L}^1 \Rightarrow \mathcal{D}_{(\mu+\varepsilon)\rho} T_{zw}(P_o, K_o)w \in \mathcal{L}^1.$$

3.2 Special Case: $N = 2$

This section constructs a controller for Theorem 3.1 in the special case where the number of families $N = 2$. This restriction results in a considerable simplification in exposition and leads to a more intuitive understanding of the nonlinear process.

Structure of the Nonlinear Compensator

Let $z(t)$ denote the compensator input. The structure of the nonlinear compensator, K_o, is given by

$$K_o z = K_1 \alpha_1 z + K_2 \alpha_2 z, \tag{1}$$

where the α_i are time-varying gains which satisfy the following. For $z \in \mathcal{L}_e^\infty$,

1. For any $t \in \mathcal{R}^+$, either $(\alpha_i z)(t) = 0$ or $(\alpha_i z)(t) = z(t)$.

2. $(\alpha_1 z)(t) + (\alpha_2 z)(t) = z(t)$.

The finite-gain $\mathcal{L}^\infty$-stability of K_o follows immediately from this definition.

The controller K_o is simply a compensator which switches between K_1 and K_2. Given this structure, the input/output dynamics of K_o are completely determined by an algorithm which specifies the switching times.

Switching Algorithm

Let μ, ε, and ρ be as in Theorem 3.1. Let T_n denote the nth switching time.

1. Choose $\gamma_1 > 0$ and $\beta_1 > 0$.

2. Set $T_0 = 0$.

3. T_{n+1} is the smallest $t > T_n$ which satisfies

$$\left\| \mathcal{D}_{(\mu+\varepsilon)\rho} z \right\|_{\mathcal{L}^1[T_n,t)} = \gamma_1 \left\| \mathcal{D}_{(\mu+\varepsilon)\rho} z \right\|_{\mathcal{L}^1[0,T_n)} + \beta_1.$$

This algorithm may be described intuitively as follows. It is unknown to which family P_o belongs. However, it *is* known that (1) each K_i achieves a disturbance rejection of μ for the family $\mathcal{F}_i$, and (2) the disturbances satisfy $w \in \mathcal{A}\mathcal{B}_{\mathcal{L}^\infty}(\rho)$. Thus,

$$P_o \in \mathcal{F}_1 \Rightarrow \mathcal{D}_{\mu\rho} T_{zw}(P_o, K_1) w \in \mathcal{L}^1,$$

$$P_o \in \mathcal{F}_2 \Rightarrow \mathcal{D}_{\mu\rho} T_{zw}(P_o, K_2) w \in \mathcal{L}^1.$$

The nonlinear compensator monitors the magnitude of the regulated output in excess of $\mu\rho$. If it appears that this excess does not decay, then it switches controllers. However, when monitoring the progress of the new controller, the compensator tolerates a larger amount of excess—thereby giving the effects of the switching itself an opportunity to decay.

The key parameters in these switching algorithms are the gains, γ, and bias terms, β. The following proposition shows how to choose these parameters to achieve the desired $\mathcal{AB}_{\mathcal{L}^\infty}(\rho)$-signal weak disturbance rejection of $\mu + \varepsilon$.

Proposition 3.1 *Under the same hypotheses as Theorem 3.1, choose any $\beta > 0$ and*

$$\gamma > \max_i \left(\sup_{P=[P_{11} \ P_{12}] \in \mathcal{F}_i} \left\| (I - P_{12}K_i)^{-1} P_{12}(K_2 - K_1) \right\|_1 \right).$$

With this choice of switching parameters, the nonlinear compensator (1) combined with the Switching Algorithm achieves an asymptotic $\mathcal{AB}_{\mathcal{L}^\infty}(\rho)$-signal weak disturbance rejection of $(\mu + \varepsilon)\rho$ for every plant in $\mathcal{F}_1 \bigcup \mathcal{F}_2$.

3.3   Transient Performance

By definition, asymptotic disturbance rejection is concerned only with the magnitude of the steady-state regulated outputs. However, it is desirable to have some idea of the magnitude of the *transient* regulated outputs. That is, how large are the regulated outputs *before* switching has ceased.

In this section, it is shown how to bound the magnitude of the transient regulated outputs. Suppose that one has the true plant $P_o \in \mathcal{F}_{i^*}$. Define

$$E[z_{tr}] \stackrel{\text{def}}{=} \mathcal{D}_{(\mu+\varepsilon)\rho} T_{zw}(P_o, K_{i^*}) w.$$

In words, $E[z_{tr}]$ denotes the (deterministic) "expected" transients—i.e., the transients which would have occurred even if the correct controller were used throughout.

Theorem 3.2 *Under the hypotheses of Theorem 3.1, let K_o be a finite-gain $\mathcal{L}^\infty$-stable compensator constructed according to Proposition 3.1. Whenever the disturbance w is such that the number of switching times (excluding T_0) is at least N, the actual transient norm $\left\| \mathcal{D}_{(\mu+\varepsilon)\rho} z \right\|_{\mathcal{L}^1}$ is bounded by an affine function of the expected transient norm $\left\| E[z_{tr}] \right\|_{\mathcal{L}^1}$.*

References

[1] ASTROM, K.J. AND WITTENMARK, B., 1989, *Adaptive Control*, Addison-Wesley, New York.

[2] DAHLEH, M.A. AND PEARSON, J.B. JR., 1988, Optimal Rejection of Persistent Disturbances, Robust Stability, and Mixed Sensitivity Minimization. *IEEE Transactions on Automatic Control*, Vol. AC–33, 722–731.

[3] DORATO, P. (ED.), 1987, *Robust Control*. IEEE Press, New York.

[4] IOANNOU, P.A. AND DATTA, A., 1991, Robust Adaptive Control: A Unified Approach. *Proceedings of the IEEE*, Vol. 79, No. 12, 1736–1767.

[5] KHAMMASH, M. AND PEARSON, J.B. JR., 1990, Robust Disturbance Rejection in ℓ^1-Optimal Control Systems. *Systems & Control Letters* **14**, 93–101.

[6] KRAUSE, J.M., KHARGONEKAR, P.P., AND STEIN, G., 1992, Robust Adaptive Control: Stability and Asymptotic Performance. *IEEE Transactions on Automatic Control*, Vol. AC–37, 316–331.

[7] MILLER, D.E. AND DAVISON, E.J., 1991, An Adaptive Controller Which Provides an Arbitrarily Good Transient and Steady-State Response. *IEEE Transactions on Automatic Control*, Vol. AC–36, 68–81.

[8] SHAMMA, J.S., 1990, Nonlinear Time-Varying Compensation for Simultaneous Performance. *Systems & Control Letters* **15**, 357–360.

Towards a General Complexity-Based Theory of Identification and Adaptive Control

G. Zames *

Abstract

This paper is based on the recent solution of two problems of control theory, which will be combined here to generate new approaches to H^∞/ℓ^1 adaptive control, as well as to produce the rudiments of a general theory of adaptation and complexity-based or information-based learning.

The problems, which have been solved by the author and his co-workers, involve,

1. optimally fast identification to reduce plant uncertainty (e.g., to a weighted ball of radius $\in$ in H^∞/ℓ^1);

2. exact computation of feedback performance under *large* plant uncertainty (including such a ball).

By combining these two solutions and using frozen-time approximations to compute optimal feedbacks under time-varying data, we obtain control laws which conform to our definition of what the term "adaptive" should mean, and in fact are in a certain sense nearly optimally adaptive.

These results are concrete and lead to algorithms. However, they also provide a paradigm of a more general theory of adaptive control, which will be outlined. We propose definitions of the notions of machine adaptation and machine learning which are independent of the internal structure of, say, the controller in the case of a feedback system; and are independent of properties such as the presence or absence of nonlinearity, time-variation, or even feedback. Instead, they are based on external performance. They enable us to address such questions as: What should the term "adaptive" and "learning" mean in the context of control? Is it possible to tell whether or not a black box is adaptive? Is adaptation synonymous with the presence of nonlinear feedback? More to the point, in design is it possible to determine beforehand whether it is necessary for a controller to adapt and learn in order to meet performance specifications, or is adaptation a matter of choice? We will claim that the answers are mostly positive.

---

*Systems and Control Group, Department of Electrical Engineering, McGill University, 3480 University Street, Montreal, Que., Canada H3A 2A7

1 Introduction

What should the terms "adaptive" and "learning" mean in the context of control? Is it possible to tell whether or not a black box is adaptive? Is adaptation synonymous with the presence of nonlinear feedback? More to the point, in design is it possible to determine beforehand whether it is necessary for a controller to adapt and learn in order to meet performance specifications, or is adaptation a matter of choice?

In this overview we shall describe recent work in the H^∞ framework which provides a means of computing certain kinds of adaptive controllers, but which also sheds some light on these more conceptual questions.

Despite the long history of research on adaptive control, and the considerable practical success of adaptive strategies associated with the names of Astrom, Ljung, Goodwin, Caines, Morse, etc., a satisfactory definition of adaptation has remained elusive (at least until the recent work to be described here). The notion that adaptation occurs when parameters vary inside a controller in response to changes in the environment has been widely regarded as deficient. Obviously, controllers with identical external behavior can have an endless variety of parameterizations; variable parameters of one may be replaced by fixed parameters of another. Indeed, in most of the recent control literature there is no clear separation between the concepts of adaptation and nonlinear feedback, or between research on adaptive control and nonlinear stability (despite the fact that existing results are very well founded mathematically; see e.g. the book by Caines [1], to whom, by the way, the author is indebted for many helpful discussions of the ideas to be presented here). This lack of clarity extends to fields other than control; e.g., in debates as to whether neural nets do or do not have a learning capacity; or in the classical 1960's Chomsky vs. Skinner argument as to whether children's language skills are learned from the environment tabula-rasa style, or to a large extent are "built-in". (How could one tell the difference anyway?).

We would like to re-examine these issues in the light of recent developments linking the theories of feedback, identification, complexity, and time-varying optimization. The perspective here is actually not new, having been outlined by the author on and off since the 1970's [14],[16], [17]. However the key mathematical details have been worked out only recently , notably in joint work with Lin, Owen and Wang [5], [21], [13], [8], [19], [9]. Other results which have a bearing on this overview have been obtained by Dahleh [12], Helmicki et al [4], Khargonekar [3], Makila [7], Poolla [11], Tse [12], to cite a few representative papers.

The objective then is to re-examine the notions of adaptation and learning, on two levels: on the conceptual level to obtain a framework of some degree of generality; on a more concrete level to get a design methodology for systems in the H^∞/l^1 "slowly time-varying" category. The results to be outlined here flesh out the following main ideas:

- Adaptation and learning involve the acquisition of information about the plant (i.e., object to be controlled).

- For feedback control the appropriate notions of information are metric, locating the plant in a metric space in one of a set of neighbourhoods of possible plants. Metric information can be quantified. The measures of *metric complexity* most frequently used for this purpose are

    1) metric dimension (inverse n-width)

    2) metric entropy.

- The object of identification is to get this metric information, which takes time to acquire. The minimum time needed to acquire it is related to the metric complexity of a priori data.

- Information obtainable at any given time that is relevant to behavior at some future target date is a monotone increasing function of time.

- Optimal performance is a monotone increasing function of relevant information.

- Nonadaptive ("robust") control optimizes performance on the basis of a priori information. On the other hand, *adaptive control is based on a posteriori information*, and uses the extra information to achieve improved performance.

To define adaptive control from this point of view, a number of mathematical constructs are needed which more or less parallel these ideas, and which we proceed to outline. It will be noted that most of them involve theory developed only since the 1980's.

2   Information (or Complexity)

What follows will be interpreted in terms of system identification experiments (although other interpretations are possible, e.g., in terms of realization, communication, etc.; at bottom one is dealing with properties of metric spaces). Identification begins with *a priori* information. A priori an uncertain plant P is known to lie in a subset S_{prior} of a metric space A. Acquiring information about the plant in the course of identification means localizing the plant in some small subset of S_{prior} of radius ϵ, and thus reducing the uncertainty.
Concretely , in the examples in this paper, A will be assumed to be an algebra of causal, stable, discrete-time systems of the form $K : \mathcal{U} \to \mathcal{U}$,

$$(Ku)(t) = \sum_{\tau=-\infty}^{\infty} k(t,\tau)u(\tau) \tag{1}$$

where the space u of inputs/outputs is either $l^{\infty}(-\infty,\infty)$ or $l^2(-\infty,\infty)$, and the norm $\|K\|_A$ is the operator norm. Initially, assume K to be time-invariant,

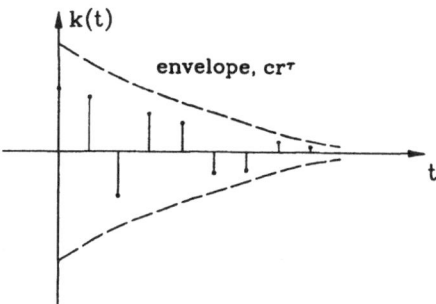

Figure 1:

in which case the weighting function $k(t, \tau)$ coincides with the impulse response $k(t - \tau)$ and has a frequency response which is assumed to be well defined by

$$\hat{k}(z) := \sum_{t=0}^{\infty} k(t)z^t, \quad |z| < 1 \tag{2}$$

and *a.e.* for $|z| = 1$.

Example 1. A is the algebra of systems with impulse responses in $l^1[0, \infty)$. The a priori information locates k in the set

$$S_{prior} = \{K \in A \ : \ \sum_{\tau=0}^{\infty} |k(\tau)|r^{-\tau} \le c\}, \ 0 < r < 1, \tag{3}$$

consisting of functions that decay exponentially in time with exponent $ln \ r$. After identification they are to lie in a band of uncertainty centered at a nominal value $\hat{k}_0$ and of radius ϵ in the H^∞ norm.

The information that k is in S_{prior} leaves K uncertain to within a tolerance ϵ. This notion of uncertainty is cruder than one giving a probability distribution, but has the virtue of being simple enough to be tractable under nonlinear transformations such as feedback, where probability distributions may be intractable. (It will be pointed out in Sect. 5 that notions of uncertainty based on neighbourhoods can be probabilistically defined).

Measures of Metric Complexity (or Information)

These are indicators of the size or something like the volume of the set of uncertainty in which the plant lies, and are variations of the original definitions of Kolmogorov. For a given set S, considered fixed, they are functions of the tolerance ϵ, which increase as ϵ decreases:

(a) *The metric Dimension* (Zames 1976 [14]), $n(\epsilon)$ of S is the dimension of the smallest subspace of A whose maximal distance from S is $\le \epsilon$.

In Example (1), $n(\epsilon) = log_r \left(\frac{c}{\epsilon}\right)$ to the nearest integer, and the optimal subspace is spanned by the first n samples of the impulse response.

(b) The *metric entropy*, $\mathcal{I}(\epsilon)$, is log_e of the smallest number of ϵ-balls needed to cover S (and is finite if S satisfies a compactness condition).

In Example 1, $\mathcal{I}(\epsilon) = (log_2 r)\left(log_r \frac{c}{\epsilon}\right)^2 + O(\text{smaller in } \epsilon)$, and the ϵ-balls are obtained by quantizing the first n samples of the impulse response.

The ϵ-dimension is the inverse of Kolmogorov's concept of n-width, which gives ϵ as a function of n. There are several other definitions of metric dimension or n-width, including the definition of time-width introduced in Lin et al [5], (see Sect. 3 below), Gelfand n-width [10], Bernstein n-width, etc. From the point of view of identification they have been shown [6] to emphasize slightly different aspects of the identification problem. They are only rough indicators of actual identification accuracy, usually off by factors of 1-3. Frequently they coincide, and can then be used interchangeably as measures of identification complexity.

The ϵ-dimension is relevant to linear identification schemes, wheras the ϵ-entropy does not depend on linearity. However, any measure of information/complexity can serve as a basis for adaptive control (see Sect. 6), and in the rest of this paper we shall use only the simpler of the above measures, namely the metric dimension, as our indicator of the quantity of what will variously be called information or complexity depending on the context.

We are particularly interested in identification in H^∞ and introduce a second, simple example, in which S_{prior} is defined entirely in terms of the frequency response.

Example 2. As in Example 1, A is the algebra of systems with impulse responses k in $l^1[0, \infty)$ and identification is to locate the frequency response in a band in H^∞ of radius ϵ, but now S_{prior} is

$$S_{prior} = \left\{ K \in A \ : \ \left|\frac{d\hat{k}(e^{i\theta})}{d\theta}\right| \leq c, \ -\pi < \theta \leq \pi \right\}$$

Here the various notions of n-width mentioned earlier give the identical result [10], [5] that

$$\epsilon(n) = \frac{c}{n}$$

or that the ϵ-dimension, $n(\epsilon) = \frac{c}{\epsilon}$ to the nearest larger integer.

Examples involving more general assumptions where $\hat{k}$ is defined on a disc of radius $r \geq 1$, and there is a bound on $\hat{k}$ or one of its derivatives, are given in [5].

3   Information vs. Time

Information takes time to acquire. There is a minimum time needed to shrink uncertainty about a plant to a tolerance ϵ, which depends on the plant's dynamics. This is a key fact for adaptive control, because it limits the speed of adaptation. We will show that the minimum time needed depends on the metric complexity of the a priori data set, and it is this fact that establishes close links between identification and complexity theories.

Another key fact will be that current information that is obtainable from identification, and that is relevant to performance at some future target date, is monotone increasing with time.

Fast identification in H^∞

Consider a plant with frequency response $\hat{k}$ a priori in a subset S_{prior} of H^∞. The input and output are observed for n instants, on an interval $[t_0, t_0 + n)$. The fast identification problem is concerned with finding an approximation to $\hat{k}$ accurate to $\epsilon > 0$ in minimum time, A) when the input is free to be optimized and, B) when it belongs to an extraneously fixed ensemble.

A) Inputs optimized, no additive noise

To extract the salient features of the problem we initially assume that there is no additive noise. (The opposite problem, where there is additive noise but no plant uncertainty has been well understood since the 1950's [1]). Initially, suppose the input to be completely free to be chosen by the controller. That will provide a benchmark against which to compare the more general case. to establish such a benchmark it is important **not to assume that the input is zero outside the observation interval**, i.e., not to rule out the possibility that prior excitation might speed up the identification. It is that possibility which makes the problem nontrivial, and there are several reasons for not excluding it, notably:

a) If observations start while the plant is in operation and the initial conditions at time t_0 are not known, the response to them acts as a noise component in the output. Until that noise decays to some small level, say at t_1, $(t_1 > t_0)$ observations of the output often convey no information about the plant. The useful output observation interval starts at t_1, whereas the input is free to be shaped earlier, after t_0.

b) If the input is designed to contain a deterministic dithering component, say an almost-periodic function, then knowledge of that component completely determines its past, which can be viewed as a known prior excitation for the purposes of the experiment. One would like to know whether such a prior excitation could be beneficial.

Suppose the input belongs to an ensemble $\mathcal{U}_{ens} \subset \mathcal{U}$ of possible inputs.

We assume throughout that S_{prior} is convex and centro-symmetric ($k \in S_{prior} \Rightarrow -k \in S_{prior}$). For an input belonging to an ensemble $\mathcal{U}_{ens} \subset \mathcal{U}$ of possible inputs, the worst-case identification error in the special case when the true system is null is

$$
\begin{aligned}
e_0(n, \mathcal{U}_{ens}) \quad &: \quad = \sup_{u \in \mathcal{U}_{ens}} \sup \{ \|K\| \\
&: \quad K \in S_{prior} \ \ and \ \ (Ku)(t) = 0 \ for \ t \in [t_0, t_0 + n) \}. \quad (4)
\end{aligned}
$$

For other systems it lies between $e_0(n, \mathcal{U}_{ens})$ and $2e_0(n, \mathcal{U}_{ens})$, so that $e_0(n, \mathcal{U}_{ens})$ is a good if imperfect indicator of error. *The optimal time-width θ^n of S_{prior} is* the error $e(n, \mathcal{U}_{ens})$ optimized over all possible (ensembles of) inputs:

$$\theta^n := \inf_{\mathcal{U}_{ens} \subset \mathcal{U}} e_0(n, \mathcal{U}_{ens})$$

The minimal time to achieve a worst-case error ϵ optimized over all inputs, which will be denoted by $T_{opt}(\epsilon)$, is the inverse of the θ^n function,

$$T_{opt}(\epsilon) = \inf \{n \; : \; \theta^n \le \epsilon\}$$

$\theta^n(\epsilon)$ or, equivalently, $T_{opt}(\epsilon)$, have been computed for a variety of convex centro-symmetric sets in [5], [18] in the H^∞ norm on frequency responses or l^1 norm on impulse responses. Examples 1 and 2 are typical. In the H^∞ case, the frequency responses or their derivatives are constrained to be analytic on a disk of radius ≥ 1 of the plane and to be bounded there; in the l^1 case the impulse responses lie in a ball after exponential weighting. It turns out that for all of these simple cases θ^n is computable exactly, and equals the norm of the largest element of S_{prior} that is null on the interval $[0, n)$. It is natural, therefore, to seek a general principle to account for the similar behaviors. This will be tied to the fact that all of these sets have a property of monotone decrease in time, which will be described next.

By the norm $\|S\|$ of any subset S of A we mean $\|S\| := \sup \{\|K\| \; : \; K \in S\}$. A subset S of A will be called *monotone decreasing* if given any fixed interval $[t_0, t_1]$, the norm of S intersected with any subspace of functions of A having support on a (variable) subinterval $[t_0', t_0' + i)$ of $[t_0, t_1]$ is monotone decreasing as t_0' increases. A somewhat more general property than monotonicity of S requires the previous statement to be true only for subintervals of length $i \le q$, in which case S will be called *q-monotone decreasing*.

Let $S|_{[a,b]}$ denote the subset of functions in S with support on the interval $[a, b]$. The following principle is derived in [18].

Monotonicity Principle. If the set $S \subset A$ is convex, centro-symmetric and q-monotone decreasing then

a) its time width θ^n satisfies

$$\|S|_{[n, n+p]}\| \le \theta^n \le \|S|_{[n, \infty)}\| \tag{5}$$

If the lower bound in (5) approaches the upper bound as $p \to q$, then:

b) $\theta^n := \|S|_{[n, \infty)}\|$ and an impulse at the start of the observation interval is an optimal input for minimizing θ^n; and

c) the minimal time $T_{opt}(\epsilon)$ equals the Gelfand metric dimension [1]

Remarks. If part b holds, then:

a) The optimal input for fast identification is null before the start of the observation interval, implying that a priori excitation can not provide any benefit.

---

[1] i.e., the inverse function of Gelfand's n-width d^n.

b) The optimized uncertainty represented by θ^n is monotonically decreasing with the time n. Equivalently, the information is increasing. This fact will be used in adaptive control.

c) The minimal time $T_{opt}(\epsilon)$ to identify a system equals the size of its a priori data set as measured by a metric complexity measure of Gelfand type. Frequently this also equals a dimension of Kolmogorov type, and is a function of the ϵ-entropy. Thus there are strong links between identification and complexity theory which make it fruitful to view these as parts of a common subject.

The optimal input is not always usable, but it does provide a benchmark for comparison.

B) Suboptimal Input Ensembles

What properties should an input have to ensure fast identification? How much speed is lost when suboptimal inputs are used, either by choice or because they are extraneously generated and fixed? To explore these issues we compare the identification performance of various ensembles to the preceeding benchmark.

Suppose then that the input u is in an ensemble $\mathcal{U}_{ens} \subset \mathcal{U}$, and that $\mathcal{U}_{ens}$ is shift-invariant. The worst-case error $e_0(u, \mathcal{U}_{ens})$ defined by (4) satisfies

$$\theta^n \leq e_0(n, \mathcal{U}_{ens}) =: \alpha(\mathcal{U}_{ens}) \, \theta^n \tag{6}$$

where θ^n is the benchmark optimized over all ensembles and $\alpha(\mathcal{U}_{ens})$ is a constant defined by (6). How big is $\alpha(\mathcal{U}_{ens})$?

Deterministic time-invariant ensembles consisting of periodic pulse strings were analysed in [5], for the various data sets mentioned previously, including those of Examples 1-2, in the l^1 and H^∞ norms. There, $\alpha(\mathcal{U}_{ens})$ lies roughly between 1 and 3. This represents a significant but not overwhelming loss in identification speed over the benchmark. The implication is that it is possible to design *time-invariant* ensembles which are reasonably efficient for identification, so long as they are deterministic.

For nondeterministic or at least aperiodic inputs, more like those which might appear inside a feedback loop in response to extraneous excitation, the deterioration can be much greater. In a forthcoming paper, we look at ensembles that satisfy a conditioning constraint on the energy present in bands of a length that is specified simultaneously in the time and frequency domains, as follows. The short-term spectrum of $u \in l^\infty(-\infty, \infty)$, of length n starting at time t is

$$\hat{u}(z; t, n) := \sum_{\tau=t}^{t+n-1} u(\tau) z^{-\tau},$$

and is defined for all z. Consider ensembles $\mathcal{U}_{ens}$ with the property that there are integers n, r, such that each $u \in \mathcal{U}_{ens}$ satisfies the conditioning inequalities

$$\underline{\beta}(n, r) \leq \left\{ \frac{r}{2\pi} \int_{\theta_0 - \frac{\pi}{r}}^{\theta_0 + \frac{\pi}{r}} |\hat{u}(e^{j\theta}, t, n)|^2 d\theta \right\}^{\frac{1}{2}} \leq \overline{\beta}(n, r) \tag{7}$$

for all $t \in \mathbb{Z}$ and all θ_0 in $-\pi < \theta_0 \leq \pi$; here $\underline{\beta}(n, r)$ and $\overline{\beta}(n, r)$ are constants for $\mathcal{U}_{ens}$ depending on n and r. Well-conditioned inputs, those for which $\beta := \overline{\beta}/\underline{\beta}$ approaches 1, (may be deterministic but) increasingly resemble white noise for large n, r.

For ensembles satisfying (7) and selected a priori data sets, the worst-case error vs. time function, $e(n, \mathcal{U}_{ens})$, satisfies the inequalities

$$\underline{\gamma} \; \beta(n, r) \; \theta^n \; \leq \; e_0(n, \mathcal{U}_{ens}) \; \leq \; \overline{\gamma} \; \beta(n, r) \; \theta^n.$$

For well conditioned inputs, the constant $\gamma(n, r)$ may remain much larger than 1. This would imply that extraneously fixed noise-like inputs may be poor for fast identification in comparison with our benchmark. A corollary would be that, where identification speed is at a premium, the use of deterministic dithering components in the input may offer considerable advantages. The results obtained also suggest ways of modulating the input to speed up identification when additive noise is present.

4   Feedback Performance vs. Identification

The purpose of feedback is to shrink uncertainty attributable either to additive disturbances or imperfect knowledge of the plant model. Feedback performance is therefore measurable by the extent of the shrinkage. That performance is itself affected by uncertainty as to the plant model, and so there is a linkage between feedback performance and plant identification. Computation of this linkage was one of the main original objectives of the H^∞ approach [16], which however remained unfulfilled despite the explosive growth of H^∞ research since the 1980's. Indeed, even the simplest problems involving nontrivial (i.e. not perfectly known at high frequencies) plant uncertainty remained unsolved, until the very recent solution [19] of the problem of optimal disturbance attenuation in H^∞ under plant uncertainty, which will be described next.
Assume that the plant P lies in a band of uncertainty in H^∞, with center P_0 and radius $V P_0$,

$$|P(e^{i\theta}) - P_0(e^{i\theta})| \leq |V(e^{i\theta})P_0(e^{i\theta})|, \quad -\pi < \theta \leq \pi, \; V \in H^\infty$$

It was shown in [2] that after parameterization of stabilizing feedbacks by a parameter $Q \in H^\infty$, the optimal sensitivity to disturbances is

$$\mu(P_0, V) := \inf_{Q \in H^\infty, \; \|V P_0 Q\|_\infty < 1} \left\| \frac{W(1 - P_0 Q)}{1 - |V P_0 Q|} \right\|_\infty \tag{8}$$

where $W \in H^\infty$ is a weight function characterising the disturbances , $\|W\|_\infty = 1$. $\mu(P_0, V)$ is an appropriate measure of feedback cost, and its reciprocal μ^{-1} a measure of performance. This performance might be expected to deteriorate as $V(e^{i\theta})$ increases. For example, if $V = \delta V_0$ for some fixed $V_0 \in H^\infty$ and variable $\delta > 0$, it might be expected that the dependence of $\mu(P_0, \delta V_0)$ on δ might look as shown in Fig. 3.

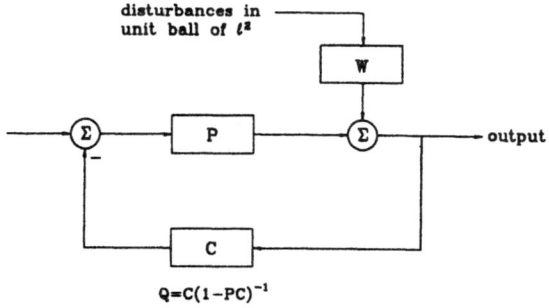

Figure 2:

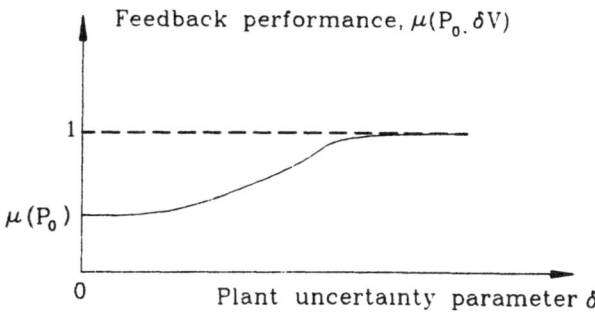

Figure 3:

Computation of $\mu(P_0, \delta V)$ remained an unsolved problem except in the limiting case $\delta \to 0$, in which case $\mu(P_0, \delta V)$ approaches the *optimal weighted sensitivity*, $\mu(P_0) := \inf_{Q \in H^\infty} \|W(1 - P_0 Q)\|_\infty$. Francis [2] had shown that (8) could be reduced to a (nonstandard) H^∞ *two-disc problem*,

$$\inf_{Q \in H^\infty} \||W(1 - P_0 Q)| + r|V P_0 Q|\|_\infty =: \chi(r) \tag{9}$$

depending on a parameter r, and then $\mu(P_0, V)$ is the smallest fixed-point r of $\chi(r) = r$. He proposed approximating the two-disc problem by a "two-block" (standard) H^∞ problem. However, it has been shown [19] that such approximation may be infinitely poor when $\chi(r)$ is highly sensitive to changes in r near the fixed points.

Solution of the two-disc problem

An exact solution for the scalar case was finally given in Owen and Zames [8], an extension to MIMO systems in [19], and a solution of the full MIMO analog to (8) in a forthcoming paper. In each case, the two-disc problem is reduced to a distance problem in Banach space and is solved using duality. The primary distance problem is convex and yields a convergent algorithm which overestimates $\chi(r)$. The dual problem underestimates $\chi(r)$, providing a stopping rule based on an error bound.

Using this approach, the computation of $\mu(P_0, V)$ as a function of V is considered in [8]. There are various conclusions concerning flatness of solutions, etc., but the ones most relevant here can be described as follows. It is assumed that the H^∞ functions V, W, UW, (where U is the inner factor of P_0) , as well as the outer spectral factor of $|W|^2 + |V|^2$, are all continuous. Also the nondegeneracy condition

$$\| \min\{|W(e^{i\theta})|, |\mu V(e^{i\theta})|\} \|_\infty < \mu$$

is imposed (which prevents W and V from being simultaneously so large at any frequency as to overwhelm the performance). The relevant results are then summarized by the following.

Theorem 1. *The feedback performance $\mu(P_0, V)$ is a monotone strictly increasing function of the plant uncertainty V, i.e., if $|V_2(e^{i\theta})| > |V_1(e^{i\theta})|$ for some $\theta \in (-\pi, \pi]$, then $\mu(P_0, V_2) > \mu(P_0, V_1)$. In the limit, $\mu(P_0, \delta V)$ approaches $\mu(P_0)$ as $\delta \to 0$, and reaches 1 for large δ. ($\mu(P_0, V)$ can be computed by a duality-based convex optimization algorithm).*

parindent=24pt

Remark. Theorem 1 was hypothesized by the author in 1976 [14], and efforts to compute $\mu(P_0, V)$ provided the initial impetus for the H^∞ approach to feedback. The mathematical tools for computing feedback performance are only now falling into place.

5   Adaptive vs. Nonadaptive Feedback

We would now like to formalize the concept that a nonadaptive feedback optimizes performance at time t on the basis of a priori information (more accurately, as though information at t were no greater than at the start of observations); whereas an adaptive feedback uses extra a posteriori information to achieve improved performance. This concept involves the two tradeoffs that were described in Sects. 3 and 4, and in fact are comptutable for selected problems in H^∞ or l^1:

a) Optimal feedback performance, such as $\mu^{-1}(P_0, V)$, is an increasing function of information, where the information is represented by sets of uncertainty S, such as the balls (P_0, δ).

b) The maximum amount of such information obtainable at a given time from an identification experiment t is a function of t representable by $S(t)$, where $S(t)$ decreases (i.e., information increases) as $t \to \infty$.

Information based on neighbourhoods

A formal definition of an adaptive system can be achieved as follows.

We shall consider an element P of the metric space A and design variable C belonging to some admissible set $\mathcal{C}$. Here, A and $\mathcal{C}$ consist of causal operators acting on functions on the time interval $[t_0, \infty)$. P and C can be thought of as plant and feedback controller, although the concepts that follow are more general.

Information about $P \in A$ will be based on a collection Σ of (not necessarily open) neighbourhoods which is assumed to be defined on A, and is ordered by containment, i.e., for $S_1, S_2 \in \Sigma$, $S_1 \subset S_2 \Rightarrow S_1 < S_2$. The collection of all balls in H^∞ of radius $\delta > 0$ is an example (but probabilistically defined neighbourhoods would also be suitable). Information about P consists of listing one or more neighborhoods $S \in \Sigma$ to which P belongs. The smaller the set S the smaller the uncertainty about P and, in our terminology, ther greater the "information represented by S". (So, small bands in the H^∞ frequency domain represent much information).

Information evolves with time, say on an interval $[t_0, \infty)$. A priori information, available at the starting time t_0, is represented by the set $S_0(0) \in \Sigma$. A posteriori information is represented by a set-valued function of time, $S(\cdot) : [t_0, \infty) \to \Sigma$. Suppose that the possible outcomes of an identification experiment belong to a collection Ω of such functions, with the following properties.

1) $S(\tau) \le S_0(0)$ for all τ, and $S(\tau)$ is monotonically nonincreasing in τ.

2) The constant function S_0, where $S_0(\tau) := S_0(0)$ $\forall \tau$, representing no increase in information, is in Ω.

A pair (S, C) determines a system which, *it should be emphasized, will always involve sets of plants rather than individual plants.*

Performance functions

The restriction of a function S on the interval $[t_0, \infty)$ to the subinterval $[t_0, t]$ will be denoted by S_t. Similarly, the restriction of an operator such as

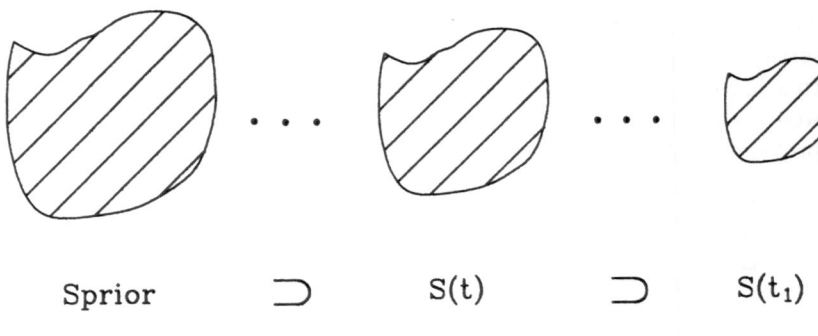

$$\text{Sprior} \quad \supset \quad \text{S(t)} \quad \supset \quad \text{S}(t_1)$$

Figure 4:

C to a domain and range consisting of functions themselves restricted to $[t_0, t]$ will be denoted by C_t

$J(S, C; t)$ is a function representing cost at or after $t \in [t_0, \infty)$, which is causally dependent on $S \in \Omega$ and $C \in \mathcal{C}$, i.e., depends only on S_t and C_t. The reciprocal function $J^{-1}(S, C; t)$ will be called a *performance function*.

For any $S \in \Omega, t \geq t_0$, let $J_{opt}^{-1}(S, t)$ be the optimal performance achievable at t for the information represented by S_t, i.e., $J_{opt}(S; t) := \inf_{C \in \mathcal{C}} J(S, C; t)$. Assume that $J_{opt}(S_t, t)$ is monotone nondecreasing in S_t; i.e. $S_t'(\tau) \geq S_t''(\tau)$ for all $\tau \in [t_0, t] \Rightarrow J_{opt}(S'; t) \geq J_{opt}(S''; t)$, in which case the information represented by S will be called *relevant* to cost J or performance J^{-1}.

Definitions

A system (S, C) is *(strictly) adaptive* if

$$J^{-1}(S, C; t) \geq J_{opt}^{-1}(S_0; t)$$

for all $S \in \Omega$ and $t \in [t_0, \infty)$, and the inequality is strict for all (for some) $S \in \Omega$ and $t \in T$ such that $S_t < S_0$ and $t \geq t_0$. Otherwise, (S, C) is *nonadaptive*.

Remark. Adaptive systems are realized by making C causally dependent on a posteriori information, whereas nonadaptive ones are dependent only on a priori information. However, a definition based on how they are realized would make some adaptive systems externally equivalent to nonadaptive ones. We prefer to base our definition on performance, which can be a purely external property. In order to do so, we need to involve optimality of performance to obtain our benchmark. For feedback systems, the (recently acquired) ability to compute such optimal performance is the key to the implementation of our definition.

Some adaptive feedbacks in H^∞

In the paper [9], feedback controllers which are adaptive in the sense of our definition are constructed in an H^∞ setting, by combing the results on performance optimization and fast identification along the lines of Sect. 3 and 4.

Two adaptive controllers are derived in [9]. In one, the a priori information is the set S_{prior} of Example 2, and the performance is $\mu(P_0(t), \delta(t))$ optimized for a ball of uncertainty $(P_0(t), \delta(t))$. That uncertainty set is derived from an identification experiment which minimizes the radius $\delta(t)$ at time t in accordance with the optimal time-width results described here. A second controller in [9] is based on a priori data in weighted l^1, and employs non-spherical uncertainty sets that provide better adaptation than balls.

The calculation of optimal performance μ, which was addressed in Sect. 3 for static uncertainty, has the new feature that now $(P_0(t), \delta(t))$ are functions of time. This is a complication which has so far been resolved only by resorting to a "frozen-time" approximation, which "slowly" takes advantage of a posteriori information to improve performance. A systematic theory of frozen-time approximations, for Volterra sum operators of the type (1), based on the concept of "double-algebras", was presented in [13], [21] by Wang et al, in which some of the ideas on adaptation were originally presented.

6 Some answers

We are now in a position to answer some of the questions raised at the outset.

When is a black box adaptive? If its performance is better than the best possible based on a priori information. This can be determined from external observations to the extent that performance is externally observable. It may require observing a complete ensemble of uncertain plants or noises or, in ergodic situations, to perform a large sequence of experiments on a single plant or disturbance. Whether this is easy or hard to do is irrelevant here.

A more important question is whether adaptation is necessary or optional if specs are to be met for an ensemble of uncertain plants. Adaptation is necessary whenever the optimal nonadaptive performance is inadequate, and adaptation provides enough improvement.

What is learning? In out context it is the improvement in performance attributable to acquired information. For the problems described at the end of Sect. 6 it can actually be measured and computed, and is represented by the difference

$$\mu(P_0(t), \delta(t)) - \mu(P_0(t_0), \delta(t_0)).$$

All of these concepts are independent of the internal realization of a system, and do not necessarily involve non-linearity or time-invariance in the controllers; for example, linear time-invariant systems can be adaptive with respect to uncertainty in the form of additive noise.

References

[1] P. E. Caines, *Linear Stochastic Systems*, Wiley, 1988.

[2] B. Francis, On Disturbance Attenuation with Plant Uncertainty, *Workshop on New Perspectives in Industrial Control System Design*, 1986.

[3] G. Gu and P. P. Khargonekar, Linear and Non-Linear Algorithms for Identification in H^∞ With Error Bounds, *IEEE Trans. Aut. contr.*, 1992, AC-37, 7, pp.953–963.

[4] A. J. Helmicki and C. A. Jacobson and C. N. Nett, Control Oriented System Identification: A Worst-Case/Deterministic Approach in H^∞, *IEEE Trans. Aut, Contr.*, AC-36, 1991, pp. 1163–1176.

[5] L. Lin, L.Y. Wang and G. Zames, Uncertainty Principles and Identification n-Widths for LTI and Slowly Varying Systems, *Proceedings of ACC 1992*, pp. 296–300.

[6] L. Lin and L.Y. Wang, "Time Complexity and Model Complexity of Fast Identification", *Proc. 32nd IEEE Conf. Dec. Control*, pp.2099-2104.

[7] P. M. Makila and J. R. Partington, Robust Approximation and Identification in H^∞, *Proc. 1991 ACC*.

[8] J.G. Owen and G. Zames, Duality Theory of Robust Disturbance Attenuation, *Automatica*, Vol 29, 3, pp. 695-705, May 1993.

[9] —— , Adaptation in the Presence of Unstructured Uncertainty, *Proc. 32nd. IEEE Conf. Dec. Control*, 1993, pp. 2110-2114.

[10] A. Pinkus, *n-Widths in Approximation Theory*, Springer-Verlag, 1985.

[11] K. Poolla and A. Tikku, On the Time Complexity of Worst -Case System Identification, *Preprint*.

[12] D. N. C. Tse and M. A. Dahleh and J. N. Tsitsiklis, Optimal Asymptotic Identification Under Bounded Disturbances, *Proc. 1991 CDC*, pp. 623-628.

[13] L.Y. Wang and G. Zames, Local-Global Double Algebras for Slow H^∞ Adaptation: Part II-Optimization of Stable Plants, *IEEE Trans.aut.contr*, AC-36(2), 1991, pp. 143-151.

[14] G. Zames, Feedback and Complexity, Plenary Lecture, *Proc. 1976 CDC*, Addenda, pp. 1-2.

[15] —— , On the Metric Complexity of Causal Linear Systems, ϵ-entropy and ϵ-dimension for Continuous time, *IEEE Trans. Aut. Contr.*, AC-38, pp. 664-667, April 1979.

[16] ——, Feedback and Optimal Sensitivity: Model Reference Transformations, Seminorms, and Approximate Inverses, *IEEE Trans. Automat.Control*, AC-23 (1981) , pp. 301-320.

[17] —— , Feedback, learning and complexity in H^∞, Plenary Lecture, *1985 Internat. Conf. Math. Th. Netw. Systems; repeated at 1989 ACC*.

[18] G. Zames, L. Lin, and L.Y. Wang, Fast Identification n-widths and Uncertainty Principles for LTI and Slowly Varying Systems, *IEEE Trans. Automatic Control*, AC-39, No. 9, Sept. 1994.

[19] G. Zames and J.G. Owen, Duality Theory of MIMO Robust Disturbance Attenuation, *IEEE Transactions on Automatic Control*, Vol 38,5, May 1993, pp. 743-751.

[20] ——— , A Note on Metric Dimension and Feedback in Discrete Time, *IEEE Trans. Aut. Contr.*, AC-38, pp. 664-667, April 1993.

[21] G. Zames and L.Y. Wang, Local-Global Double Algebras for Slow H^∞ Adaptation: Part I-Inversion and Stability, *IEEE Trans.aut.contr*, AC-36(2), 1991, pp. 130-142.

[22] ———— , "What is an Adaptive System?", *Proc. 29th. IEEE Conf. Dec. Control*, Vol. 5, Dec. 1990, pp. 2861-2864.

Focusing on the Knowable

Controller Invalidation and Learning

Michael G. Safonov*†

The task of science is to stake out the limits of the knowable and to center the consciousness within them.

Rudolf Virchow — Berlin, 1849

Abstract

Modern and classical control theories fail to clearly distinguish conclusions based on data from those that are a consequence of assumptions. The "unfalsified control concept" is offered as a possible solution to this dilemma. The concept provides a set-theoretic characterization of control-relevant information which is based solely on experimental data and is untainted by plant modeling assumptions or related prejudices. It forms the basis for an objective analysis of adaptation and learning, leading directly to practical model-free adaptive control algorithms. The basic idea is to invalidate, or falsify, controllers off-line using previously collected experimental plant data. The theory may be viewed as complementing traditional plant-model-based theories of control by providing needed additional guidance when real-time data fails to corroborate modeling assumptions.

Motivation and Issues

Describing the scientific method, Nobel laureate mathematician-philosopher Bertrand Russell [1] once said,

"The essential matter is an intimate association of hypothesis and observation. The Greeks were fruitful in hypotheses, but deficient in observation. Aristotle, for example, thought that women have fewer teeth than men, which he could not have thought if he had had a proper respect for observation."

Like Aristotle [2] and many theoreticians since, we control theorists are too easily tempted to build grand, but fanciful theories based on assumptions which may fail to hold. We may rightly claim to prove stability, robustness and even

*Phone 1-213-740-4455. FAX 1-213-740-4449. Email safonov@bode.usc.edu.

†Research supported in part by AFOSR grants F49620-92-J-0014 and F49620-95-I-0095.

optimality, but only subject to assumptions that our models are valid to within conjectured tolerances or that our uncertainties conform to assumed probability distributions. Unfortunately, proofs of stability and optimality based on mathematical models all too often have little predictive value when applied to physical systems. The problem here is that mathematical theories of control typically give insufficient attention to the implications of possible future observations which may be at odds with assumptions. The Achilles heel of modern control theory has been this habit of "proof by assumption."

Even robust control theory fails to adequately address the question of what to do when subsequently obtained experimental data does not corroborate assumptions about uncertainty bounds, though model-validation theory provides a conservative first step in the right direction (see [3] and references therein). An essential question is:

> *How can we separate the control-relevant information in experimental data from that in our assumptions?*

Answering this question is key to understanding feedback control and learning. It turns out that it is not necessarily difficult to formulate this question in precise mathematical terms, though curiously traditional control theory problem formulations have apparently failed to do so.

We need to focus on the data. Unlike mathematics which is constrained only by one's imagination in creating self-consistent theorems and assumptions, the theories of science are more rigidly constrained to conform to experimental data — or at least they should be. While control theory has brought the full power of mathematics to bear on the question of how to proceed from assumptions about the physical world to conclusions about stability robustness and even optimality, the same attention has not been brought to bear on the critical question of how experimental data and performance goals constrain the choice of controller. The answer to this question is precisely what is needed to put feedback control theory on a sound scientific footing. To find the answer, it will be necessary to develop problem formulations that focus attention squarely on what of control relevance can be deduced from experimental data alone.

> *To start, put aside the dogma that control analysis must begin with a model.*

Of course, there can be no denying that models are useful in control design. They help us to formulate hypotheses about suitable controller structures by distilling apparently simple patterns from complex data. But we must recognize that models may also be misleading, since they can cause us to draw conclusions which we could not draw from the data alone. Models are essentially an aggregate of our a priori knowledge and prejudices. These prejudices are only partially removed when we augment a plant model with bounded uncertainties (Δ's), stochastic noises, or unknown parameters. Uncertain models are still models. And they still reflect prejudices.

A proper understanding of feedback and learning requires us to carefully examine the control-relevant information in the data so that we may better recognize situations in which our modeling assumptions are mistaken. To do so, we must temporarily cleanse our minds of models and other prejudicial assumptions about the plant to be controlled, for only then can we hope to develop a scientifically sound analysis of the proper role of experimental data. That is, we need to find mathematical problem formulations which allow us to clearly distinguish experimentally derived information from that which is a consequence of our modeling assumptions.

But, we must reconcile ourselves to the fact faced by all scientists that there are severe limitations on what can be logically concluded from data alone. To quote Sir Karl Popper [4, p. 114], noted philosopher and author of numerous works on the logical basis of scientific discovery,

> *The Scientist ... can never know for certain whether his theory is true, although he may sometimes establish ... a theory is false.*

No matter how many past experiments have corroborated a scientific theory, there always remains the possibility that a new experiment may disprove the theory. Consequently, a scientific theory remains forever a tentative conjecture as scientists seek, but never fully succeed, to validate the theory through diligent efforts to devise experiments which may falsify the theory. The role of mathematics in science is to test the logical self-consistency of various components of a scientific theory, as well as to assist in performing the computations needed to test whether a theory is consistent with experimentally observed facts. And, we must be prepared to abandon theories and models which are inconsistent with data. As Isaac Newton said when confronted with experimental data which seemed to refute his theory [5],

> *"It may be so, there is no arguing against facts and experiments."*

Like a scientific theory, a control law's ability to meet a given performance specification may be regarded as a conjecture to be subjected to validation by experimental data. And, as with a scientific theory, one must accept that in the absence of prejudicial assumptions the most that one can hope to learn about a control law from experimental data alone is that the conjecture is false, i.e., that the control law cannot meet a given performance specification. For example, without prejudicial assumptions about the plant it is not logically possible from data alone to deduce that a given controller stabilizes the plant. But, using only past data it turns out that it is sometimes possible to show that a candidate feedback controller is *not* capable of meeting certain quantitative specifications that are closely related to closed-loop stability, even without actually trying the controller on the real plant. It also turns out to be true that even a very small amount of open-loop experimental plant data can often be used to draw sweeping conclusions about the unsuitability of large classes of feedback controllers.

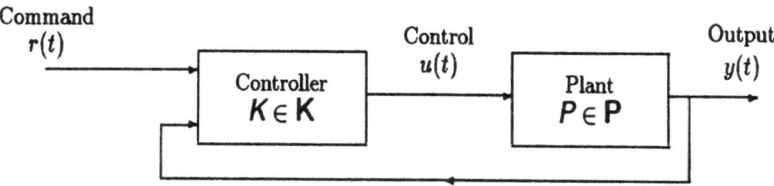

Figure 1: Feedback control system.

A Simple Example

Consider the system in Figure 1. Suppose that nothing is known about the plant P, save that an experiment has been performed during which some otherwise unknown nonlinear dynamical feedback controller was in place. Further, suppose that during this experiment the control input to the plant was observed to be $u_{data}(t) = 1$ and the experimentally observed output was $y_{data}(t) = -1$ over the same time-interval $t \in [0,1]$. Although this particular closed-loop data is consistent with the simple plant model $y = -u$, another simple model matching the data is $y \equiv -1$. The actual plant that generated this data might even be a complicated time-varying, nonlinear, dynamical system and subject to disturbances and noise such as

$$y + \dot{y} = -2u^3(t) + 1 - \dot{u}(t)sin(t)cos(u^2(t)),$$

though a good scientist would never opt for such a complex model without some corresponding complexity in the data u_{data}, y_{data}. We know only the experimental data. With this and nothing more, we may still conclude that if another feedback law $u = k(r - y)$ were to have been in the loop at the time of the experiment and if the hypothetical command input

$$r(t) = \tilde{r}(t,k) \triangleq -1 + \frac{1}{k} \tag{1}$$

had been applied, then this would have resulted in the previously observed experimental data being exactly reproduced and, moreover, the resultant closed-loop error $\tilde{e}(t) \triangleq \tilde{r}(t,k) - y(t)$ would have been $\tilde{e}(t) = 1/k$. Furthermore,

$$\sup_{\substack{r \neq 0 \\ r}} \frac{\|e\|_{L_2[0,\tau)}}{\|r\|_{L_2[0,\tau)}} \geq |\tilde{e}|/|\tilde{r}| = 1/|k - 1|. \tag{2}$$

We thus see that even without a plant model or any assumptions whatsoever about the plant, experimental data alone is sufficient to allow one to logically conclude that if $|k - 1| < 0.1$ then the closed-loop system fails to meet a quantitative stability specification requiring closed-loop L_{2e}-gain less than 10. With additional experimental data, one might be able to identify other values of k that are inconsistent with this L_{2e}-gain specification, if there are any. And, if one collected further data while the plant was being subjected to a rich class of "persistently exciting" inputs, then one might hope eventually to arrive at a situation in which practically all inconsistent values of k had been eliminated or "falsified." Through this process of elimination, the set of "good" values of the feedback gain k would be identified.

The foregoing example suggests that by setting control system performance goals that are only slightly less ambitious than stability or robustness, it still may be possible to develop scientifically sound algorithms for designing good feedback control laws directly from past data — without recourse to a priori plant models or other prejudicial assumptions about the plant. To do this scientifically, we must focus squarely on what is knowable from the data alone. The key is to ask not that stability and robustness be proved, but to ask only that the experimental data not be inconsistent with stability and robustness — i.e., ask only that the system be *not-provably-unrobust*.

Unfalsified Control

In 1620 Sir Francis Bacon [6] identified the problem of scientific discovery as follows:

> *"Now what the sciences stand in need of is a form of induction which shall analyse experience and take it to pieces, and by a due process of exclusion and rejection lead to an inevitable conclusion."*

While nearly four centuries of scientific thought have not fulfilled Bacon's wish for the certainty of inevitable conclusions about the laws that govern the physical world, the scientific method continues to revolve around a process of exclusion and rejection of hypotheses that are falsified by observations [4]. In collaboration with my students T. Tsao and T. Brozenec, I have recently examined the application of the falsification principle to the problem of identifying the set of not-provably-unrobust controllers in the absence of plant models or related assumptions [7, 8, 9, 10]. Our work may be regarded as an attempt to simplify, streamline and reduce the conservativeness associated with works (e.g., [11, 12, 3]) which strive to achieve this same goal indirectly via a marriage of model validation and robust control theory.

We have considered systems of the general form in Figure 1, adopting the input-output perspective introduced by George Zames and Irwin Sandberg in their celebrated papers on stability [13, 14, 15, 16, 17]. And, like Zames [16], Rosenbrock [18], Safonov [19, 20], Willems [21] and others, we cut directly to the core of the problem by throwing away the plant state and thinking of the plant as simply a collection of input-output pairs $(u, y) \in \mathcal{U} \times \mathcal{Y}$; i.e., a "graph" consisting of the points (u, y) in the "$\mathcal{U} \times \mathcal{Y}$-plane". Then, treating each candidate controller's ability to meet specifications as a hypothesis to be validated against past data, we may perform a computation to test the candidate control laws for inconsistencies between the three components that define our data-only control design problem, viz.,

I. experimental input-output data from the plant u_{data}, y_{data},

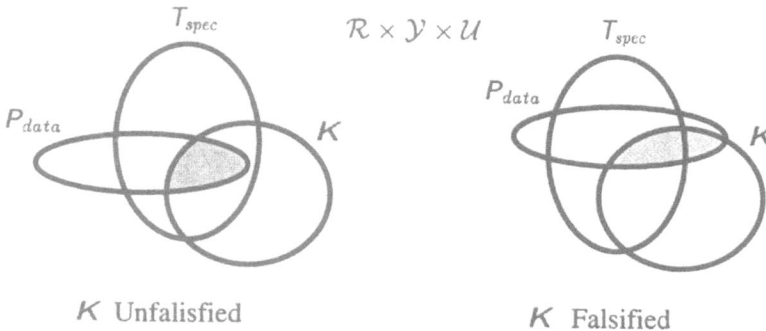

K Unfalisified $\qquad\qquad\qquad\qquad$ K Falsified

Figure 2: A controller $K(s)$ is *unfalsified* by the data if $P_{data} \cap K \subset T_{spec}$, otherwise K is falsified.

II. constraints on (r, y, u) imposed by the control law $u(s) = K(s) \begin{bmatrix} y(s) \\ r(s) \end{bmatrix}$, and

III. a performance specification $(r, y, u) \in T_{spec}$.

We call this idea the *unfalsified control concept*, since the "validated" control laws that emerge from this process are precisely those whose ability to meet the performance specification is not falsified by experimental data.

These three components of unfalsified control problem each define a corresponding subset of the space $\mathcal{R} \times \mathcal{Y} \times \mathcal{U}$, viz.,

$$P_{data} \left\{ \ (r, y, u) \ \middle| \ u = u_{data}, y = y_{data} \ \right\} \subset \mathcal{R} \times \mathcal{Y} \times \mathcal{U},$$

$$K = \left\{ \ (r, y, u) \ \middle| \ u(s) = K(s) \begin{bmatrix} y(s) \\ r(s) \end{bmatrix} \ \right\} \subset \mathcal{R} \times \mathcal{Y} \times \mathcal{U}$$

$$T_{spec} \subset \mathcal{R} \times \mathcal{Y} \times \mathcal{U}.$$

The controller $K(s)$ is *invalidated* (i.e., its ability to meet the performance specification T_{spec} for all r is falsified) by the open-loop data $u_{data}, y_{data})$ if and only if the following condition fails to hold (cf. Fig. 2):

$$P_{data} \cap K \subset T_{spec}. \tag{3}$$

The unfalsification condition (3) is simply a restatement of what it means mathematically for experimental data and a performance specification to be consistent with a particular controller. Yet, simple though it may be, it does not seem to have been incorporated in either classical or modern control problem formulations. It has some interesting and, I think, important implications:

- Condition (3) is nonconservative; i.e., it gives "if and only if" conditions on $K(s)$. It uses all the information in the data — and no more.

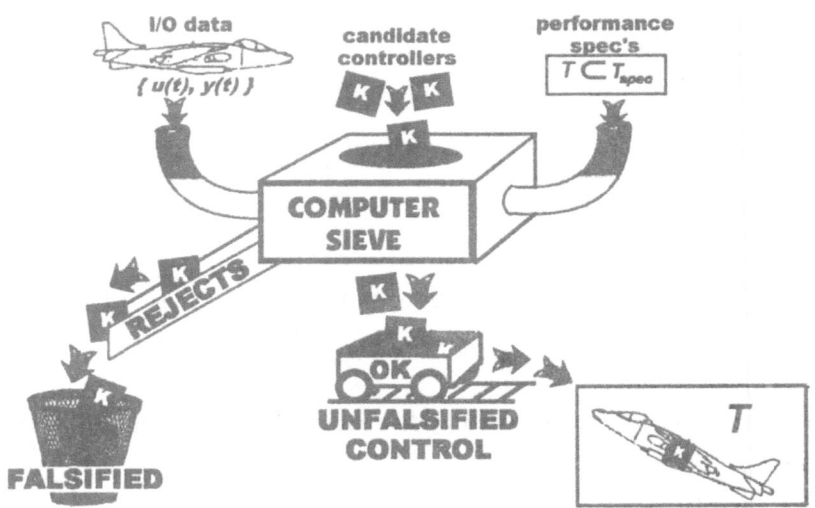

Figure 3: Computer "sieve" checks controllers for consistency with experimental data and performance specifications.

- Condition (3) provides a "plant-model free" test for controller invalidation. No plant model is needed to test its conditions. It depends only on the data, the controller and the specification.

- In (3), data (u, y) which invalidates a controller $K(s)$ need not have been generated with the controller K in the feedback loop; it may be open loop data or data generated by some other control law (which need not even be in K).

- The unfalsification condition (3) accommodates noise and plant uncertainty. In the case that the system to be controlled is subject to random disturbances, sensor noise or uncertain nonlinear or dynamical "parasitics," then the performance criterion T_{spec} must be be chosen to reflect the fact that achievable performance will be correspondingly limited. But the terms P_{data} and K on the left-hand side of the falsification condition (3) are unaffected.

- The relation (3) provides the mathematical basis for constructing a computer "sieve" for screening controllers. This sieve non-conservatively identifies controllers which are consistent with both performance goals and past observations of plant input-output data. See Fig. 3.

Learning control is possible using the controller sieve provided by (3). If one has a class, say **K**, of candidate controllers, then one may evaluate condition (3) for each candidate controller $K \in \mathbf{K}$ and thereby "identify" the subset $\mathbf{K}_{OK} \subset \mathbf{K}$ of unfalsified controllers, i.e., those controllers whose ability to meet the performance specification T_{spec} is not invalidated by the data u_{data}, y_{data}. With minor modifications described in [7, 8, 9, 10], this invalidation process may be carried out in real-time by recursively evaluating certain functions of the plant data u_{data}, y_{data}. The result is a practical "unfalsified adaptive control" theory which, in apparent contrast to previous theories, is focused squarely on what is knowable from the experimental data alone, independent of any assumptions on the plant.

Of course, if the theory is to lead to a non-empty set of unfalsified controllers, then it is essential that it begin with a sufficiently rich class of ideas for candidate controllers **K**. As Linus Pauling put it [22],

"Just have lots of ideas and throw away the bad ones."

Conclusions

Because they fail to clearly distinguish information derived from new data from that which follows from previous knowledge, both classical and modern control problem formulations provide inadequate guidance when experimental results fail to corroborate a priori expectations based on plant models or old data. To address this concern, a non-traditional plant-model-free formulation of the control problem has been proposed in which the rather modest goal of "unfalsification" replaces traditional goals such as stability or optimality. Admittedly, the proposed goal deprives us of the ability to make grand claims about the plant's future behavior (e.g., asymptotic stability), but we console ourselves with the fact that such limits on our powers of clairvoyance would seem to be inherent in any unprejudiced scientific analysis of feedback control problems based solely on data. On the other hand, unfalsified control theory does give a precise mathematical characterization of how the available experimental data can be used to eliminate from consideration broad classes of controllers and, through this process of elimination, identify controllers which insofar as is possible to ascertain from data alone are at least not provably incapable of achieving traditional performance goals. The results complement traditional results based on a priori assumptions about the plant by providing a sharp characterization of how the set of suitable control laws shrinks when experimental data falsifies a priori expectations. And, when implemented in real time, the results lead directly to scientifically sound adaptive control algorithms.

The salient feature of our theory of learning control is that, for a given specification T_{spec} and given class of admissible controllers **K**, plant models and other prejudicial assumptions about the plant or its uncertainty do not enter into the equations. Instead prejudices and models are relegated to the more appropriate role of influencing initial choices for the performance specification T_{spec} and for the class of candidate controllers **K**. As Bertrand Russell said,

"The essential matter is an intimate association of hypothesis and observation."

References

[1] Bertrand Russell. Nature and origin of scientific method. In British Broadcasting Corp., editor, *The Western Tradition*, pages 23–28, London, England, 1949. Vox Mundi.

[2] Aristotle. *Historia Animalium*, volume II. c. 300BC. In *The Works of Aristotle*, Vol. IV. Oxford, England: Clarendon, 1910. Trans. by D'Arcy W. Thompson under editorship of J. A. Smith and W. D. Ross.

[3] R. L. Kosut. Uncertainty model unfalsification: A system identification paradigm compatible with robust control design. In *Proc. IEEE Conf. on Decision and Control*, volume 1, pages 3492–3497, New Orleans, LA, December 13–15, 1995. IEEE Press, New York.

[4] K. R. Popper. *Conjectures and Refutations: The Growth of Scientific Knowledge*. Routledge, London, 1963.

[5] David Brewster. *Memoirs of the life, writings, and discoveries of Sir Isaac Newton*. T. Constable and Co., Edinburgh, Scotland, 1855.

[6] Francis Bacon. *Novum Organum*. 1620. English translation by R. L. Ellis and J. Spedding (eds), in *The Philosophical Works of Francis Bacon*, Routledge, 1905.

[7] M. G. Safonov and T. C. Tsao. The unfalsified control concept and learning. In *Proc. IEEE Conf. on Decision and Control*, pages 2819–2824, Lake Buena Vista, FL, December 14–16, 1994. IEEE Press, New York.

[8] M. G. Safonov and T. C. Tsao. The unfalsified control concept and learning. *IEEE Trans. on Automatic Control*, submitted 10/94, revised 7/95.

[9] T. F. Brozenec and M. G. Safonov. Controller invalidation. In *Proceedings IFAC World Congress*, San Francisco, CA, USA, July 19–23, 1996. Submitted June 1995.

[10] T. F. Brozenec and M. G. Safonov. Controller invalidation. *Automatica*, submitted 8/95.

[11] R. L. Kosut. Adaptive calibration: An approach to uncertainty modeling and on-line robust control design. In *Proc. IEEE Conf. on Decision and Control*, pages 455–461, Athens, Greece, December 10–12, 1986. IEEE Press, New York.

[12] R. Smith and J. Doyle. Model invalidation — a connection between robust control and identification. *IEEE Trans. on Automatic Control*, AC-37(7):942–952, July 1992.

[13] G. Zames. Functional analysis applied to nonlinear feedback systems. *IEEE Trans. on Circuit Theory*, CT-10:392–404, September 1963.

[14] I.W. Sandberg. On the L_2-boundedness of solutions of nonlinear functional equations. *Bell System Technical Journal*, 43(4):1581–1599, July 1964.

[15] I.W. Sandberg. A frequency-domain condition for the stability of feedback systems containing a single time-varying nonlinear element. *Bell System Technical Journal*, 43(4):1601–1608, July 1964.

[16] G. Zames. On the input–output stability of time-varying nonlinear feedback systems — Part I: Conditions derived using concepts of loop gain, conicity, and positivity. *IEEE Trans. on Automatic Control*, AC-15(2):228–238, April 1966.

[17] G. Zames. On the input–output stability of time-varying nonlinear feedback systems — Part II: Conditions involving circles in the frequency plane and sector nonlinearities. *IEEE Trans. on Automatic Control*, AC-15(3):465–467, July 1966.

[18] H. H. Rosenbrock. *State Space and Multivariable Theory*. NY:Wiley, 1970.

[19] M. G. Safonov. *Robustness and Stability Aspects of Stochastic Multivariable Feedback System Design*. PhD thesis, Dept. Elec. Eng., MIT, 1977. Supervised by Michael Athans.

[20] M. G. Safonov. *Stability and Robustness of Multivariable Feedback Systems*. MIT Press, Cambridge, MA, 1980.

[21] J. C. Willems. Paradigms and puzzles in the theory of dynamical systems. *IEEE Trans. on Automatic Control*, AC-36(3):259–294, March 1991.

[22] J. Angier (exective producer). Linus Pauling: Crusading scientist. *Transcript of broadcast of NOVA*, 417, 1977.

[23] R. Virchow. *Der Mensch (On Man)*. Berlin, 1849. In *Disease, Life and Man Selected Essays of Rudolf Virchow* (trans. L. J. Rather), Stanford University Press, Stanford, CA, pp. 67–70, 1958.

An Approach to Switching Control: Theory and Application

Daniel E. Miller*
Department of Electrical and
Computer Engineering
University of Waterloo
Waterloo, Ontario, N2L 3G1
Canada
Phone: (519) 885-1211 ext. 5215
Fax: (519) 746-3077

Michael Chang and Edward J. Davison[†]
Department of Electrical and
Computer Engineering
University of Toronto
Toronto, Ontario, M5S 1A4
Canada
Phone: (416) 978-6342
Fax: (416) 978-0804

Abstract

In this paper we briefly summarize a switching approach to adaptive control which has been investigated by the authors over the last ten years. In this approach, we do not carry out plant estimation, but rather we switch between a pre-defined set of controllers in accordance with an auxiliary switching signal. We consider both the stabilization problem, and the tracking and disturbance regulation problem. The advantage of this approach is that the amount of plant uncertainty which can be tolerated is typically larger than that tolerated by traditional estimator-based adaptive controllers. In this paper we give a brief overview of our work in this area as well as a brief sketch of the historical development.

1   Introduction

Uncertainty in the plant model is an unavoidable problem in systems control, and adaptive control is one way to deal with it. In this approach, the controller is nonlinear and possibly time-varying; it typically consists of a linear time-invariant compensator together with a tuning mechanism which adjusts the compensator gains to match the plant model.

Our work has focussed on a switching control approach to adaptive control. Here, each of the controllers typically consists of a linear time-invariant compensator whose gains switch in accordance with an auxiliary switching signal; it is a non-parametric approach – no plant parameter estimation is carried out. This approach has the advantage that various control objectives can be

---

*This work was supported by the Natural Sciences and Engineering Research Council of Canada via a research grant.

[†]This work was supported by the Natural Sciences and Engineering Research Council of Canada under Grant A4396.

achieved in the face of *significant* plant uncertainty; in particular, it is capable of dealing with unstructured uncertainty in the plant model, and an upper bound on the plant order is almost never required. The drawback is that the transients may often be large; an exception is the controller of [26], [32].

Our interpretation of the history of this approach goes as follows. In 1982 it was conjectured in [36] that there does not exist a "smooth" LTI controller which can stabilize a first-order system without knowledge of the high-frequency gain. This was disproved in [37], where the author presented a controller which stabilizes a system without this knowledge, thereby introducing the notion of a "Nussbaum gain". Following this, results on higher order LTI systems were developed. In [2], it was shown that to stabilize a LTI multivariable plant, it is enough that the plant be minimum phase with invertible high-frequency gain. In a surprising result by Mårtensson [38], it was shown that to stabilize a multivariable LTI system it is sufficient that the order of a LTI stabilizing compensator be known; this was shown to be almost necessary information if a "smooth" nonlinear time-invariant controller is to be used [1].

At about the same time, our work in the area was just beginning, and with the following impetus. In a paper on "tuning regulators" [8], the objective was to design a controller to provide error regulation for a small class of signals when the plant is stable but the plant parameters are unknown. For the case of step inputs, it was shown that using only knowledge of the dc gain, which can be obtained experimentally, it is possible to determine if the objective can be achieved, and if so, the form of the controller; however the controller is parametrized by a scalar which must be tuned manually on-line. The goal of the MASc thesis of the first author was to develop a method to tune this parameter automatically on-line. This goal was achieved, and a switching controller was developed which consists of a multivariable integral compensator together with a switching mechanism to tune the scalar parameter; this work, as well as extensions of it, were reported in [24], [23], [30], [34].

In 1986, we became aware of the work of [2] and [38], which spurred us to use our approach to solve other problems. In the remainder of the paper we briefly discuss our results, and present an experimental application study of one of the controllers. For a survey of the area of non-identifier based adaptive control, see [14].

2 Stability

Here our plant has the form

$$\dot{x} = Ax + Bu \qquad (1)$$
$$y = Cx, \qquad (2)$$

with $x(t) \in \mathbf{R}^n$ the state, $u(t) \in \mathbf{R}^m$ the control input, and $y(t) \in \mathbf{R}^r$ the output.

The control objective here is that of *asymptotic stabilization*: we want $x(t) \to 0$ as $t \to \infty$ using a bounded control input $u(t)$.

Fix m and r. Let Σ_l denote the set of all controllable and observable systems (of arbitrary order n) which have an lth order lti stabilizing compensator. In [38] it is shown that for each $l \in \mathbf{Z}^+$ there exists an $(l+1)^{th}$ order controller which stabilizes every plant in Σ_l. Furthermore, it is shown in [1] that a "smooth" finite dimensional nonlinear time-invariant controller of order l can asymptotically stabilize at most a set slightly larger than Σ_l. In [38] an infinite-dimensional controller is given which stabilizes every system in $\cup_{i=0}^{\infty}\Sigma_l$.

Our contributions to this problem are as follows:

- **Asymptotic stability**: [25]

  We present a switching controller which has the same capabilities as that of [38], but which can be modified to be noise tolerant at a small expense. A modified version of this is presented in [16].

- **Asymptotic stability with minimal knowledge**: [17], [19]

  By switching between time-varying compensators rather than lti ones, a smooth adaptive switching controller of dimension $r+1$ is constructed which asymptotically stabilizes every system in $\cup_{i=0}^{\infty}\Sigma_l$.

The drawback of the above results are that transients can be extremely large, even when the initial condition is small. In [13] this problem is considered, and a switching controller is given which provides "Lyapunov-like" exponential stability for all plants in a compact set. Our results on this problem are as follows:

- **Exponential stability**: [27], [28]

  For each $l \in \mathbf{Z}^+$, a switching controller is given which provides "Lyapunov-like" exponential stability for all plants in Σ_l. The advantages over that of [13] are: no pre-computation of a finite sub-covering of an open covering is required, and the set of uncertainty is larger. A modified version which is easier to implement is given in [16].

3   Tracking/Disturbance Regulation

Here our system has the form

$$\dot{x} \;=\; Ax + Bu + Ew \tag{3}$$
$$y \;=\; Cx + Fw \tag{4}$$
$$e \;=\; y_{ref} - y(t), \tag{5}$$

with x, y, and u defined as above, $w(t) \in \mathbf{R}^q$ an unmeasurable disturbance signal, y_{ref} the measurable reference signal, and e the tracking error.

Here the goal is to provide error regulation for a class of reference signals. We partition this section into two parts: when the set of reference signals is a finite dimensional vector space, and when it is given by the output of a forced reference model.

3.1 The Servomechanism Problem

With $\alpha(s) = \sum_{i=1}^{p} \alpha_i s^i$ a polynomial with simple zeros, each of which lies on the imaginary axis, let

$$Y_{ref} := \{y_{ref} \in C_\infty : \sum_{i=1}^{p} \alpha_i y_{ref}^{(i)}(t) = 0, \ t \geq 0\},$$

$$W := \{w \in C_\infty : \sum_{i=1}^{p} \alpha_i w^{(i)}(t) = 0, \ t \geq 0\}.$$

Our control objective is to have $e(t) \to 0$ as $t \to \infty$ for every $y_{ref} \in Y_{ref}$ and $w \in W$, while ensuring that x and u are bounded. Our results for this problem are:

- **Step inputs (integral control):** [24], [30]

  Here we consider the case of $\alpha(s) = s$, i.e. we are concerned with step inputs. With m and r fixed, we design a switching controller which achieves the above objective for every plant satisfying the following constraints: it is stable, it has no transmission zeros at zero, and $m \geq r$.

- **Step inputs (PI/PID control):** [3], [4], [5]

  The results of [24], [30] are extended to include PI self-tuning controllers in [3], [4], and PID self-tuning controllers in [5]. In the latter case, switching is now partially based upon a direct measurement of the error signal $e(t)$.

- **Step inputs with a control input constraint:** [23], [34]

  Here we extend the results of [24], [30] to the case when there is a bound on the control input.

- **General Case:** [22], [33], [7]

  In [22], [33], with $m \geq r$ fixed, for each $l \in \mathbf{Z}^+$ we design a controller which achieves the stated control objective for every system in Σ_l which has no transmission zeros at the zeros of $\alpha(s)$. By restricting the class of uncertainty so that there are only a finite number of candidate controllers, the results of [7] attempt to improve the closed loop transient response.

- **Family of Plants Case:** [31]

  Here we examine the special case in which the plant model lies in a finite set. For each possible plant model, we design a satisfactory controller to achieve the control objective. We then switch between them, trying each one at most once; under a modest technical requirement, we prove that the control objective is achieved for every admissible plant model. Two distinct approaches to solving the problem, one of which is observer based, are presented.

3.2 The Model Reference Adaptive Control Problem

In this setup, we have $w = 0$, and a reference model

$$\dot{x}_m(t) = A_m x_m(t) + B_m u_m(t) \tag{6}$$

$$y_{ref}(t) = C_m x_m(t) \tag{7}$$

which embodies the desired behaviour to the exogenous input u_m; we assume that A_m is stable. Here the goal is the following: for every bounded reference model input u_m, we would like $e(t) \to 0$ as $t \to \infty$ while maintaining u and x bounded. Our results for this problem are:

- **Necessary conditions in MRAC**: [29], [35], [20], [15]

  In the siso case, we show that the control objective can be achieved by a causal nonlinear time-varying controller for an uncertain system *only if* there is an upper bound on the plant relative degree, and the set of possible non-minimum phase zeros is finite.

- **"Almost Perfect" Control**: [26], [32]

  The original goal of exact asymptotic tracking is demanding, so we modify our control objective. Here we consider the siso case and design a controller which achieves "almost perfect" control. For each $r \in \mathbf{N}$, we present a controller, parametrized by $\varepsilon > 0$, $T > 0$, and $\delta > 0$, which, for every bounded u_m, yields u and x bounded,

  $$|e(t)| \leq \max\{\varepsilon, |e(0)| + \delta\}, \quad t \in [0, T],$$

  and

  $$|e(t)| < \varepsilon, \quad t > T,$$

  for every minimum phase plant of relative degree less than or equal to r.

- **MRAC for non-minimum phase systems**: [18], [21]

  Here we consider the discrete-time multivariable case. Given a set of possible plant models $\mathcal{P}$ together with a feasible performance index α, we present a controller which provides, for every plant in $\mathcal{P}$, performance of the form

  $$\limsup_{t \to \infty} |e(t)| \leq \alpha \limsup_{t \to \infty} |u_m(t)|.$$

4 Experimental Apparatus Studies

In order to evaluate the behaviour of the previous "switching controllers" when applied to "real world" industrial systems, an experimental apparatus called MARTS (Multivariable Apparatus for Real Time Control Studies), located at the Systems Control Group, Department of Electrical and Computer Engineering, University of Toronto, was used. This experimental system consists of a collection of industrial actuators and sensors, which gives a highly interacting multivariable hydraulic system. A brief description of this system is given as follows.

4.1 Experimental Apparatus

The system consists of an interconnection of industrial commercial actuators and sensors used to monitor and control a nonlinear hydraulic system as described in figure 1. In this system, the by-pass, drainage, and interconnecting valves are all adjustable manually to enable the selection of desired equilibrium column heights and to control the degree of interaction existing between both columns. Actuator valves for both columns also enable one to individually apply positive constant system disturbances w_1 and w_2. A Texas Instruments (TM 990/101/MA-1) real-time digital computer using the PDOS operating system with a 12 bit A/D, D/A board and a variable sampling time ≥ 10 msec. (not shown) is also used to control the system. For a detailed description of the apparatus, the reader is referred to [9].

The control objective using MARTS is to regulate the level of both column heights (if possible) for all initial conditions and disturbances applied; we note, however, that the plant is marginally open loop stable, and as the interconnecting valve angle θ increases from $0°$ to $45°$, the plant becomes increasingly difficult to control.

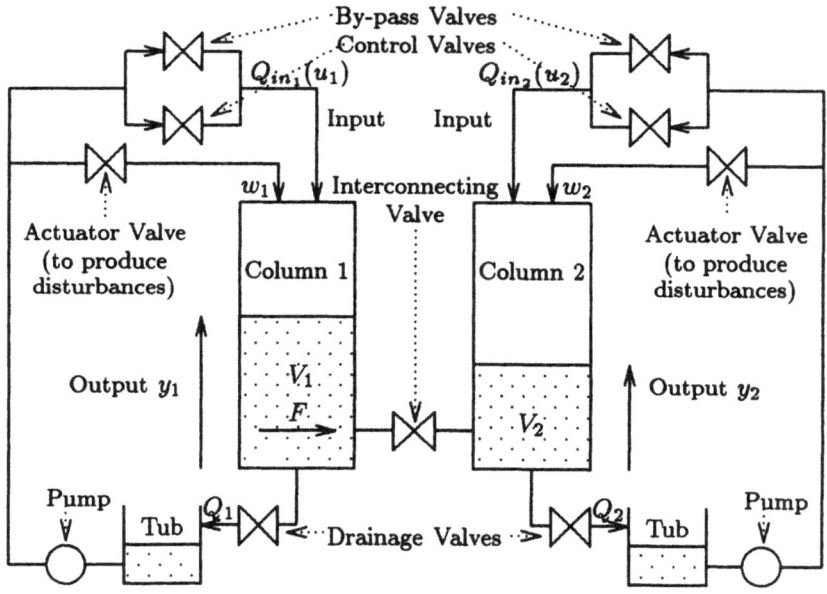

Figure 1: Schematic diagram of the MARTS setup (not to scale).

4.2 Conventional Controller Design Results

In this section, experimental results obtained using the high performance controller design method given in [12] and [11], which require that a mathematical model of the system be available, are presented. In [6], a LTI model of MARTS for the case when $\theta = 30°$ is developed, and using the performance index [12]

$$J_\epsilon := \int_0^\infty (e^T e + \epsilon \dot{u}^T \dot{u}) dt$$

with $\epsilon = 1$, the optimal controller is

$$u = -K_0^1 e + K_1^1 \int_0^t e(\tau) d\tau \tag{8}$$

with gains of

$$[\ K_0^1\ |\ K_1^1\ ] = \left[\begin{array}{cc|cc} -1.3776 & -0.0131 & 1.0000 & 0.0000 \\ -0.0131 & -1.3753 & 0.0000 & 1.0000 \end{array} \right]. \tag{9}$$

4.2.1   Experimental Results Obtained

When controller (8)-(9) is digitally implemented on the MARTS system using a sampling period of $T = 0.4$ seconds, the experimental results of figure 2 are obtained for the case in which $\theta = 30°$ and

$$(y_{ref}^1(t), y_{ref}^2(t)) := \left\{ \begin{array}{ll} (3000, 2500), & 0 \leq t < 500 \\ (3500, 2000), & 500 \leq t < 1000 \\ (3000, 2500), & 1000 \leq t < 1500 \\ (3500, 2000), & 1500 \leq t < 2000 \text{ seconds} \end{array} \right.$$

is the reference input signal. In this instance, it is seen that the controller provides excellent performance. For comparison purposes, the case of $\epsilon = 1000$ is also implemented, and presented in figure 2.

When an "unexpected" event occurs using the conventional controller (8)-(9), however, catastrophic and undesirable results may, and most probably will, occur, as demonstrated in figure 3. Here, the following gross change in the MARTS configuration was made at $t = 1200$ seconds: with the controller (8)-(9) implemented on the MARTS system (which is designed for the case when $\theta = 30°$), and with $T = 0.75$ seconds, the plant's configuration was suddenly changed at $t = 1200$ seconds by reversing the output leads with the reference input given by

$$(y_{ref}^1(t), y_{ref}^2(t)) := (2500, 2250), \quad t \geq 0$$

applied. In this case, the response shown in figure 3 was experimentally obtained, which shows that the controller (8)-(9) *fails* to bring about tracking and/or regulation for such a severe configuration change. Unfortunately, such a failure of the controller (8)-(9) is not unexpected, since a *drastic* change in the plant has occurred at $t = 1200$ seconds.

It will now be shown that one class of the self-tuning servomechanism controllers previously described do not have this limitation, i.e. they can readily adjust to severe plant configuration changes.

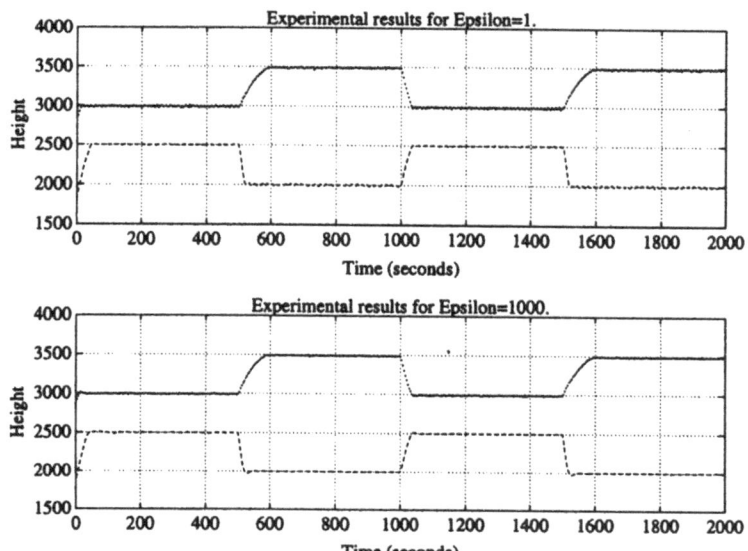

Figure 2: Experimental proportional-integral results of y_1 (dotted) and y_2 (dashed) using $T = 0.4$ seconds and $\theta = 30°$ with conventional controller (8)-(9) applied.

4.3 Switching Controller Output Results

Now consider the following self-tuning PID controller [5]. With $c > 0$, $\rho > 0$, $\lambda > 0$, and $f_1(k)$, $f_2(k)$ and $g(k)$ controller parameters, set

$$
\begin{aligned}
\dot{\eta}(t) &= \varepsilon(t)e(t), \quad \eta(t_k^+) \equiv 0, \quad t \in (t_k, t_{k+1}], \\
\dot{z}(t) &= -cz(t) - y(t), \quad z(t_k^+) \equiv 0, \quad t \in (t_k, t_{k+1}], \\
d(t) &= -c^2 z(t) - cy(t), \\
u(t) &= K(t)\left(\eta(t) + \rho\varepsilon(t)e(t) + \varepsilon(t)d(t)\right), \\
\dot{e}_f(t) &= -\lambda e_f(t) + \lambda e(t), \quad e_f(t_k^+) \equiv 0, \quad t \in (t_k, t_{k+1}].
\end{aligned}
\tag{10}
$$

Here, we set

$$
\varepsilon(t) = g(k), \quad t \in (t_k, t_{k+1}],
$$

$t_1 := 0$, and for each $k \geq 2$ such that $t_{k-1} \neq \infty$, the switching time t_k is defined by

$$
t_k := \begin{cases}
\begin{aligned}
&\min(t) \text{ such that} & \text{if this minimum exists} \\
&\quad \text{i)} \quad t > t_{k-1}, \text{ and} \\
&\quad \text{ii)} \quad \|[\eta(t)^T \; z(t)^T]^T\| = f_1(k-1) \text{ and/or} \\
&\quad \quad \|e_f(t)\| = f_2(k-1)
\end{aligned} \\
\infty & \text{otherwise.}
\end{cases}
$$

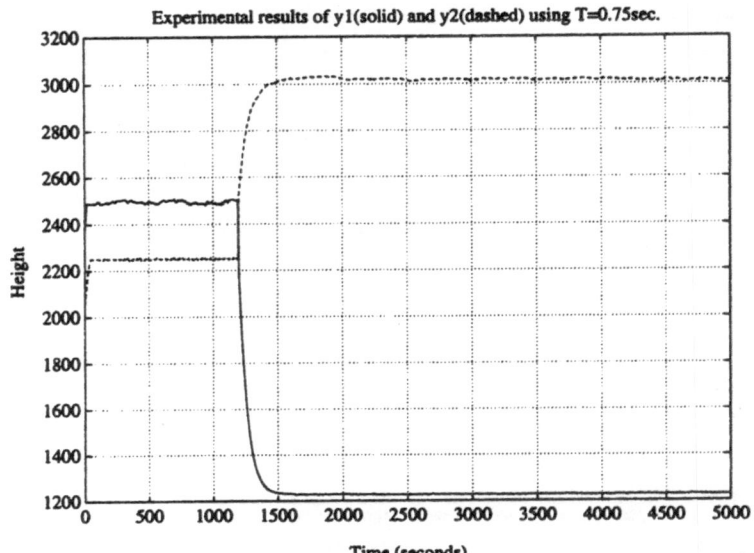

Figure 3: Experimental proportional-integral results for y_1 (solid) and y_2 (dashed) with $T = 0.75$ seconds, $\theta = 20°$ and the outputs reversed at $t = 1200$ seconds, showing failure of conventional controller (8)-(9).

Finally, with

$$W_1 = \begin{bmatrix} 1 & 0 \\ 0 & 1 \end{bmatrix}, \quad W_2 = \frac{1}{2}\begin{bmatrix} -1 & -\sqrt{3} \\ \sqrt{3} & -1 \end{bmatrix}, \quad W_3 = \frac{1}{2}\begin{bmatrix} -1 & \sqrt{3} \\ -\sqrt{3} & -1 \end{bmatrix},$$

$$W_4 = \begin{bmatrix} 1 & 0 \\ 0 & -1 \end{bmatrix}, \quad W_5 = \frac{1}{2}\begin{bmatrix} -1 & -\sqrt{3} \\ -\sqrt{3} & 1 \end{bmatrix}, \quad W_6 = \frac{1}{2}\begin{bmatrix} -1 & \sqrt{3} \\ \sqrt{3} & 1 \end{bmatrix},$$

we set

$$K(t) := W_{((k-1) \bmod 6)+1}, \quad t \in (t_k, t_{k+1}].$$

Using the parameters in table 1, together with

$$
\begin{aligned}
\theta &:= 20°, \\
\lambda &:= 10, \\
T &:= 0.75 \text{ seconds}, \\
\rho &:= 20, \\
c &:= 10,
\end{aligned}
$$

and with two choices of $y_{ref}(t)$ given as

$$(y_{ref}^1(t), y_{ref}^2(t)) := \begin{cases} (3000, 2500), & 0 \le t < 600 \\ (2500, 2000), & 600 \le t < 1200 \\ (3000, 2500), & t \ge 1200 \text{ seconds}, \end{cases} \quad (11)$$

$$(y_{ref}^1(t), y_{ref}^2(t)) := (2500, 2250), \quad t \ge 0, \quad (12)$$

the output results shown in figures 4 and 5[1] are obtained.

Fig.	$f_1(k)$	$f_2(k)$	$g(k)$	$k(\infty)$	$K(\infty)$	$\varepsilon(\infty)$	y_{ref}
4	$5\alpha^\alpha$	$210\alpha + 35\beta$	$10/5^\alpha$	19	W_1	0.016	(11)
5	$5\alpha^\alpha$	$210\alpha + 35\beta$	$10/5^\alpha$	24	W_6	0.016	(12)

Table 1: Summary of the parameters used for figures 4 and 5, where $\alpha(k) := \operatorname{int}\left(\dfrac{k+5}{6}\right)$ and $\beta(k) := ((k-1) \bmod 6) + 1$.

It is observed that in both instances, reference tracking and disturbance rejection occurs, and that the proposed self-tuning PID controller is quite successful in achieving its mandate, despite the severe configuration change which occurs in figure 5 at $t = 1200$ seconds, and despite the relatively little *a priori* plant information required. References [10] and [6] provide additional experimental results regarding this study.

5 Conclusions

This paper presents a brief overview of the results that we have obtained on a non-identifier based approach to adaptive control. The main advantage of this method is that the set of plant uncertainty which is tolerated is much larger than that required for the more traditional identifier based approach; in fact, the results of [30] have formed the basis of a controller for a third generation large flexible space structure. The main disadvantage is that in some of our results the transient behaviour can be very poor, which is not surprising given the large amount of plant uncertainty allowed.

A potential application of this theory is that of providing a backup to an existing LTI controller, thereby yielding an extra degree of robustness.

References

[1] C. I. Byrnes, U. Helmke, and A. S. Morse. Necessary Conditions in Adaptive Control. In *Modeling, Identification, and Robust Control*, pages 2–14. North-Holland, New York, 1986.

[2] C. I. Byrnes and J. C. Willems. Adaptive Stabilization of Multivariable Linear Systems. In *Proceedings of the 23'rd IEEE Conference on Decision and Control*, pages 1574–1577, 1984.

[3] M. Chang and E. J. Davison. Control of Unknown Systems using Switching Controllers: an Experimental Study. In *Proceedings of the 1994 American Control Conference*, pages 2984–2989, 1994.

---

[1]In figure 5, the outputs were reversed at $t = 1200$ seconds.

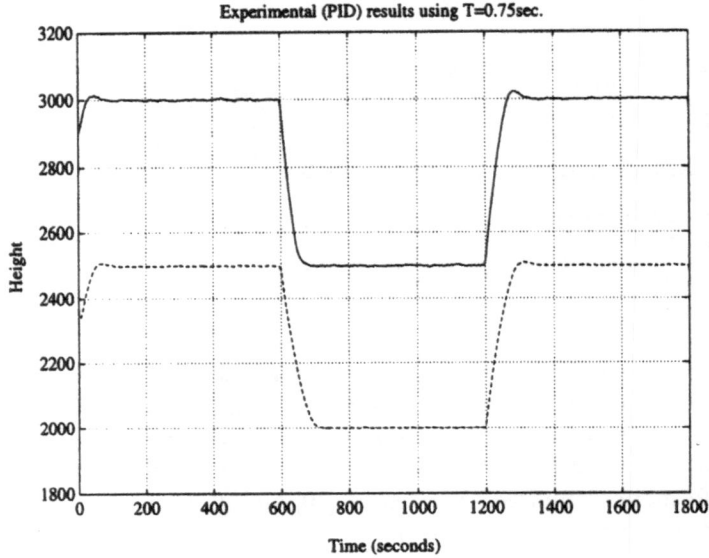

Figure 4: Experimental results of y_1 (solid) and y_2 (dashed) with the self-tuning PID controller applied to the MARTS system.

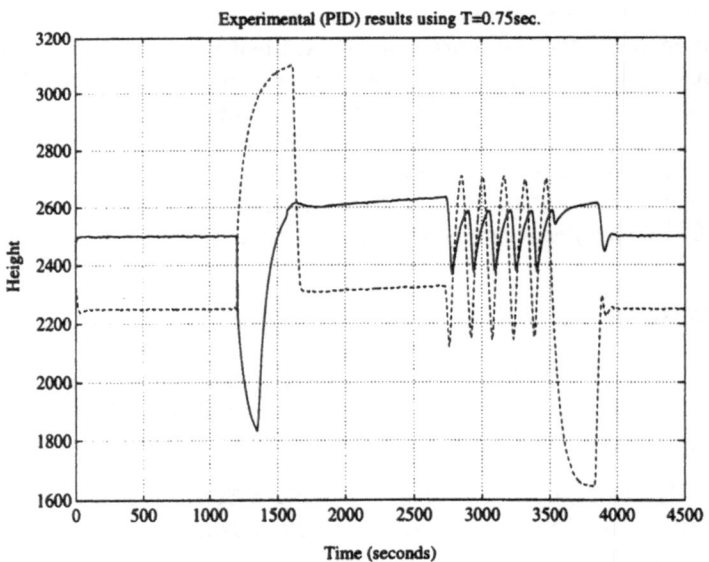

Figure 5: Experimental results of y_1 (solid) and y_2 (dashed) with the self-tuning PID controller applied to the MARTS system, and with both outputs reversed at $t = 1200$ seconds, showing successful control of the perturbed system.

[4] M. Chang and E. J. Davison. Control of Unknown Systems using Switching Controllers: The Self-Tuning Robust Servomechanism Problem. In *Proceedings of the 33'rd IEEE Conference on Decision and Control*, pages 2833–2838, 1994.

[5] M. Chang and E. J. Davison. Control of Unknown MIMO Systems using Self-Tuning PID Controllers. In *Proceedings of the 1995 American Control Conference*, pages 564–568, 1995.

[6] M. Chang and E. J. Davison. New Directions in Industrial Control: Intelligent Control – An Experimental Study Applied to a Multivariable Hydraulic System. In I. Lasiecka and B. Morton, editors, *Control Problems in Industry: Proceedings from the SIAM Symposium on Control Problems, San Diego, California, July, 1994*, volume 21 of *Progress in Systems and Control Theory*, pages 65–95. Birkhäuser Press, Boston, 1995.

[7] M. Chang and E. J. Davison. Switching Control of a Family of Plants. In *Proceedings of the 1995 American Control Conference*, pages 1015–1020, 1995.

[8] E. J. Davison. Multivariable Tuning Regulators: The Feedforward and Robust Control of a General Servomechanism Problem. *IEEE Transactions on Automatic Control*, 21(1):35–47, February 1976.

[9] E. J. Davison. Description of Multivariable Apparatus for Real Time Control Studies (MARTS). Systems Control Report 8514, University of Toronto, Department of Electrical Engineering, November 1985.

[10] E. J. Davison and M. Chang. Intelligent Control: Some Preliminary Results. In B. A. Francis and A. R. Tannenbaum, editors, *Feedback Control, Nonlinear Systems, and Complexity*, volume 202 of *Lecture Notes in Control and Information Sciences*, pages 67–87. Springer-Verlag, London, 1995.

[11] E. J. Davison and I. J. Ferguson. The Design of Controllers for the Multivariable Robust Servomechanism Problem Using Parameter Optimization Methods. *IEEE Transactions on Automatic Control*, 26(1):93–110, February 1981.

[12] E. J. Davison and B. M. Scherzinger. Perfect Control of the Robust Servomechanism Problem. *IEEE Transactions on Automatic Control*, 32(8):689–702, August 1987.

[13] M. Fu and B. R. Barmish. Adaptive Stabilization of Linear Systems Via Switching Control. *IEEE Transactions on Automatic Control*, 31(12):1097–1103, December 1986.

[14] A. Ilchmann. Non-Identifier-Based Adaptive Control of Dynamical Systems: A Survey. *IMA Journal of Mathematical Control & Information*, 8:321–366, 1991.

[15] D. E. Miller. On Necessary Assumptions in Discrete-Time Model Reference Adaptive Control. *International Journal of Adaptive Control and Signal Processing*, to appear.

[16] D. E. Miller. *Adaptive control of Uncertain Systems*. PhD thesis, University of Toronto, 1989.

[17] D. E. Miller. A Second Order Nonlinear Time-Varying Controller Which Provides Asymptotic Stabilization. In *Proceedings of the 1992 American Control Conference*, pages 2332–2336, 1992.

[18] D. E. Miller. Model Reference Adaptive Control for Nonminimum Phase Systems. In *Proceedings of the 1993 European Control Conference*, pages 190–194, 1993.

[19] D. E. Miller. Adaptive Stabilization Using a Nonlinear Time-Varying Controller. *IEEE Transactions on Automatic Control*, 39(7):1347–1359, July 1994.

[20] D. E. Miller. On Necessary Assumptions in Discrete Time Model Reference Adaptive Control. In *Proceedings of the First Asian Control Conference*, volume 3, pages 65–68, 1994.

[21] D. E. Miller. Model Reference Adaptive Control for Nonminimum Phase Systems. *Systems & Control Letters*, 26:167–176, 1995.

[22] D. E. Miller and E. J. Davison. A New Self-Tuning Controller to Solve the Servomechanism Problem. In *Proceedings of the 26'th IEEE Conference on Decision and Control*, pages 843–849, 1987.

[23] D. E. Miller and E. J. Davison. The Self-Tuning Robust Servomechanism Problem with Control Input Constraints. In *Proceedings of the 26'th IEEE Conference on Decision and Control*, pages 837–842, 1987.

[24] D. E. Miller and E. J. Davison. The Self-Tuning Robust Servomechanism Problem. In *Proceedings of the 26th IEEE Conference on Decision and Control*, pages 384–390, 1987.

[25] D. E. Miller and E. J. Davison. An Adaptive Controller which can Stabilize any Stabilizable and Detectable LTI System. In C. I. Byrnes, C. F. Martin, and R. E. Saeks, editors, *Analysis and Control of Nonlinear Systems*, pages 51–58. Elsevier Science Publishers B. V., Amsterdam, 1988.

[26] D. E. Miller and E. J. Davison. An Adaptive Controller to Achieve an Arbitrarily Good Transient and Steady-State Response. In *Proceedings of the 26th Allerton Conference on Communications, Control, and Computing*, pages 907–916, 1988.

[27] D. E. Miller and E. J. Davison. An Adaptive Controller to Provide Lyapunov Stability. In *Proceedings of the 27th IEEE Conference on Decision and Control*, pages 1934–1939, 1988. This paper subsequently appeared in *Advances in Adaptive Control*, IEEE Press, pp. 21-26, 1991.

[28] D. E. Miller and E. J. Davison. An Adaptive Controller Which Provides Lyapunov Stability. *IEEE Transactions on Automatic Control*, 34(6):599–609, June 1989.

[29] D. E. Miller and E. J. Davison. On Necessary Assumptions in Continuous Time Model Reference Adaptive Control. In *Proceedings of the 28'th IEEE Conference on Decision and Control*, pages 1573–1578, 1989. This paper subsequently appeared in *Advances in Adaptive Control*, IEEE Press, pp. 14-20, 1991.

[30] D. E. Miller and E. J. Davison. The Self-Tuning Robust Servomechanism Problem. *IEEE Transactions on Automatic Control*, 34(5):511–523, May 1989.

[31] D. E. Miller and E. J. Davison. Adaptive Control of a Family of Plants. In D. Hinrichsen and B. Mårtensson, editors, *Control of Uncertain Systems: Proceedings of an International Workshop, Bremen, West Germany, June 1989*, volume 6 of *Progress in Systems and Control Theory*, pages 197–219. Birkhäuser Press, Boston, 1990.

[32] D. E. Miller and E. J. Davison. An Adaptive Controller Which Provides an Arbitrarily Good Transient and Steady-State Response. *IEEE Transactions on Automatic Control*, 36(1):68–81, January 1991.

[33] D. E. Miller and E. J. Davison. An Adaptive Tracking Problem. *International Journal of Adaptive Control and Signal Processing*, 6(1):45–63, January 1992.

[34] D. E. Miller and E. J. Davison. An Adaptive Tracking Problem with a Control Input Constraint. *Automatica*, 29(4):877–887, July 1993.

[35] D. E. Miller and E. J. Davison. On Asymptotic Model Matching. *Mathematics of Control, Signals, and Systems*, 6(4):322–340, 1993.

[36] A. S. Morse. Recent Problems in Parameter Adaptive Control. In *Proc. CNRS Colloquium on Development and Utilization of Mathematical Models in Automatic Control*, pages 733–740, Belle-Isle, France, 1982.

[37] R. D. Nussbaum. Some Remarks on a Conjecture in Parameter Adaptive Control. *Systems & Control Letters*, 3(5):243–246, November 1983.

[38] B. Mårtensson. The Order of any Stabilizing Regulator is Sufficient a priori Information for Adaptive Stabilization. *Systems & Control Letters*, 6(2):87–91, July 1985.

On the Performance and Complexity of a Class of Hybrid Controller Switching Policies*

Sanjeev R. Kulkarni and Peter J. Ramadge
Department of Electrical Engineering
Princeton University, Princeton NJ 08544
email: {kulkarni,ramadge}@princeton.edu

Abstract

Based on observations of the past inputs and outputs of an unknown system Σ, a countable set of predictors $O_p, p \in \mathcal{P}$, is used to predict the system output sequence. Using performance measures derived from the resultant prediction errors a decision rule is to be designed to select a $p \in \mathcal{P}$ at each time k. We study the structure and memory requirements of decision rules that converge to some $q \in \mathcal{P}$ such that the qth prediction error sequence has desirable properties, e.g., is suitably bounded or converges to zero. In a very general setting we give a positive result that there exist stationary decision rules with countable memory that converge (in finite time) to a 'good' predictor. These decision rules are robust in a sense made precise in the paper. In addition, we demonstrate that there does not exist a decision rule with finite memory that has this property. This type of problem arises in a variety of contexts but one of particular interest is the following. Based on the decision rule's selection at time k a controller for the system Σ is chosen from a family $\Gamma_p, p \in \mathcal{P}$, of predesigned control systems. We show that for certain mimo linear systems the resultant closed loop controlled system is stable and can asymptotically track an exogenous reference input.

1 Introduction

Predictor based controller switching policies have recently been examined in a variety of contexts, e.g., [9], [7], [10], [11], [1] and [13]. Of particular relevance here is the work of [7]. In [7] a family of concurrently operating LTI predictors are used to predict the output of the plant and the resulting prediction errors are used to form a performance measure for each predictor. Then, at a sequence of sample times, one compares the performance of the predictors and selects the controller corresponding to the best predictor at that time. The sequence

*This work was supported in part by the National Science Foundation under grants IRI-9457645 and ECS-9216450, by EPRI under grant RP8030-18, and by the Army Research Office under grant DAAL03-92-G-0320.

of selected predictors is not required to converge and in general will not do so. Nevertheless, the system variables remain bounded and the output of the siso system asymptotically tracks a constant set-point. In [9] and [13] switching is used to select a controller structure matched to the similarity invariants of the plant, and in [1],[10], [11] it is used to improve the transient performance of stable adaptive control schemes.

Alternative controller search methods, e.g., [12], [5], [3], [2], [6], have used a 'pre-routed' search through the controllers. The 'pre-routed' search procedures fall into the following basic pattern. A scheduling or routing sequence $\pi: \mathbb{N} \to \mathcal{P}$ is designed so that it has the 're-visitation property' that every $p \in \mathcal{P}$ appears infinitely often in π. The sequence π is the predetermined order in which the candidate controllers will be tried. Then a switching logic is designed that uses the plant inputs and outputs to determine if and when the next element of π should be tried. So the actual time spent 'dwelling' on controller $\pi(i)$ will depend on the observed plant variables but the order in which the candidate controllers are examined does not. For obvious reasons this form of search is usually impractical unless $\mathcal{P}$ is a small finite set. However, these methods have been useful for showing the existence of convergent adaptive control algorithms under weak a priori plant assumptions. A good review of this work is given in [6].

Here we discuss some results that combine attributes of both predictor performance based search with the finite time convergence property of a pre-scheduled search. Our results are complementary to those of [7] where convergence is not required. The advantage of a convergent policy is that it is possible to examine the limiting controlled system. However, convergence does not imply that the decision rule terminates; if the plant is changed the rule can again begin switching.

Our approach to this problem is to first study the basic capabilities and limitations of decision rules in a general framework. We obtain results on the existence and complexity of convergent decision rules for several generic problems. We then apply the results on the decision rules to the predictor selection problem as a special case. This in turn leads to a natural switching policy for the controller switching problem for which we can obtain certain performance guarantees. In particular, for mimo linear time-invariant systems we show that there exist convergent controller switching rules based on the predictor performance measures such that if the individual controllers have been designed to adequately track an admissible exogenous signal, then within the limits imposed by disturbances so will the supervised system. A complete version of the present paper with proofs of the results stated here can be found in [4].

2   Predictor Selection

We are given a fixed but unknown discrete-time system Σ described by:

$$
\begin{aligned}
x(k+1) &= A(k)(x(k), u(k)), \quad x(k_0) = x_0 \\
y(k) &= C(k)(x(k))
\end{aligned}
\tag{1}
$$

where $x(k) \in \mathbb{R}^n$, $u(k) \in \mathbb{R}^m$, $y(k) \in \mathbb{R}^l$ and $A(k): \mathbb{R}^{n \times m} \to \mathbb{R}^n$, $C(k): \mathbb{R}^n \to \mathbb{R}^l$. The initial condition x_0 is an unknown point in $\mathbb{R}^n$. We make no assumptions on the input sequence to Σ. For simplicity we often assume that the initial time k_0 is zero.

A predictor O_p for Σ is a sequence of functions $O_k^p: \mathbb{R}^{kl} \times \mathbb{R}^{km} \to \mathbb{R}^l$ with the prediction at time k given by $\hat{y}_p(k) = O_k^p(y(k-1), \ldots, y(0), u(k-1), \ldots, u(0))$. The associated prediction error is $e_p(k) \triangleq \hat{y}_p(k) - y(k)$, $k \geq 0$. In practice it may be convenient to let $\hat{y}_p(k)$ be a function of a state variable $w_p(k)$ taking values in a finite dimensional vector space, i.e., a system O_p with

$$
\begin{aligned}
w_p(k+1) &= M_p(k)(w_p(k), y(k), u(k)), \quad w(0) = w_0 \in \mathbb{R}^{n}(p) \\
\hat{y}_p(k) &= C_p(k)(w_p(k))
\end{aligned}
\tag{2}
$$

where $w_p(k) \in \mathbb{R}^{n(p)}$, $u(k) \in \mathbb{R}^m$, $y(k) \in \mathbb{R}^l$ and $M_p(k): \mathbb{R}^{n(p) \times m \times l} \to \mathbb{R}^{n(p)}$, $C_p(k): \mathbb{R}^{n(p)} \to \mathbb{R}^l$. A system of the form (2) is easy to design, for example, if the systems of interest are linear and time-invariant.

It is not our task here to deal with the existence or design of the predictors O_p. Rather, we focus on decision rules that at each time utilize the output errors of the predictors to select a model from $\mathcal{P}$.

Specifically, let $J^k: \mathbb{R}^{l \times (k+1)} \to \mathbb{R}^+$ be a sequence of real-valued performance functions and for each $p \in \mathcal{P}$ let $J_p(k) \triangleq J^k(e_p(0), \ldots, e_p(k))$. The real number $J_p(k)$ measures the performance of predictor p at time k, with a smaller value indicating better performance. At each time k we observe the performance indices of all the predictors in $\mathcal{P} = \{p_1, p_2, \ldots\}$, i.e., we observe $\overline{J}(k) \triangleq (J_{p_1}(k), J_{p_2}(k), \ldots)$ and then select a model from $\mathcal{P}$. In general, the performance measures $J_p(k)$ will not converge and may not be consistently ordered. Nevertheless the sequence of selections is required to converge to a fixed element of $\mathcal{P}$ (in finite time since $\mathcal{P}$ is discrete) with the corresponding performance measure being acceptable.

For example, assume that for some $p^* \in \mathcal{P}$, $\|e_{p^*}(k)\| \leq L g_1(k) + g_2(k)$ for known nonnegative sequences g_1, g_2 and some $L \in \mathbb{R}$. The term $L g_1(k)$ models the effect of unknown initial conditions and $g_2(k)$ models asymptotic behavior. We assume throughout that $g_1 \in l^1$.

To form a predictor performance measure let $\{a(k), k \geq 0\}$ be a nonnegative sequence with not all terms zero and set $J^k(e(0), \ldots, e(k)) = \sum_{j=0}^{k} a(k-j)\|e(j)\|$. If $h_i(k) = \sum_{j=0}^{k} a(k-j)g_i(j)$, $i = 1, 2$, then

$$
J_{p^*}(k) = \sum_{j=0}^{k} a(k-j)\|e_{p^*}(j)\| \leq L h_1(k) + h_2(k)
$$

There are several possibilities depending on the assumptions we place on g_2 and a. First, if $g_2 \equiv 0$ and $a \in l^\infty$, then $h_2 \equiv 0$ and h_1 is bounded. Thus $J_{p^*}(k)$ has a finite limit supremum. In this case, we might seek a decision rule that will converge to some $q \in \mathcal{P}$ with $\limsup_{k \to \infty} J_q(k) < \infty$. Now if a is persistent in the sense that there exists $\nu > 0$ and an integer N such that for every $k \geq 0$, $\sum_{j=k}^{k+N-1} a(j) > \nu$, then it is easily shown that $e_q \in l^1$ and hence $\lim_{k \to \infty} e_q(k) = 0$.

Alternatively, if $a \in l^1$, then $h_1 \in l^1$ and $J_{p^*}(k)$ converges to zero at rate $h = h_1$. In this case, we might seek a decision rule that converges to $q \in \mathcal{P}$ with $J_q(k) \leq Lh(k)$ for some $L > 0$. Since $h \in l^1$, it would then follow that $J_q \in l^1$ and hence that $e_q \in l^1$.

Finally, if g_2 is a bounded sequence and $a \in l^1$, then h_2 is a bounded sequence and $J_{p^*}(k)$ has a finite limit supremum. If a bound M on $\limsup_{k \to \infty} g_2(k)$ is known, then for a given $\varepsilon > 0$, we might seek a decision rule that will converge to $q \in \mathcal{P}$ with $\limsup_{k \to \infty} J_q(k) \leq M + \varepsilon$. Since a is nonnegative with not all terms zero, it would then follow that e_q is a bounded sequence with $\limsup_{k \to \infty} \|e_q(k)\| \leq K(M + \varepsilon)$, where $K = 1/\|a\|_\infty$.

The existence and complexity of decision rules to solve the above problems is examined in the next section. It is shown in Theorem 3.1 below, for example, that there exist decision rules based on the performance functions discussed above that converge to a predictor with good performance.

3 Decision Rules

In this section, we pose the design of decision rules in a general framework, which will then allow us to treat the predictor selection problem outlined in the previous section as a special case. Our goal is to obtain results on the basic limitations and complexity requirements of convergent decision rules.

A data sequence $\overline{J} \triangleq \{\overline{J}(k), k \geq 0\}$, with $\overline{J}(k) \in (\mathbb{R}^+)^{|\mathcal{P}|}$, is given and at each time $k \geq 0$ we are told the value of $\overline{J}(k)$. Based on the presented data we must select a parameter $d(k) \in \mathcal{P}$ such that the sequence $\{d(k), k \geq 0\}$ converges and the limit satisfies a given criterion of success. To complete the problem specification we need to define how a decision rule is to work, what constraints are placed on the data sequence $\overline{J}$, and what is the criterion of success.

We require each admissible data sequence $\overline{J}$ to be an element of some given collection $\mathcal{F}$ of data sequences. Associated with $\mathcal{F}$ is a criterion of correct selection. This is a map S that maps a pair $\overline{J} \in \mathcal{F}$ and $p \in \mathcal{P}$ to the value 1 if p is a suitable limit for our decision rule under the input sequence $\overline{J}$ and the value 0 otherwise.

Let $\mathcal{P}^{\mathbb{N}}$ denote the set of all sequences of elements of $\mathcal{P}$. Then a decision rule is defined to be a causal map from $\mathcal{F}$ into $\mathcal{P}^{\mathbb{N}}$. We regard $\overline{J} \in \mathcal{F}$ as the input to the rule and the sequence of decisions $d(k), k \geq 0$, as the output.

A decision rule *with memory* may depend on the entire past history of observations and explicitly on the time k. Such rules can be written as a sequence of mappings $\phi_k \colon \mathbb{R}^{|\mathcal{P}| \times (k+1)} \to \mathcal{P}$ where $\phi_k(\overline{J}(0), ..., \overline{J}(k))$ represents the decision at time k. A natural restriction is to insist that past observations be summarized by an element of a set V. In this case, a decision rule can be realized as a sequence of state transition mappings $s_k \colon \mathbb{R}^{|\mathcal{P}|} \times V \to V$ and an initial state $v_0 \in V$, together with a sequence of mappings $\phi_k \colon \mathbb{R}^{|\mathcal{P}|} \times V \to \mathcal{P}$. The decisions are then formed by setting

$$v(k+1) = s_k(\overline{J}(k), v(k)), \quad v(0) = v_0 \tag{3}$$
$$d(k) = \phi_k(\overline{J}(k), v(k)) \tag{4}$$

Here $v(k+1)$ denotes the state of the rule after the kth observation and $d(k)$ denotes the decision at time k using the new observation $\overline{J}(k)$. If for some $s \colon \mathbb{R}^{|\mathcal{P}|} \times V \to V$ and $\phi \colon \mathbb{R}^{|\mathcal{P}|} \times V \to \mathcal{P}$ we have $s_k = s$ and $\phi_k = \phi$, then the decision rule is said to be stationary. If V is a finite set then the decision rule is implemented by a finite state automaton. In this case, we say that the decision rule has finite memory. If V is a countable set, then we say that the decision rule has countable memory.

We now formalize the concept of a successful decision rule as follows.

Definition 1 *We call a decision rule* successful *for* $(\mathcal{F}, \mathcal{S})$ *if for every observation sequence* $\overline{J} \in \mathcal{F}$ *the corresponding sequence of decisions* $d(k)$ *satisfies*

(i) $q \triangleq \lim_{k \to \infty} d(k)$ *exists, and*

(ii) $\mathcal{S}(\overline{J}, q) = 1$

Clearly, as the size of the class $\mathcal{F}$ of admissible sequences is increased the more demanding becomes the task of the decision rule. Similarly, for each $\overline{J}$ as the set of $p \in \mathcal{P}$ for which $\mathcal{S}(\overline{J}, p) = 1$ decreases, the requirements for a successful decision rule become more stringent. Thus, the complexity of successful decision rules may increase as the size of $\mathcal{F}$ increases, or, for a fixed $\mathcal{F}$, as the set on which $\mathcal{S}(\overline{J}, p) = 1$ decreases. On the other hand, the larger the size of $\mathcal{F}$ the more robust, in an intuitive sense, is a successful rule, and the smaller the set on which $\mathcal{S}(\overline{J}, p) = 1$, the stronger are the properties of the limit.

We now give three specific examples of admissible data sequences and their corresponding success criteria. These classes are sufficiently general to cover a variety of interesting situations. They will be the focus of attention in the sequel.

• **Finite limit supremum observations:** This set, denoted $\mathcal{F}_{\text{FLS}}$, consists of all data sequences $\overline{J}$ with the property that there exists at least one $p \in \mathcal{P}$ such that $\limsup_{k \to \infty} J_p(k) < \infty$. The success criterion $\mathcal{S}_{\text{FLS}}$ for $\mathcal{F}_{\text{FLS}}$ is simply $\mathcal{S}_{\text{FLS}}(\overline{J}, p) = 1 \iff \limsup_{k \to \infty} J_p(k) < \infty$. So, a decision rule is successful for $(\mathcal{F}_{\text{FLS}}, \mathcal{S}_{\text{FLS}})$ if the decisions converge to some $q \in \mathcal{P}$ with finite limsup (but not necessarily the minimum limsup).

• **Finite limsup observations with known bound:** for a known $M < \infty$, let $\mathcal{F}_{\text{FLS}}^M$ be the subset of $\mathcal{F}_{\text{FLS}}$ for which there exists at least one $p \in \mathcal{P}$

such that $\limsup_{k\to\infty} J_p(k) \leq M$. For a fixed $\epsilon > 0$, let $\mathcal{S}_{\text{FLS}}^{M+\epsilon}$ be defined as $\mathcal{S}_{\text{FLS}}^{M+\epsilon}(\bar{J}, p) = 1 \iff \limsup_{k\to\infty} J_p(k) \leq M + \epsilon$. Hence, for this problem, we know that at least one sequence has limsup less than or equal to M, but to succeed we need only find one that has limsup less than or equal to $M + \epsilon$.

• **Finite limsup observations with known bound and known rate:** Given a known $M < \infty$ and a known positive sequence $\bar{g} = \{g(k)\}$ with $\lim_{k\to\infty} g(k) = 0$, let $\mathcal{F}_{\text{FLS}}^{M,\bar{g}}$ consist of all data sequences $\bar{J}$ for which there exists at least one $p \in \mathcal{P}$ and some constant $L < \infty$ (that is unknown and can depend on $\bar{J}$) such that $J_p(k) \leq M + Lg(k)$. The success criterion corresponding to this class will be denoted $\mathcal{S}_{\text{FLS}}^{M,\bar{g}}$ and is defined by $\mathcal{S}_{\text{FLS}}^{M,\bar{g}}(\bar{J}, p) = 1 \iff \exists L < \infty$ such that $J_p(k) \leq M + Lg(k)$. For this class we know that for some $p \in \mathcal{P}$, the tails of $J_p(k)$ decay to less than or equal to M at some known rate, but we may not know the constant. This will be useful for problems in which we have guarantees on the rate at which data sequences converge, but the constant may depend on some unknown parameters. The success criterion requires that we converge to some $q \in \mathcal{P}$ for which $J_q(k)$ also decays to less than or equal to M at the same rate as for p (although the constants may differ).

The following theorem gives positive results for each of the three decision problems.

Theorem 3.1 *Let $\mathcal{P}$ be a finite or countable index set.*

1. *There exists a stationary decision rule with countable memory that is successful for $(\mathcal{F}_{\text{FLS}}, \mathcal{S}_{\text{FLS}})$.*

2. *For every $M < \infty$ and every $\epsilon > 0$, there exists a stationary decision rule with memory of size $|\mathcal{P}|$ that is successful for $(\mathcal{F}_{\text{FLS}}^M, \mathcal{S}_{\text{FLS}}^{M+\epsilon})$.*

3. *For every $M < \infty$ and every positive sequence $\bar{g} = \{g(k)\}$ with $g(k) \to 0$, there exists a decision rule with countable memory that is successful for $(\mathcal{F}_{\text{FLS}}^{M,\bar{g}}, \mathcal{S}_{\text{FLS}}^{M,\bar{g}})$.*

Note that part 2 has a smaller class of admissible sequences but a stronger performance criterion than part 1. The resulting decision rule is less complex in that a memory of only size $|\mathcal{P}|$ is used as opposed to a countable memory. On the other hand, part 3 again has a smaller class of admissible sequences and a stronger performance criterion than part 2, but the decision rule for part 3 is more complex in that it uses a countable memory and is non-stationary.

The result of part 3 can actually be extended to the case in which each $p \in \mathcal{P}$ may have a different rate of convergence to its limsup (assuming the limsup of the sequence $J_p(k)$ is finite). In this case, the class of admissible sequences consist of those for which for some $p \in \mathcal{P}$, $J_p(k) \leq M + Lg_p(k)$ where the convergence rate $g_p(k)$ can depend on $p \in \mathcal{P}$. In this case, the success criterion $\mathcal{S}$ can be modified so that if the decision rule converges to $q \in \mathcal{P}$, then we will be guaranteed that the sequence $J_q(k)$ has limsup bounded by M with corresponding convergence rate $g_q(k)$. The decision rule for this case simply uses $M + L_{v(k-1)}g_{\pi_{v(k-1)}}(k)$ in place of $M + L_{v(k-1)}g(k)$ in the memory state transition rule for $v(k)$.

The proof given in [4] is constructive but is neither unique nor optimal. The decision rules constructed contain, as a component, a pre-routed path. However, this can be used in conjunction with the performance data to determine the actual search path through the candidate predictors. Thus the actual trajectory $d(k)$ through $\mathcal{P}$ is, in general, 'sample path' dependent.

It is possible to construct, in an ad hoc fashion, many successful rules for the problems of Theorem 3.1. Some of these will be more efficient than others. For example, common sense suggests that the indices to the sequences π and h in the proof of Theorem 3.1 should be updated separately. In this case the state $v(k)$ consists of a pair of integers (n, m) where n is the index into h and m is the index into π. However, good characterizations of efficiency and the degree to which the selections of successful rules can be sample path dependent are not available at this point, although the 'hysteresis switching rule' of [9], [13] can be formulated as a special case of the constructions given in [4].

Finally, as a counterpoint to Theorem 3.1, we state the following result which shows that the memory requirements of the decision rules presented in Theorem 3.1 cannot be simplified.

Theorem 3.2 *Let $\mathcal{P}$ be a finite or countable index set with $|\mathcal{P}| \geq 2$.*

1. *There is no finite memory decision rule that is successful for $(\mathcal{F}_{\mathrm{FLS}}, \mathcal{S}_{\mathrm{FLS}})$.*

2. *For any $M < \infty$ and $\epsilon > 0$, there is no decision rule with memory of size less than $|\mathcal{P}|$ that is successful for $(\mathcal{F}^M_{\mathrm{FLS}}, \mathcal{S}^{M+\epsilon}_{\mathrm{FLS}})$.*

3. *For any $M < \infty$ and positive sequence $\bar{g} = \{g(k)\}$ with $g(k) \to 0$, there is no finite memory decision rule that is successful for $(\mathcal{F}^{M,\bar{g}}_{\mathrm{FLS}}, \mathcal{S}^{M,\bar{g}}_{\mathrm{FLS}})$.*

4   Controller Switching Policies

In this section we consider the design of suitable controller switching policies for the plant Σ where the controllers are selected from a family of predesigned controllers $\Gamma_p, p \in \mathcal{P}$. As our strongest results are for linear time-invariant mimo systems we focus attention to this case. We begin by defining the families of predictors, models, and controllers.

Assume that each predictor $O_p, p \in \mathcal{P}$, is a linear time-invariant system with state space representation:

$$O_p: \begin{aligned} w_p(k+1) &= M_p w_p(k) + B_p u(k) + K_p y(k), \quad w_p(0) = w_{p0} \\ \hat{y}_p(k) &= C_p w_p(k) \\ e_p(k) &= C_p w_p(k) - y(k) \end{aligned} \tag{5}$$

Similarly, each controller Γ_p is described by

$$\Gamma_p: \begin{aligned} z_p(k+1) &= F_p z_p(k) + G_p y(k) + R_p r(k), \quad z_p(0) = z_{p0} \\ u(k) &= H_p z_p(k) \end{aligned} \tag{6}$$

where r is an exogenous reference input in an admissible class.

The predictor O_p and controller Γ_p will be constrained through the following model:

$$\Sigma_p : \quad \begin{aligned} x_p(k+1) &= (M_p + K_p C_p) x_p(k) + B_p u_p(k), \quad x_p(0) = x_{p0} \\ y_p(k) &= C_p x_p(k) \end{aligned} \tag{7}$$

This is obtained by setting $y(k) = \hat{y}_p(k)$ in (5). If we let $A_p \triangleq M_p + K_p C_p$, then (C_p, A_p) is a detectable pair and O_p is an observer for Σ_p. One could regard the detectable model (7) as given and derive $M_p = A_p - K_p C_p$ by stabilizing A_p using output injection. Either point of view is valid; the first is more convenient for our purposes here.

The equations for the closed loop system that result when Γ_q is connected in feedback with Σ_q are

$$\begin{aligned} \begin{pmatrix} z_q \\ x_q \end{pmatrix}(k+1) &= \begin{pmatrix} F_q & G_q C_q \\ B_q H_q & A_q \end{pmatrix} \begin{pmatrix} z_q \\ x_q \end{pmatrix}(k) + \begin{pmatrix} R_q \\ 0 \end{pmatrix} r(k) \\ y_{qq}^r(k) &= (0 \ C_q) \begin{pmatrix} z_q \\ x_q \end{pmatrix}(k) \end{aligned} \tag{8}$$

We let

$$A_{qq} \triangleq \begin{pmatrix} F_q & G_q C_q \\ B_q H_q & A_q \end{pmatrix} \tag{9}$$

The predictors and controllers are required to satisfy the following basic constraints. For each $p \in \mathcal{P}$:

(L1) The matrices M_p, $p \in \mathcal{P}$, have their eigenvalues inside the circle of radius $\sigma < 1$; and

(L2) A_{pp} is stable, i.e., Γ_p stabilizes Σ_p.

We will assume that in addition to satisfying (L2), Γ_q has been designed so that the controlled system (Γ_q, Σ_q) gives good tracking performance over the admissible class of reference signals. For example, Γ_q might be designed to stabilize Σ_q and to keep an appropriate weighted induced norm of the mapping from r to the tracking error $y_{qq}^r - r$ of (8) small.

It is also reasonable to expect that Γ_q has been designed so that the closed loop system (Γ_q, Σ_q) has good disturbance and noise attenuation properties. Of particular interest in this regard is the system:

$$\begin{aligned} \begin{pmatrix} z_q \\ x_q \end{pmatrix}(k+1) &= \begin{pmatrix} F_q & G_q C_q \\ B_q H_q & A_q \end{pmatrix} \begin{pmatrix} z_q \\ x_q \end{pmatrix}(k) + \begin{pmatrix} G_q \\ K_q \end{pmatrix} \delta(k) \\ \varepsilon_{qq}(k) &= (0 \ C_q) \begin{pmatrix} z_q \\ x_q \end{pmatrix}(k) + \delta(k) \end{aligned} \tag{10}$$

This is obtained from (8) by setting r to zero and adding a disturbance δ as an additive term in the output. See Figure (1). Let β_q denote the induced

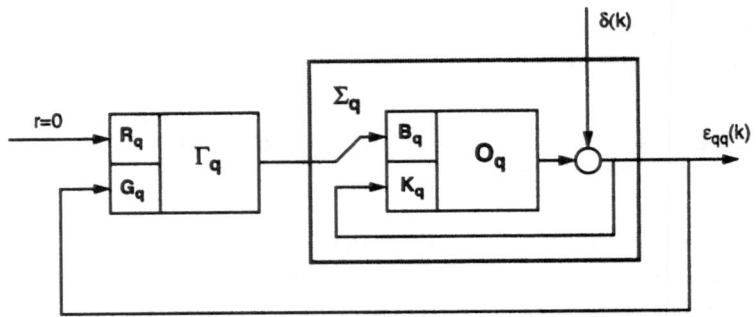

Figure 1: Closed loop system (Γ_q, Σ_q) with a disturbance.

∞-norm of the mapping from δ to ε_{qq}. Since A_{qq} is stable, β_q is finite. For reasons to be seen shortly, a desirable design requirement of Γ_q is that β_q is small.

The plant Σ is assumed to be a linear time-invariant system with state space representation:

$$\Sigma: \quad \begin{aligned} x(k+1) &= Ax(k) + Bu(k) + Dv(k), \quad x(0) = x_0 \\ y(k) &= Cx(k) + n(k) \end{aligned} \tag{11}$$

Here v and n are exogenous disturbance and sensor noise signals respectively.

We assume that the plant satisfies:

(P1) For some $p^* \in \mathcal{P}$, (Γ_{p^*}, Σ) is stable.

A consequence of (P1) is that (C, A) is detectable, i.e., $A - KC$ is stable for suitable K; and (A, B) is stabilizable.

We now relate the performance of Γ_q when connected in feedback with the plant Σ, to the concurrent performance of predictor O_q. The framework of this result is quite general and includes unmodeled dynamics, disturbances and sensor noise. The result applies without change to continuous-time systems and easily extends to time-varying linear systems.

Proposition 4.1 *Let Σ be the plant specified by (11) and Σ_q, O_q and Γ_q be specified by (7), (5) and (6) respectively. Assume that (P1), (L1) and (L2) are satisfied. Let r, v and n be elements of the admissible classes of exogenous reference, disturbance and noise signals respectively.*

Let the system Σ, driven by v and n, be connected in feedback with the control system Γ_q driven by r. Let y denote the output of Σ and e_q denote the prediction error sequence of O_q along the resultant trajectory. Then

$$y(k) = y_{qq}^r(k) + \varepsilon_{qq}(k), \quad k \geq 0$$

where y_{qq}^r is an output sequence of the disturbance and noise free stable closed loop system (8), and ε_{qq} is the output of the stable system (10) with $\delta(k) = -e_q(k)$. In particular:

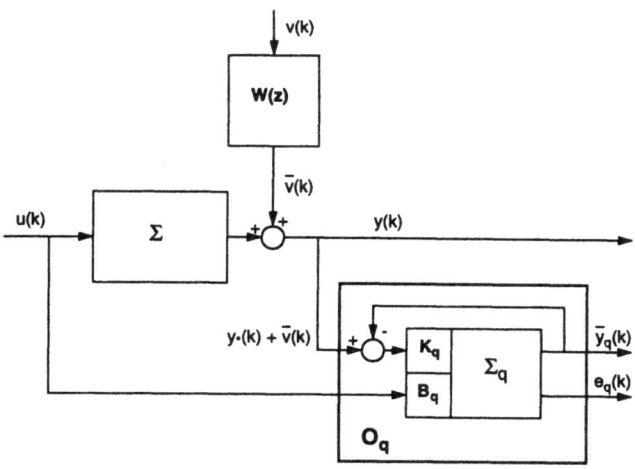

Figure 2: The set-up for additive output disturbances with the details of the predictor O_q.

1. If $e_q \to 0$ *exponentially, then* $\varepsilon_{qq} \to 0$ *exponentially. In addition, if* $v = n = r \equiv 0$, *then* $x(k), z_q(k), w_q(k) \to 0$ *exponentially, and if* r, v, n *are bounded, then* $x(k), z_q(k), w_q(k)$ *are bounded.*

2. If $\|e_q\|_{\bar{p}} < \infty$, *for some* $1 \le \bar{p} \le \infty$, *then* $\|\varepsilon_{qq}\|_{\bar{p}} < \infty$ *and*

$$\limsup_{k \to \infty} \|\varepsilon_{qq}(k)\| \le \beta_q \limsup_{k \to \infty} \|e_q(k)\|.$$

In addition, if r, v, n *are bounded, then* x, z_q, *and* w_q *are bounded.*

The results of Section 3 and Proposition 4.1 can now be used to prove a result for prediction error based controller switching policies. We first define and comment on several special situations of interest.

• **Exact matching:** When $\Sigma = \Sigma_{p^*}$ for some unknown $p^* \in \mathcal{P}$, we follow [7] and say we have exact matching. Ideally we would like to have exact matching with $v = n \equiv 0$. However, this is clearly a mathematical idealization.

• **Output disturbances:** If the mapping from the disturbance v to the output y of Σ is stable, then the effect of the disturbance can be modeled as an additive term $\hat{v}$ at the output of the plant. A commonly used model in this situation is that $\hat{v}$ is the output of a known stable LTI filter $W(z)$ with the input v satisfying $\|v\|_\infty \le 1$. For simplicity we lump n with $\hat{v}$ (or equivalently assume $n \equiv 0$). To assess the effect of this disturbance on the prediction errors let $G_p(z) = C_p(zI - M_p)^{-1} K_p$ and α_p denote the induced ∞-norm of the stable mapping defined by the transfer matrix $G_p(z)W(z)$ from v to e_p. Since the predictors and $W(z)$ are given, we assume that the α_p are known. The set-up for additive output disturbances is shown in Figure 2.

• **SISO set point regulation:** In this case assume that all systems are siso, the disturbances can be modeled as an additive term at the output, the reference signals and disturbance signals are constants, and $n \equiv 0$. To ensure asymptotic set-point regulation we assume that the plant has a pole at 1 and no zeros at 1. If necessary this can be achieved by the inclusion of a summer in the forward path. Since this structure is known, we also include a summer in the observer dynamics by ensuring that each model A_p has an eigenvalue of 1 and that the transfer function of the triple (C_p, A_p, K_p), does not have a zero at 1. If in addition, $\Sigma = \Sigma_{p^*}$ for some $p^* \in \mathcal{P}$, then we have an exact matching siso set point problem. This is the equivalent tracking problem to that considered by Morse in [7].

We are now ready to state the main result of this section.

Theorem 4.2 *Let Σ_p, Γ_p, and O_p, $p \in \mathcal{P}$, be linear systems specified by (7), (5), and (6) where r, v, n are elements of the admissible classes of exogenous reference, disturbance and noise signals respectively. Assume that for each $p \in \mathcal{P}$ conditions (L1) and (L2) are satisfied. Let Σ be an unknown plant of the form (11) satisfying (P1) and with unknown initial state $x(0)$. Then:*

1. *Ideal Case: There exists a convergent controller selection rule such that if $\Sigma = \Sigma_{p^*}$ for some $p^* \in \mathcal{P}$, and $v = n \equiv 0$, then for some $q \in \mathcal{P}$ (depending on the initial condition and r), $y(k) - y_{qq}^r(k) \to 0$ exponentially. In addition, if $r \equiv 0$, then $x, w_q, z_q \to 0$ exponentially, and if r is bounded, then x, w_q, and z_q are all bounded.*

2. *Exact Matching with Output Disturbances: For each $\epsilon > 0$ and $\rho \geq 1$ there exists a convergent controller selection rule such that if $\Sigma = \Sigma_{p^*}$ for some $p^* \in \mathcal{P}$, r is bounded, and the assumptions of output disturbances are satisfied, then for some $q \in \mathcal{P}$ (depending on the initial condition, r, and v) $\limsup_{k \to \infty} \|y(k) - y_{qq}^r(k)\| \leq \beta_q(\rho\alpha_q + \epsilon) < \infty$, and x, w_q, and z_q are bounded.*

3. *Exact Matching SISO Set Point Problem: There exists a convergent controller selection rule such that if $\Sigma = \Sigma_{p^*}$ for some $p^* \in \mathcal{P}$, and the assumptions of the siso set point problem are satisfied, then for some $q \in \mathcal{P}$, $\lim_{k \to \infty} d(k) = q$, $y(k) - r \to 0$ exponentially, and the sequences x, w_q, and z_q are all bounded. In addition, if $v = r \equiv 0$, then $x, w_q, z_q \to 0$ exponentially.*

Part 1 of Theorem 4.2 is included to indicate what can be achieved in an idealized case. The output of the supervised control system is eventually the sum of an output trajectory of the closed loop system (Γ_q, Σ_q), for some $q \in \mathcal{P}$, and a term that converges exponentially to zero. Hence if the controlled systems (Γ_p, Σ_p) all do an adequate job of asymptotically tracking r, then under ideal assumptions so will the supervised control system for Σ. A similar statement can be made when the signal ε_{qq} is bounded except that to be useful the bound must be small compared to the magnitude of the signal r. This can be assured even if e_q does not converge to zero but an asymptotic bound on e_q is possible.

Part 2 indicates that under the common additive output disturbances model it is still possible to achieve convergence. If the disturbances are small, a reasonable assumption in many cases, the quality of the resultant tracking will still be good. Instead of using a worse case disturbance model we could also have used a finite power model. Part (3) is a discrete-time version of the problem considered in [7]. The new contribution is that we show that set-point regulation is possible even when controller selection is required to converge. It is possible to relax the rule used in the proof of part 3 so that in the limit $J_q(k) < \epsilon$. This simplifies the rule but sacrifices tracking performance.

We end this section with a brief discussion showing how our results can also be applied to continuous-time LTI systems.

Suppose we have a set of continuous-time LTI predictors O_p, $p \in \mathcal{P}$, and corresponding controllers Γ_p with state space representations analogous to those in (5) and (6), respectively. We also assume that we have a continuous-time performance measure that is keeping track of how well predictor p is performing. That is, for each p we have a real-valued performance measure $J_p(t)$ that depends on $e_p(\tau)|_{0 \le \tau \le t}$. For example, in the case of siso systems we might fix $\lambda \in \mathbb{R}$ and take $J_p(t) = \int_0^t e^{-\lambda(t-\tau)} e_p^2(\tau) \, d\tau$.

Now, let $t_1 < t_2 < \cdots$ be a sequence of times with $t_n \to \infty$. The t_n denote the potential switching times. That is, at time t_n, we will apply an appropriate decision rule based on the performance indices sampled at time t_n: $\overline{J}(t_n) = \{J_p(t_n)\}_{p \in \mathcal{P}}$. Based on the output of the decision rule, the corresponding controller will be applied during the interval $t_n < t \le t_{n+1}$. Since the controller can be switched only at a discrete set of times, convergence of a switching policy, etc., can be reduced to a discrete-time analysis. Then, by suitable choices for the performance functions we can apply our results from Section 3. Of course, to conclude properties of the resulting continuous-time closed loop system like those in Theorem 4.2, assumptions analogous to those in Theorem 4.2 must be made.

5 Conclusion

We have analyzed in a general framework the existence and memory requirements of discrete decision rules based on real valued data sequences. We have shown that it is possible to combine the two extremes of performance based search and pre-routed search to obtain some of the desirable characteristics of both methods. The rules contain a mechanism to generate a potential search through the candidate predictors with the re-visitation property. However, this is combined with the available performance measure so that only those candidates which are classified as exhibiting 'acceptable' performance at time k are considered.

Open problems that remain to be addressed include:

- A good characterization of the efficiency of convergent selection rules.

- A characterization of the sample path dependence of the selection sequence, e.g., when is a built in pre-routed search path necessary?

- A treatment of more general classes of plant uncertainty.

References

[1] K. Ciliz and K. Narendra, "Multipole model based adaptive control of robotic manipulators," *Proc. of the 33rd IEEE Conf. on Decision and Control*, pp. 1305-1310, Lake Buena Vista, FL, Dec. 1994.

[2] S. J. Cusumano and K. Poolla, "Adaptive control of uncertain systems: a new approach," *Proc. 1988 American Control Conference*, pp. 355-359, Atlanta, Georgia, June 1988.

[3] M. Fu and B. R. Barmish, "Adaptive stabilization of linear systems via switching controls," *IEEE Trans. on Automatic Control*, 31 (12), pp. 1079-1103, Dec. 1986.

[4] S.R. Kulkarni and P.J. Ramadge, "Model and Controller Selection Based on Output Prediction Errors," to appear in *IEEE Transactions on Automatic Control*.

[5] B. Mårtensson, "The order of any stabilizing regulator is sufficient a priori information for adaptive stabilization," *Systems and Control Letters*, 6 (2), pp. 87-91, July 1985.

[6] D. E. Miller, "Adaptive stabilization using a nonlinear time-varying controller," *IEEE Trans. on Automatic Control*, 39 (7), pp. 1347-1359, July 1994.

[7] A. S. Morse, "Supervisory control of families of linear set-point controllers – part 1: exact matching," preprint, March 1993.

[8] A. S. Morse, "Supervisory control of families of linear set-point controllers – part 2: robustness," preprint, Nov. 1994.

[9] A. S. Morse, D. Q. Mayne, and G. C. Goodwin, "Applications of hysteresis switching in parameter adaptive control," *IEEE Trans. Automatic Control*, 37 (9), pp. 1343-1354, Sept. 1992.

[10] K. Narendra and J. Balakrishnan, "Improving transient response of adaptive control systems using multiple models and switching," *Proc. of the 32rd IEEE Conf. on Decision and Control*, San Antonio, Texas, Dec. 1993.

[11] K. Narendra and J. Balakrishnan, "Intelligent control using fixed and adaptive models," *Proc. of the 33rd IEEE Conf. on Decision and Control*, pp. 1680-1685, Lake Buena Vista, Florida, Dec. 1994.

[12] R. D. Nussbaum, "Some remarks on a conjecture in parameter adaptive control," *System and Control Letters*, 3 (5), pp. 243-246, Nov. 1983.

[13] S. R. Weller and G. C. Goodwin, "Hysteresis switching adaptive control of linear multivariable systems," *IEEE Trans. Automatic Control*, 39 (7), pp. 1360-1375, July 1994.

Parallel Algorithms for Adaptive Control: Robust Stability

Felipe M Pait and Fuad Kassab Jr.
pait,fuad@lac.usp.br
Universidade de São Paulo
Laboratório de Automação e Controle
Av. Prof. Luciano Gualberto trav 3 158
São Paulo SP 05508-900 Brasil

Abstract

A class of parallel algorithms for adaptive control of siso linear systems is described. The systems considered are assumed to belong to one among a finite number of classes of admissible process models, and each class is robustly stabilizable by some linear time-invariant controller. The control used is chosen in real time — from among the outputs of a finite set of linear time-invariant candidate controllers — by a tuner or supervisor, according to observations of suitably defined "identification errors." The method preserves the robustness properties of the linear control design in an adaptive context. We expect that parallel algorithms of the type discussed here will be useful tools to exploit the compromise between performance of an adaptive control system and the computational power of the hardware in which it is implemented. Another application is to fault-tolerant control.

Keywords: Adaptive control; robust control; parallelism; hybrid control.

1 Introduction

Stability of adaptive control schemes for linear time-invariant processes is the subject of a large literature. By now it is reasonably well understood how to bring the tools of stability analysis of linear systems into an adaptive context, so that, combining ideas from parameter estimation theory with those of linear control synthesis, stable adaptive systems can be designed. The main thrust of research in linear adaptive control nowadays comes from the equally important and challenging questions of robustness and performance.

Robustness issues arise out of the inherently inaccurate nature of plant models, which makes it impossible for any model to explicitly account for the full range of possible behaviors of a process when a control loop is closed.

Traditional robust control theory furnishes techniques to deal with these uncertainties provided they are somehow "small," but for larger uncertainties it is frequently advisable to use some sort of adaptation to close the control loop. We have in mind, for instance, the sort of uncertainty which results from faults in the process, including changes in dynamics and failures of actuators or sensors. Conventional parameter adaptive control is one idea to consider in this case. The difficulties are that the transient performance of an adaptive systems is often poor when the parameter uncertainty is large or when the use of probing signals is not desirable.

This paper presents a class of parallel algorithms for robust adaptive control. The main idea behind the use of parallel algorithms is to split the task of computing a feedback control input into a (possibly very large) number of subtasks, which can be performed independently and simultaneously, establishing a trade-off between the specifications of the control algorithm and the processing capabilities of the computer hardware in which it is implemented. The significance of parallel algorithms with respect to this compromise is further discussed in [12] and in §2. Our purpose here is to define what is required of the parameterized controllers and of the tuners (or "supervisory control algorithms") in order to establish robust stability of the overall adaptive system. Algorithms satisfying those requirements are exemplified.

In [7, 8] Morse introduces a scheme of robust supervisory control of a family of set-point controllers. By using tuning algorithms (or *supervisors*) possessing the properties described in §6 we are able to obtain tighter results concerning the robust stability of the overall adaptive system. In fact the uncertainty radii are limited only by our ability to do a certainty-equivalence robust control synthesis, thus answering one of the questions posed in [8]. The idea of using switching between multiple models to improve the transient performance of an adaptive system has also been exploited in [11]. The parallel schemes developed here also owe to the literature on multiple model adaptive control [1, 2], which is concerned primarily with stochastic convergence. In contrast, our main interest is robust deterministic stability.

In the robust adaptive control literature it is usual to introduce modifications to standard adaptive laws so as to preserve stability in the presence of sufficiently small unmodeled dynamics [4]. We approach the problem by specifying from the start that the tuner and the certainty-equivalence controllers have properties that are conducive to robust adaptive stability, and are thus able to prove less conservative results. A different approach to robustness, which combines concepts from variable-structure systems with model-reference adaptive control, is presented in [3]. The kind of switching discussed in this contribution involves sliding modes and is very different from the type of hybrid systems we have in mind.

This paper is organized as follows: We first state the assumptions made about the class of processes we attempt to control and describe the structure of the overall adaptive system, and then show how standard identification techniques (§4) and robust control design tools (§5) are used to construct parallel adaptive control systems. Tuner properties that ensure closed-loop stability are stated in §6, and a supervisory algorithm which has them is described. In §7 we explain an analysis method based on detectability of the process + controller parameterized linear system, an input-output property of slowly time-varying systems, and a small-gain argument. In §2 we try to motivate, in an informal and somewhat speculative manner, the study of parallel algorithms for adaptive control; the proofs are omited from this version.

2   Speculation: Why Parallel Algorithms?

Control Theory has passed by the revolution in digital computation like a ship in the night. This statement may appear ludicrous in face of the widespread usage of electronic computers and software tools which perform all sorts of tasks related to control engineering: digital control systems, computer-aided design and analysis software, simulation packages, text processors... However from one point of view the observation is quite appropriate: the way control algorithms are formulated has not changed. The conceptual framework of control theory — dynamical system analysis and synthesis — makes scant reference to the processing power available. Quite unlike the theory of Computer Science, largely devoted to algorithmic complexity and computational efficiency.

When the strategy to deal with a specific problem is defined and implemented, one always pays attention to the type of hardware used. The more demanding the plant, and the more complex the algorithms, the higher is the processing power required. That often calls for fast computers and for specialized control hardware architectures where each processing unit is responsible for a specific task. Also, questions related to numerical analysis have received a great deal of attention, particularly in the context of control design. However, with regard to the algorithms themselves, theory rarely mentions the role of the available computing power. Bibliographical research through journals specializing in control theory will show that references to computational capacity in the *formulation* of control algorithms are infrequent and have localized impact.

That leads one to ask: Is it possible to translate higher processing power of a control system into better overall closed-loop performance? Can one devise algorithms that make explicit use of more powerful hardware, for instance of parallel processors, to better achieve the control goals? Can one develop a framework to exploit a trade-off between computational power and control system performance?

Questions like those led us to pursue the development of parallel algorithms for adaptive control. By that we mean algorithms in which one basic task is divided in many subtasks which can be performed independently and simultaneously. The method chosen is to use several subsystems — each corresponding to a feedback control design model and each capable of generating both an identification error and a candidate control signal. Adaptation is accomplished by switching the candidate controls into the feedback loop, based on observations of the respective error signals. Thus the task which conventional adaptive control approaches using "local" methods, like least-squares parameter estimation, can here be treated by a supervisor of a "global" character — global in that its logic can take into account all candidate controls via their respective error signals. The computations can easily be split among independent computer tasks, making them natural candidates for implementation using parallel computer systems. Presumably such methods can lead to better performance than conventional adaptive control, as the number of design models is increased. Concluding this digression, we remark that adaptive systems — conceived to adjust and optimize the behavior of a controller according to real-time data — are among the control techniques most needy of firm results concerning robustness and performance.

3   Formulation

Our interest is in the adaptive control of a siso linear time-invariant process Σ_P

$$
\begin{aligned}
\dot{x}_P(t) &= A_P x_P(t) + b_P\big(u(t) + w_u(t)\big) \\
y(t) &= c_P x_P(t) + w_y(t)
\end{aligned}
\tag{1}
$$

belonging to a known class $\mathcal{C}_P$ of admissible process models. Here $x_P : \mathbb{R}^+ \to \mathbb{R}^{n_P}$ is the process' state, $u \in \mathbb{R}$ is the control input, $y \in \mathbb{R}$ the measured output, and $w_u, w_y \in \mathbb{R}$ are respectively input and output bounded noise signals. We assume $\mathcal{C}_P$ is contained in the union of a finite number of classes $\mathcal{C}_k$, $k \in \mathcal{K} = \{1, 2, \ldots, n\}$. Each class $\mathcal{C}_k$ is a ball centered around a known *nominal process model* Σ_{N_k} whose strictly proper transfer function is $\pi_{N_k}(s)$. More precisely, we make the following assumption:

Assumption 1 The transfer function of each element of $\mathcal{C}_k$ can be written as

$$
\big(1 + \Delta(s)\Omega_k(s)\big)\pi_{N_k}(s)
\tag{2}
$$

where

- Ω_k are preestablished stable, proper perturbation weighting transfer functions;

- Δ is a stable, proper transfer function such that $\|\Delta(s+\lambda)\|_\infty < \delta$, where δ and $\lambda \geq 0$ are real numbers; and

- $\big(1 + \Delta(s)\Omega_k(s)\big)\pi_{N_k}(s)$ has the same number of unstable poles as $\pi_{N_k}(s)$, and there is no unstable pole-zero cancelation in (2).

We call all such transfer functions Δ, and by extension all elements of $\mathcal{C}_k$ thus defined, *admissible*.   ♣♣

To control Σ_P, we employ a family of linear time-invariant controllers Σ_{C_k}, each of them a dynamical system whose inputs are u, y, and a reference signal r, and whose outputs are u_k and e_k. The n *identification errors* e_k serve as surrogate "performance indicators," based upon which we will select at each instant which of the candidate control signals u_k is to control Σ_P. That is to say, the input of the tuning algorithm is the identification error vector $e_I = [e_1 \quad e_2 \quad \cdots \quad e_n]'$, and its output is a switching signal $\sigma : \mathbb{R}^+ \to \mathcal{K}$. Selecting $u = u_\sigma$ closes the adaptive control loop.

The nominal process models play the role of design models, upon which certainty-equivalence controllers will be designed. The requirements we make on each Σ_{C_k} are twofold:

- First, Σ_{C_k} must *robustly stabilize* the nominal process model Σ_{N_k}. In other words, the feedback interconnection of Σ_{C_k} with each element of $\mathcal{C}_k$ is internally stable in the linear system theory sense; and

- Second, e_k must be "small" in some well-defined sense if $\Sigma_P \in \mathcal{C}_k$.

The first requirement arises out of certainty-equivalence control; the second out of the idea of analysing the overall adaptive system employing the certainty-equivalence stabilization theorem [6]. Notice that by using fixed stabilizable nominal systems as design models we have circumvented the problem of loss of stabilizability of the design model, which appears in conventional parameter adaptive control when tuning is done continuously over a parameter set.

Remark 2 At this point those who are familiar with the adaptive control literature may wonder about the relevance of the problem under consideration. Identifier-based parameter adaptive control usually employs smoothly parameterized design models, rather than finite sets of nominal models. The set of exactly matchable transfer functions considered here is finite, as opposed to uncountable. For the present scheme to be a viable alternative to standard adaptive control, we need to answer the following question: Is it possible to make $\mathcal{C}_P$ large via an appropriate choice of the nominal models Σ_{N_k}? In other words, are there 1) an "interesting" class $\mathcal{C}_P$; 2) a number n; and 3) sets $\mathcal{C}_{P_k}$, each of them robustly stabilizable, such that $\mathcal{C}_P \subset \bigcup_k \mathcal{C}_{P_k}$?

To answer this question, consider a metric space of linear systems $\mathcal{M}$, for example the set of systems with real-rational transfer functions endowed with the gap metric. $\mathcal{M}$ is not necessarily a compact space; it may include distributed-parameter systems for instance. So choose a *compact* subset $\mathcal{X} \subset \mathcal{M}$, which could be a family of parameterized linear systems (design models) of McMillan degree not exceeding a given number.[1] Note that $\mathcal{X}$ need not have any convexity property or even be connected, as is usually required when recursive estimation techniques are to be used. Clearly it is not reasonable to take $\mathcal{C}_P$ equal to $\mathcal{M}$, and it is too restrictive to take it equal to $\mathcal{X}$.

A plausible candidate is the set $\mathcal{Y}$ defined as follows. For each point $x \in \mathcal{X}$, take an open ball $B(x, \delta(x))$ centered in x with radius $\delta(x)$, where $\delta(x)$ is some bounded function of x. Then define the set $\mathcal{Y}$ as the union of all such balls. In general, $\mathcal{Y}$'s closure is not compact unless $\mathcal{M}$ is finite-dimensional. Thus a finite

---

[1]Finitely parameterized design models for systems with McMillan degree not exceeding a given number are constructed in [10].

covering for $\mathcal{Y}$ cannot be constructed as a subset of an arbitrary open covering. We are nonetheless interested in constructing a finite covering for $\mathcal{Y}$, as a means of ensuring that each element of $\mathcal{Y}$ is an element of at least one among a finite number of sets. Additionally, those sets cannot be too "large," because there are limits to what can be achieved via robust control. For instance, a trivial covering could have just one set — $\mathcal{Y}$'s closure. But controllers that stabilize each process in $\mathcal{Y}$ may not exist, or may result in unacceptable performance compromises. The following proposition shows that the construction of a finite covering is possible, at least in principle.

Proposition 3 *It is possible to construct finite covering for* $\mathcal{Y}$.

Practical contructions are a subject for future research. Some applications to fault-prone systems are treated in [5]. ♣ ♣

4 Identification

The first component of the controller Σ_C is the identifier Σ_I, consisting of subsystems Σ_{I_k} given by

$$
\begin{aligned}
\dot{x}_I &= A_I x_I + b_I u + d_I y \\
e_k &= c_{I_k} x_I - y
\end{aligned}
\tag{3}
$$

where A_I is a stability matrix. Note that all identifiers are constrained to share the same state, so (3) together with $e_I = [\, e_1 \quad e_2 \quad \cdots \quad e_n \,]'$ describes the identifier $\Sigma_I{}^2$. One way to define the matrices above is as follows: represent each transfer function π_{N_k} as the ratio of two stable rational transfer functions β_k and $1 - \alpha_k$, where α_k and β_k are strictly proper and the denominators of all α_k and β_k, $k \in \mathcal{K}$, are the same Hurwitz polynomial. Then pick an observable pair (c, A) whose characteristic polynomial is precisely this common denominator, and choose

$$
A_I = \begin{bmatrix} A' & 0 \\ 0 & A' \end{bmatrix}, \qquad b_I = \begin{bmatrix} c' \\ 0 \end{bmatrix}, \qquad d_I = \begin{bmatrix} 0 \\ c' \end{bmatrix}.
$$

Finally, let $c_{I_k} = [\, b_k' \quad d_k' \,]$ and pick b_k and d_k such that

$$
\begin{aligned}
c(sI - A)^{-1} b_k &= \beta_k(s) \\
c(sI - A)^{-1} d_k &= \alpha_k(s).
\end{aligned}
$$

Note that A, c, b_k, and d_k define the following "natural" realization of the kth nominal process model Σ_{N_k}

$$
\begin{aligned}
\dot{x}_{N_k} &= (A + d_k c) x_{N_k} + b_k u_{N_k} \\
y_{N_k} &= c x_{N_k},
\end{aligned}
$$

---

[2] In order for the identifier to "scale well" as n increases, it is important to construct it in the manner described above rather than as a collection of n minimal state observers. In the latter case Σ_I's state would have dimension $n(n_P - 1)$ as opposed to $2n_P$.

and that the linear system above can be represented as

$$y_{N_k}(t) = c(sI - A)^{-1} \left(b_k u_{N_k}(t) + d_k y_{N_k}(t) \right)$$
$$= \alpha_k(s) y_{N_k}(t) + \beta_k(s) u_{N_k}(t).$$

This expression is in a familiar form often encountered in the design of identifiers.

From Assumption 1, there exists some $k_* \in \mathcal{K}$ such that $\Sigma_P \in \mathcal{C}_{k_*}$. Thus there exists an admissible Δ such that

$$y(t) = \alpha_{k_*} y(t) + \beta_{k_*} u(t) + \Delta \Omega_{k_*} \beta_{k_*} u(t) + w_{k_*}(t),$$

where for each k we write $w_k(t) = \beta_k(1 + \Delta \Omega_k) w_u(t) + (1 - \alpha_k) w_y(t)$, up to a term depending on the initial states of Σ_P and Σ_I and decaying exponentially fast with $e^{A't}$. Since w_u and w_y are bounded, $w \equiv w_{k_*}$ is also bounded, namely, $|w(t)| \leq \bar{w}$ for some constant $\bar{w}$. This results in the following

Identifier Property:

$$e_{k_*} = -\Delta(s) \Omega_{k_*}(s) \beta_{k_*}(s) u(t) - w(t). \tag{4}$$

Of course k_* is a priori unknown. This expression for e_{k_*} can be derived from the plant uncertainty model used.   ♣ ♣

5 Robust Control

The other component of Σ_C is the internal regulator Σ_R; the subsystem Σ_{R_k} is given by

$$\dot{x}_R = A_R x_R + b_R u + d_R y + h_R r$$
$$u_k = -c_{R_k} x_R - f_{R_k} y + g_{R_k} r \tag{5}$$

where $r(t)$ is a reference signal bounded on any compact set. As with the identifiers, we require the n regulators to share the same state x_R, so (5) together with $u_C = [u_1 \ u_2 \ \cdots \ u_n]'$ describes Σ_R. Often some of the elements of the vector x_R above are the elements of the identifier's state x_I, used to construct an "adaptive state observer;" an explicit characterization of this decomposition is not needed for the points we wish to make here. In most applications A_R is chosen, without loss of generality, as a stability matrix; in the present case this has the important practical advantage of ensuring that, if the process input and output and the reference signal stay bounded, then all signals u_k will be bounded, whichever is connected in feedback with the process.

Each certainty-equivalence controller Σ_{R_k} will be designed so that the feedback interconnection Σ_{cl_k} of Σ_{C_k} with the nominal process model Σ_{N_k} is robustly stable. In order to make this statement in a more precise way, let us assume that the triple $(\bar{c}, \bar{A}, \bar{b})$ belongs to the class $\mathcal{C}_k$ and write the matrix triple corresponding to the closed loop system formed by its interconnection with Σ_{R_k} as

$$
\left(\begin{bmatrix} \bar{c} & 0 \end{bmatrix}, \begin{bmatrix} \bar{A} - \bar{b}f_{R_k}\bar{c} & -\bar{b}c_{R_k} \\ (d_R - b_R f_{R_k})\bar{c} & A_R - b_R c_{R_k} \end{bmatrix}, \begin{bmatrix} \bar{b}g_{R_k} \\ b_R g_{R_k} + h_R \end{bmatrix} \right).
$$

Defining $\pi_{R_k}(s) = f_{R_k} + c_{R_k}(sI - A_R + b_R c_{R_k})^{-1}(d_R - b_R f_{R_k})$, it can be verified that the necessary and sufficient conditions for the closed loop system above to be stable are the

Robust Stabilization Requirements: .

1. There are no unstable pole-zero cancelations when $\psi_k(s)$ is formed;

2. The zeros of $1 + \pi_{R_k}(s)\pi_{N_k}(s)$ lie all in the open left half plane; and

3. $\|\Omega_k(s)\eta_k(s)\|_\infty < 1/\delta$ where η_k, sometimes called the closed-loop complementary sensitivity function, is

$$
\eta_k(s) = \frac{\pi_{R_k}(s)\pi_{N_k}(s)}{1 + \pi_{R_k}(s)\pi_{N_k}(s)}.
$$

In fact, we impose the stronger requirements that $1 + \pi_{R_k}(s+\lambda)\pi_{N_k}(s+\lambda)$ be a Hurwitz polynomial and $\|\Omega_k(s+\lambda)\eta_k(s+\lambda)\|_\infty = \varepsilon_k < \varepsilon_*$, where $\varepsilon_* < 1/\delta$ and λ is the positive constant appearing in Assumption 1. ♣ ♣

$\mathcal{H}_\infty$ control theory furnishes techniques to design controllers which satisfy the requirements above, for a given value of δ. Alternatively, a convenient robust control technique can be used to design Σ_R, and the robustness margin δ can be computed a posteriori. Since our main concern is with the supervisory adaptive loop, we take as given a family of controllers Σ_{R_k}, $k \in \mathcal{K}$, each of them satisfying conditions 1,2, and 3 above when connected in feedback with a plant belonging to $\mathcal{C}_k$, respectively.

6 Tuning algorithm

In our scheme, the tuner plays the role of a supervisor, whose purpose is to generate, at each time t, a switching function $\sigma(t)$ which determines the signal to be used to control the process. The essential feature of the supervisor is that "it must be able to evaluate the potential closed-loop performance of each controller without actually trying it out in the feedback loop" [7]. We explicitly rule out prerouted supervisory schemes which try out controllers by switching each of them into the feedback loop in order to evaluate their performance. Our reason is that we are afraid such schemes will result in unacceptable performance compromises. The tuner is thus a hybrid system composed of continuous-time dynamical systems together with switching logic, and it takes as input the identification error vector e_I. For a discussion of adaptive control using logic-based switching, see [9].

The requirements we make of the candidate tuner are formulated in terms of n auxiliary signals $v_k(t)$, defined out of considerations about the uncertainty model. With the quadruple $(c_{\Omega_k}, A_\Omega, b_\Omega, d_{\Omega_k})$ a detectable, stabilizable realization of the stable, proper transfer function $\Omega_k(s)$, construct the subsystem Σ_Ω as follows:

$$\begin{aligned}
\dot{x}_\Omega(t) &= A_\Omega x_\Omega(t) + b_\Omega \begin{bmatrix} b'_k & 0 \end{bmatrix} x_I \\
v_k(t) &= c_{\Omega_k} x_\Omega(t) + d_{\Omega_k} \begin{bmatrix} b'_k & 0 \end{bmatrix} x_I.
\end{aligned} \tag{6}$$

We call the overall system Σ_Ω. Since by definition the transfer function from the process input u to $\begin{bmatrix} b'_k & 0 \end{bmatrix} x_I$ is equal to $\begin{bmatrix} b'_k & 0 \end{bmatrix} (sI - A_I)^{-1} b_I = c(sI - A)^{-1} b_k = \beta_k(s)$ (§4), in operator notation we can write

$$v_k(t) = \Omega_k(s) \beta_k(s) u(t).$$

Given an arbitrary time function $\sigma(t)$, we define the signals e_σ, u_σ, and v_σ as the composition of the maps from $\mathcal{K}$ to the kth element of the vectors e_I and u_C, respectively. We can state the required

Tuner Properties:

1. σ's rate of change, denoted $\dot{\sigma}(t)$, is nondestabilizing with sufficiently small growth rate; and

2. The switching function $\sigma(t)$ is such that

$$\begin{aligned}
\|e^{-\lambda(t-\tau)} e_\sigma(\tau)\|_t^2 < \\
\delta_*^2 \|e^{-\lambda(t-\tau)} v_\sigma(\tau)\|_t^2 + a_0 + a_1(t) + a_2(t),
\end{aligned} \tag{7}$$

where

- $\delta_* < 1/\varepsilon_*$ is a nonnegative constant;

- a_0 is a positive number;

- $a_1(t)$ is a function of time which depends causally on $w(\cdot)$ and is either bounded (if $\lambda > 0$) or grows at most as fast as t (if $\lambda = 0$), e.g. $a_1(t) = \bar{a}_1 \|e^{-\lambda(t-\tau)} w\|_t^2$, $a_1(t) = \bar{a}_1 \left(\sup_{\tau \leq t} |w(\tau)| \right) t$ (with $\bar{a}_1$ a constant); and

- $a_2(t)$ is bounded by $\|\bar{a}_2(\tau) e^{-\lambda(t-\tau)} \begin{bmatrix} x_I & x_\Omega \end{bmatrix}\|_t^2$, where $\bar{a}_2$ is nondestabilizing with sufficiently small growth rate. ♣♣

The tuner properties are formulated in a more general setting than is required because we do not want to commit ourselves to any particular algorithm. We would like to leave open the possibility of using some form continuous tuning, by itself or in a hybrid arrangement with the kind of discrete tuning discussed in the present paper. The intuition behind the properties lies in the idea that if Σ_{N_k} is the "correct" nominal process model, then the error e_k must be "small," taking into account the admissible uncertainties and the noise. If identification is good then e_σ must be small as well. Tuners with the properties above allow us to construct adaptive systems and prove their closed-loop stability, making explicit use of the quantitative robustness characteristics intrinsic to the linear control design.

Under the not very realistic assumption that w_{k_*} is square integrable, or that an upper bound on the power of the noise signals is known, it becomes straightforward to construct a tuner with the required properties. The algorithm described below has tuner properties 1 and 2 and requires no such prior knowledge, and it can be applied when the constant λ appearing in Assumption 1 and in the robust stability requirements is positive or even zero.

Step 1: Let $\{t_1, t_2, \ldots, t_i, \ldots\}$ be an unbounded, infinite sequence of candidate switching times. σ is held constant in the intervals $[t_i, t_{i+1})$. This sequence may be determined a priori or in real time; all we require is that $\lim_{i \to \infty} t_i = \infty$ (which implies that the number of switching times is finite in any finite interval) and that $t_{i+1} - t_i$ is bounded.

Step 2: Choose δ_* such that $\delta < \delta_* < 1/\varepsilon_*$. For each $k \in \mathcal{K}$ define the *performance signal* $J_k(t)$ by $J_k(0) = 0$ and

$$\frac{d}{dt} J_k(t) = e_k^2(t) - \delta_*^2 v_k^2(t). \tag{8}$$

Step 3: Let $\bar{J}_k(t_i) = \max_{0 < t_i \le t}(J_k(t_i)/t_i)$. Pick an arbitrary nonnegative number a and let $\underline{k}(t_i)$ be the least positive integer such that $\bar{J}_{\underline{k}(t_i)}(t_i) \le \bar{J}_k(t_i)$, $k = 1, 2, \ldots, n$. Then choose an arbitrary value for $\sigma(t_0)$ and let

$$\sigma(t_{i+1}) = \begin{cases} \sigma(t_i) & \text{if } \bar{J}_{\sigma(t_i)}(t_{i+1}) \le \\ & \quad \bar{J}_{\underline{k}(t_{i+1})}(t_{i+1}) + a; \\ \underline{k}(t_{i+1}) & \text{otherwise.} \end{cases} \tag{9}$$

The idea of this algorithm is to pick the nominal model which has the least maximum associated cost, using hysteresis in order to ensure slow tuning. The algorithm is somewhat naive but has what is required of a tuner.

Proposition 4 *Consider a process $\Sigma_P \in \mathcal{C}_P$, and an identifier possessing the identifier property (4) in §3. Suppose that Assumption 1 holds and that $\lambda = 0$. The algorithm described in steps 1, 2, and 3 ensures that tuner properties 1 and 2 are satisfied.*

A proof of this proposition is given in the complete version of this paper, together with an alternative algorithm.

7 Analysis

The first aim of this section is to study the linear time-varying system obtained when we close the supervisory control loop making $u = u_\sigma$, where $\sigma(\cdot)$ is a time function which varies slowly, though not necessarily continuously (i.e., σ has tuner property 1). The resulting closed-loop system formed by Σ_P, Σ_{I_σ}, Σ_{R_σ}, and Σ_{Ω_σ}, called Σ_σ, admits a model with state $x = [\, x_P'\;\; x_I'\;\; x_R'\;\; x_\Omega'\,]'$ and dynamics of the form:

$$
\begin{aligned}
\dot{x}(t) &= A_\sigma x(t) + b_\sigma r(t) + d_\sigma w(t) \\
e_\sigma(t) &= c_\sigma x(t) + f_\sigma w(t) \\
v_\sigma(t) &= \bar{c}_\sigma x(t) + \bar{f}_\sigma w(t).
\end{aligned}
\tag{10}
$$

Here $w' = [\, w_u\;\; w_y\,]'$ and all matrices are defined in the obvious way. Besides x, the states of the overall adaptive, nonlinear system are the tuner's states; in our case, the performance indices (J_k or $J_k^\sharp$), the switching signal σ, and a timing signal which is reset at every candidate switching time. A complete state-space description is not needed for the points we wish to make here, but it is useful to observe that Σ_σ is linear, piecewise time-invariant, and that the right-hand side of the equation above satisfies Lipschitz conditions on $[0, \infty)$ because $\mathcal{K}$ has finite cardinality and all coefficient matrices are bounded. Hence the dynamical system above will admit solutions in $[0, \infty)$, which are bounded in any compact subinterval in which σ does not commute an infinite number of times. Tuner property 1 ensures that this condition holds in any interval, and is therefore sufficient for global existence of solutions.

Because Σ_{R_k} was constructed to stabilize Σ_{N_k}, which is the kth design model, the certainty-equivalence stabilization theorem [6] guarantees that Σ_σ is *tunable*; that is to say, the pair (c_k, A_k) is detectable for each fixed value of the parameter k. In the present work stronger demands on Σ_{R_k} were made — robust stabilization — hence we can expect to draw stronger conclusions about the pair (c_k, A_k).

Theorem 5 *With Σ_σ as defined in (10), for each k there exists an output injection h_k such that $A_k - h_k c_k$ is a stability matrix and*

$$
\|\bar{c}_k(sI + \lambda I - A_k + h_k c_k)^{-1} h_k\|_\infty < \varepsilon_*.
$$

The proof, hanging on the robust stabilization requirements, is given in the appendix. The second result we need is

Theorem 6 *Let the matrix triple $(C(p), A(p), B(p))$ be a continuously differentiable function of a parameter vector p such that, for each p, $A(p) + \lambda I$ is a stability matrix and*

$$
\|C(p)(sI + \lambda I - A(p))^{-1} B(p)\|_\infty < \gamma.
$$

If $\hat{p}(t)$ takes values in a compact set and $\dot{\hat{p}}$ is nondestabilizing in $[0, t)$, there exist positive definite matrices $P(\hat{p})$ and $R(\hat{p})$ and a bounded, nondestabilizing matrix $Q(t)$ such that, along the trajectories of

$$
\begin{aligned}
\dot{x}(t) &= (A(\hat{p}) - Q(t))x(t) + B(\hat{p})u(t) \\
y(t) &= C(\hat{p})x(t),
\end{aligned}
\tag{11}
$$

the function

$$V(t) = x'P(\hat{p})x + \int_0^t e^{-\lambda(t-\tau)}(y'(\tau)y(\tau) - \gamma^2 u'(\tau)u(\tau))d\tau$$

obeys the differential inequality

$$\dot{V}(t) \leq -2\lambda V(t) - x'(t)R(t)x(t).$$

This theorem extends a well-known result: a parameterized linear time-invariant system which is stable for every fixed value of the parameters remains exponentially stable if the parameters change slowly, that is, with nondestabilizing derivative. A proof is given in the appendix. For related results with an operator-algebraic point of view, see [15]. The significance of Theorem 6 may become more readily apparent through

Corollary 7 *Under the conditions of Theorem 6, system (11) is stable and has induced $\mathcal{L}^2$ norm smaller than γ, that is, $\|y\|_t \leq \gamma\|u\|_t$ under zero initial conditions.*

We are now in a position to sketch an analysis of the overall system. According to Theorem 5, there exists an output injection matrix h_σ such that $(\bar{c}_\sigma, A_\sigma - h_\sigma c_\sigma, h_\sigma)$ satisfies the conditions of Theorem 6. Thus there exist a positive definite P_σ and a matrix $Q(t)$ such that the derivative of

$$V(t) = x'P_\sigma x + \|e^{-\lambda(t-\tau)}\bar{c}_\sigma x\|_t^2 - \varepsilon_*^2\|e^{-\lambda(t-\tau)}e_\sigma\|_t^2$$

is negative definite along the trajectories of the differential equation

$$\dot{x} = (A_\sigma - h_\sigma c_\sigma - Q(t))x + h_\sigma e_\sigma.$$

The homogeneous part of (10) can be written as

$$\begin{aligned}\dot{x} &= (A_\sigma - h_\sigma c_\sigma - Q(t))x + h_\sigma e_\sigma + Q(t)x \\ v_\sigma &= \bar{c}_\sigma x\end{aligned}$$

where $Q(t)$ is nondestabilizing because σ is (tuner property 1). Thus

$$\dot{V}(t) \leq -2\lambda V(t) + x'(-R(t) + P_\sigma Q(t) + Q'(t)P_\sigma)x$$

which integrated gives

$$x'(t)P_\sigma x(t) + \|e^{-\lambda(t-\tau)}v_\sigma\|_t^2 - \varepsilon_*^2\|e^{-\lambda(t-\tau)}e_\sigma\|_t^2 \leq$$
$$e^{-2\lambda t}V(0) + \int_0^t e^{-2\lambda(t-\tau)}x'(-R(\tau) + P_\sigma Q(\tau) + Q'(\tau)P_\sigma)x d\tau.$$

Recall tuner property 2 which states that

$$\frac{1}{\delta_*^2}\|e^{-\lambda(t-\tau)}e_\sigma\|_t^2 - \|e^{-\lambda(t-\tau)}v_\sigma\|_t^2 < \frac{a_0 + a_1(t) + a_2(t)}{\delta_*^2}.$$

The crucial step in the analysis is to make a small-gain argument combining the previous two equations:

$$x'P_\sigma x + \frac{1 - \delta_*^2 \varepsilon_*^2}{\delta_*^2} ||e^{-\lambda(t-\tau)} e_\sigma||_t^2 <$$

$$e^{-2\lambda t} x_0' P_0 x_0 + \frac{a_0 + a_1(t) + a_2(t)}{\delta_*^2} -$$

$$\int_0^t e^{-2\lambda(t-\tau)} x'(R(\tau) - P_\sigma Q(\tau) - Q'(\tau)P_\sigma) x \, d\tau. \tag{12}$$

The coefficient multiplying $||e^{-\lambda(t-\tau)} e_\sigma||_t^2$ is positive because $\delta_* \varepsilon_* < 1$ (tuner property 2), and $R(t)$ remains positive-definite. It is a textbook exercise to complete the analysis using norm inequalities and a Bellman-Gronwall type inequality. One concludes that the homogeneous part of (10) behaves like an exponentially stable system driven by a bounded term depending on the noise. The analysis can be extended to take into account the forcing terms r and w: if r and w are bounded then x is bounded. If moreover r and $w \to 0$, then x decays exponentially to the interior of a ball whose radius depends on the constant a_0.

8   Concluding Remarks

We have described a class of adaptive systems which can be constructed in a modular fashion, with an identifier, an internal regulator, and a tuner as main components. The first two form a parameterized controller, which can be implemented as a number of linear dynamical systems running in parallel, and they can be designed using standard linear system design techniques. The number of parallel controllers is limited only by the hardware in which the feedback control system is implemented, and can in principle be turned into a tool to achieve specified performance and robustness measures for plants about which there can be a considerable amount of uncertainty. The type of tuner discussed consists of a switching algorithm in charge of selecting the most appropriate signal to be fed back into the process, with basis on comparison between the identification errors.

The analysis using Theorems 5 and 6 together with a small-gain argument is a natural extension of the method employed in [6, 14]. It is straightforward and independent of tuner and controller particulars. Moreover the provable robustness margins are "tight" in that they make full use of the robustness properties built into the linear control schemes used to implement the internal regulators. We emphasize that the method can be used to analyze a variety of robustly stable adaptive systems, including those with infinite or even uncountable parameter space. The main difficulty is solving the minimization problem of constructing a supervisor possessing the required properties. We do not at present know how to do without finiteness.[3]

---

[3] If continuous tuning is carried out by means of traditional recursive parameter estimation

Perhaps the main lesson of the present work is that, in order to obtain robust stability of an adaptive system, the admissible unstructured uncertainties ought to be taken into account when closing *both* the feedback loop and the adaptive loop. If either the certainty-equivalence controller or the tuner are chosen without due regard to unmodeled dynamics, the robustness margins of the overall adaptive system will tend to be overly conservative and difficult to compute. We would like to proceed in the direction of quantitative bounds on the performance of the adaptive systems described, and particularly towards a compromise between performance and the degree of parallelism n of the control system. One possible application of algorithms of the type discussed here, which would particularly benefit from quantitative performance bounds, is as an alternative to estimation-based adaptive control schemes; another is in designing control systems to accommodate unpredictable faults into modes which are a priori only imperfectly known.

8.1 Acknowledgements

The authors are grateful to A. S. Morse, Oswaldo L. V. Costa, João Hespanha, and Paulo Sérgio P. Silva for helpful discussions which have contributed to this work. The first also made available an advance copy of [7].

This research was supported by FAPESP — State of São Paulo Research Council, under grants 91/0508-3 and 93/2464-9. The first author is also supported by CNPq — Brazilian Research Council, grant 520961/93-5 and by the Rosa G. and Henrique S. Pait Foundation for the Advancement of the Arts and Sciences. A previous version of this paper has appeared in the *1994 IFAC Symposium on Robust Control Design* in Rio de Janeiro [13].

The bibliography of the complete paper is given in the sequel; the proofs are omitted.

References

[1] B. D. O. Anderson and J. B. Moore. *Optimal Filtering*. Prentice-Hall, Englewood Cliffs, 1979.

[2] M. Athans, D. Castañon, K.-P. Dunn, C. S. Greene, W. H. Lee, N. R. Sandell, Jr., and A. S. Willsky. The stochastic control of the F-8C aircraft using a multiple model adaptive control (MMAC) method — Part I: Equilibrium flight. *IEEE Trans. Automatic Control*, 22(5):768–780, Oct. 1977.

[3] L. Hsu, A. D. de Araújo, and R. R. Costa. Analysis and design of I/O based variable structure adaptive control. *IEEE Trans. Automatic Control*, 39(1):4–21, Jan. 1994.

(least-squares, gradient-like, etc.), the analysis can still be employed if tuner property 2 is shown to hold. We reckon that it is not easy to establish it without being quite conservative with regard to admissible uncertainties.

[4] P. A. Ioannou and J. Sun. *Stable and Robust Adaptive Control*. Prentice-Hall, Englewood Cliffs, 1995.

[5] F. Kassab Jr. *Estabilização Adaptativa Robusta: Uma Aplicação de Algoritmos Paralelos a Controle Tolerante a Falhas*. PhD thesis, Universidade de São Paulo, São Paulo, Brazil, 1995. (In Portuguese).

[6] A. S. Morse. Towards a unified theory of parameter adaptive control — Part 2: Certainty equivalence and implicit tuning. *IEEE Trans. Automatic Control*, 37(1):15–29, Jan. 1992.

[7] A. S. Morse. Supervisory control of families of linear set-point controllers — Part 1: Exact matching. Technical report, Yale University, Mar 1993. submitted for publication.

[8] A. S. Morse. Supervisory control of families of linear set-point controllers — Part 2: Robustness. Technical report, Yale University, Nov 1994. submitted for publication.

[9] A. S. Morse. Control using logic-based switching. In A. Isidori, editor, *Trends in Control*, pages 69–113. Springer-Verlag, 1995.

[10] A. S. Morse and F. M. Pait. MIMO design models and internal regulators for cyclically-switched parameter-adaptive control systems. *IEEE Trans. Automatic Control*, 39(9):1809–1818, Sept. 1994.

[11] K. S. Narendra and J. Balakrishnan. Performance improvement in adaptive control systems using multiple models and switching. In *Proc. Seventh Yale Workshop on Adaptive and Learning Systems*, pages 27–33, New Haven, Connecticut, May 1992.

[12] F. M. Pait and F. Kassab Jr. Algoritmos paralelos para controle adaptativo: Projeto de pesquisa. In 10º *Congresso Brasileiro de Automática*, pages 677–682, Rio de Janeiro, Brazil, Setembro 1994. (In Portuguese).

[13] F. M. Pait and F. Kassab Jr. Parallel algorithms for adaptive control: Robust stability. In *IFAC Symposium on Robust Control Design*, pages 406–411, Rio de Janeiro, Brazil, Setembro 1994.

[14] F. M. Pait and A. S. Morse. A cyclic switching strategy for parameter adaptive control. *IEEE Trans. Automatic Control*, 39(6):1172–1183, June 1994.

[15] G. Zames and L. Y. Wang. Local-global double algebras for slow $\mathcal{H}_\infty$ adaptation: Part I — Inversion and stability. *IEEE Trans. Automatic Control*, 36(2):130–142, Feb. 1991.

Lecture Notes in Control and Information Sciences

Edited by M. Thoma

1993–1996 Published Titles: